AF545085

AN INTRODUCTION TO

Electrical Wiring

SECOND EDITION

AN INTRODUCTION TO

Electrical Wiring

SECOND EDITION

John M. Doyle

RESTON PUBLISHING COMPANY
A Prentice-Hall Company
Reston, Virginia 22090

Library of Congress Cataloging in Publication Data

DOYLE, JOHN M.
An Introduction to electrical wiring.

Includes index.
1. Electric wiring, Interior. I. Title.
TK3271.D69 1980 621.319'24 79–16004
ISBN 0–8359–3185–4

A Prentice-Hall Company
Reston, Virginia 22090

10 9 8 7 6 5 4 3 2 1

Printed in the United States of America

To My Daughter Tamara Elaine

Contents

V
ELECTRIC MOTORS AND FARM INSTALLATIONS 265

12
Motors 267

13
Farm Installations 307

Appendix: Reference Tables 321

Index 329

Preface

Like its predecessor, this edition is written for electrician trainees in vocational-technical schools, technical institutes, and industrial apprenticeship programs; attention is centered on urban and rural single-family dwellings. Once the reader has acquired experience in the wiring of such structures, the changeover to commercial and/or industrial systems should proceed smoothly.

This is an entirely practical text. The only mathematics used—and sparingly—is arithmetic and the most elementary forms of algebra.

All methods discussed are in accordance with the 1978 edition of the National Electrical Code (NEC) published by the National Fire Protection Association (NFPA), 470 Atlantic Avenue, Boston, Massachusetts 02210. All students, trainees, and apprentices should either have a copy or have easy access to one.

Some sections of the original work have been completely rewritten for improved clarity—for example, Section 3–4, which deals with the zonal-cavity method of determining the illumination requirements for good lighting. Several new sections, on such things as residential lightning-protection systems and "long life" and "extended life" lamps, have been added. All additions, deletions, and modifications needed to make the book conform to the 1978 edition of the NEC are included.

The book is arranged in five parts. Part 1 is a simplified but entirely adequate treatment of the basic concepts of electricity, magnetism, and illumination. Part 2 deals with the sizing, characteristics, and permissible uses of various types of electrical conductors, and introduces the reader to common circuit components, such as switches, receptacles, and fuses. Part 3 covers the general considerations involved in any wiring job, such as wiring methods, grounding, the sizing of service entrances, and the provision of adequate lighting. Part 4 gives detailed instructions for the installation of complete wiring systems in buildings either under construction (new work) or already constructed (old work). Part 5 covers the various types of motors and methods for farm installations.

Because actual hardware is handled in any worthwhile training program, photographs of such hardware are not included in this book. Illustrations are used liberally, however, including the pictorial type, where such usage clarifies the discussion.

The reader may note some repetition in the book. This is intentional and is done to reinforce certain basics that the electrician must remember at all times.

Practical examples are scattered throughout the book and questions and/or problems appear at the end of each chapter. In addition, numerous tables, reprinted either in whole or in part by permission of the National Fire Protection Association, are included; the reader should also make use of the more extensive data given in the NEC.

My sincere thanks to the NFPA for granting permission to quote portions of the NEC.

Practical suggestions for improvement of either the content or the presentation of this book will be most welcome. Mail your comments to me, in care of Reston Publishing Company, 11480 Sunset Hills Road, Reston, Virginia 22090. Please write the book title across the top of your correspondence.

To those who used the first edition, my special thanks. I hope you will find this edition even more to your liking.

AN INTRODUCTION TO

Electrical Wiring

SECOND EDITION

I

BASIC PRINCIPLES AND CONCEPTS

1

Electricity and Magnetism

1–1 ELECTRIC CHARGES

When two clean, small glass balls are suspended near each other by separate threads, they hang straight down next to each other. See Figure 1–1(a). If the balls are rubbed with a silk cloth, they no longer hang straight down, but *repel* each other, as shown in Figure 1–1(b).

Several important deductions can be made from this experiment. (1) Rubbing the balls with a silk cloth produces a force of repulsion between the balls. This force is called an *electric charge*. (2) When the balls are uncharged, no force exists between them, and they are said to be *electrically neutral*. (3) Since both balls are rubbed with the same material, they must acquire the same kind of charge. (4) That is, *like charges repel*.

Let a small rubber ball replace one glass ball. When both balls are again rubbed with a silk cloth, they move toward each other, as shown in Figure 1–1(c). What does this indicate? Again the experiment leads to important deductions. (1) The rubber ball must acquire a different kind of charge than the glass ball. (2) *Unlike charges attract*.

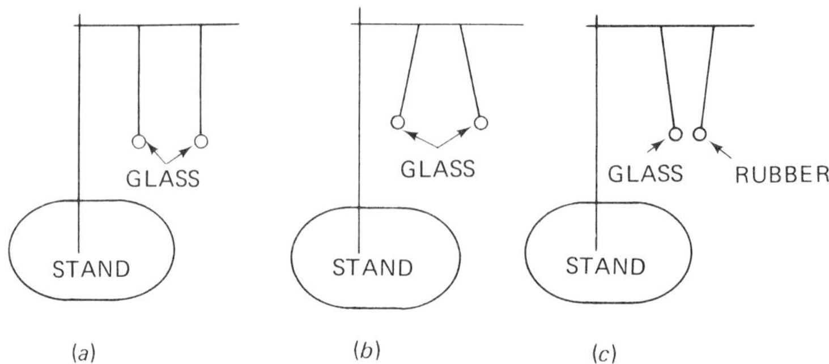

FIGURE 1–1. Demonstrating the laws of attraction and repulsion between electric charges: (*a*) electrically neutral; (*b*) like charges repel; (*c*) unlike charges attract.

The two opposite kinds of electric charge are called *positive* and *negative*. The condition of positiveness or negativeness is called *polarity*. Positive polarity is indicated by a plus sign (+) and negative by a minus sign (−).

Because the form of electricity demonstrated by the above experiments is motionless, it is called *static* electricity.

Finally, it should be noted that a charged (electrified) body retains its charge for a certain length of time, but eventually the electrification leaks off. If the body comes into contact with earth (ground), it discharges quickly.

Now that you know the meaning of charge (electrification), the two kinds of charges, and the laws of attraction and repulsion between charges, you are ready to study the *electron theory*, which is so named because it explains how electric energy in the form of electrons moves from one point to another in an electric circuit.

1–2 THE ELECTRON THEORY

The invisible and smallest whole elemental component, called the *atom*, is the basic building block of all matter, whether solid, liquid, or gas. It is often pictured as shown in Figure 1–2. The central portion, the *nucleus*, has a positive electric charge. Spinning around that nucleus in egg-shaped orbits are elementary particles called *electrons*. Each electron has a negative charge; the sum of these negative charges, under normal conditions, is exactly equal to the positive charge of the nucleus. Since the two charges are equal to one another, the *net* electrical charge of a normal atom (one containing a specified stable number of electrons) is zero.

FIGURE 1–2. Basic structure of an atom.

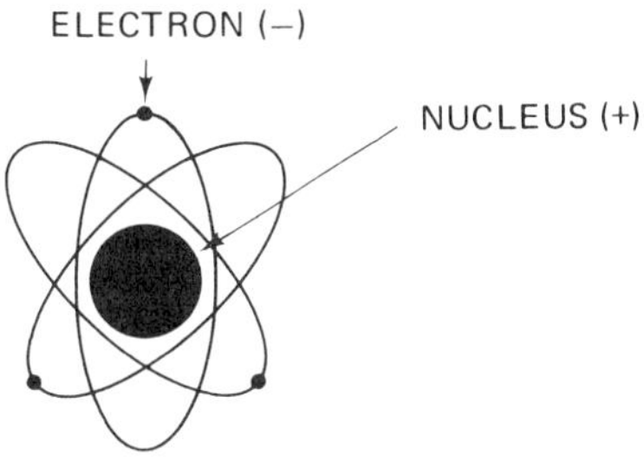

Because they have unlike charges, a force of attraction exists between the nucleus and its orbiting electrons. This force is exactly balanced, however, by a *centrifugal force* that tends to push the electrons outward from the center of rotation. As a result, the electrons assume *fixed* orbits around their nucleus.

For instance, there are 29 electrons in one atom of copper. Of these, 28 are bound permanently—at least, for our purposes—to the nucleus, but only a very weak binding force exists between the nucleus and the one remaining electron. This twenty-ninth electron, located in the outermost orbit, is called a *valence* electron.

When a valence electron finds itself midway between two nuclei, as shown in Figure 1–3, it is attracted equally by each nucleus. In this position of balanced attraction, a valence electron often leaves its own orbit and moves into the outermost orbit of a neighboring atom. For this reason, valence electrons are often called *free* electrons. *The movement of free electrons is the means by which electrical energy is carried from one point to another.*

FIGURE 1–3. When a valence electron finds itself midway between two nuclei, it is attracted equally by each nucleus.

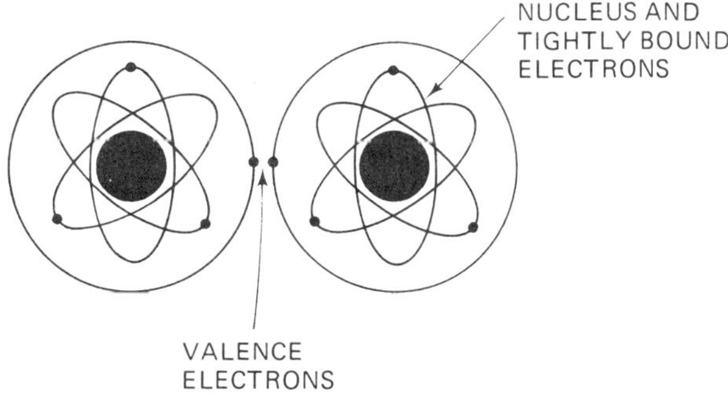

1–3 CONDUCTORS AND INSULATORS

When a material contains many free electrons *per unit volume* (for example, per cubic inch), it is called a *conductor*. Although any metal may be used as an electric conductor, copper is the most frequent choice. Aluminum is another very popular conductor.

In electric circuits it is often necessary to confine the free electrons to a given path or to prevent contact either *between* or *with* conductors. Materials containing very few free electrons per unit volume are used for this purpose; they are called *insulators*. Plastic, rubber, porcelain, glass, and air are the most common insulators encountered in electrical wiring.

The operation of most switches depends on the insulating properties of air. Simply by opening (breaking) the conducting path at a selected point, a switch prevents the transfer of free electrons and, therefore, the transfer of electric energy from one point in a conductor to another.

When the electric *pressure* (see Section 1–6) applied across an insulator is greater than that specified by the manufacturer, the insulator breaks down and becomes a conductor. The arcing of electricity across an open switch is one such example of insulator breakdown. For this reason, insulators must always be operated within their specified ratings.

1–4 THE PARTS OF AN ELECTRICAL CIRCUIT

All practical electric circuits, regardless of how simple or how complex they may be, require four basic parts: (1) a *source* of electrical energy that forces free electrons to flow through the circuit; (2) *conductors* for carrying the free electrons around the complete circuit; (3) a *load*, that is, a device or devices to which electrical energy is supplied; and (4) a *control device* to turn the circuit "on" or "off."

The *schematic* diagram of a simple electric circuit is shown in Figure 1–4. The source may be an electric outlet in a home, a battery, an electric generator, or some other device. As you will learn later, either of two basic types of sources are used: *alternating current* (ac) or *direct current* (dc). The symbol shown in Figure 1–4 is for a direct-current source. A different symbol is used to indicate an alternating-current source.

The conductors are usually copper wire; however, aluminum may also be used.

The control device in Figure 1–4 is a *switch*. When the switch is open, as shown, free electrons cannot move around the circuit. A complete path that permits the flow of free electrons is formed when the switch is closed.

The load may be any one of a wide variety of devices, such as a light

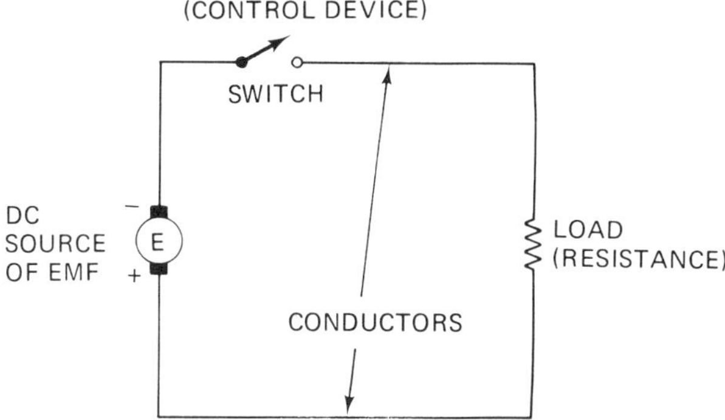

FIGURE 1–4. Schematic diagram of a simple electric circuit.

bulb, a motor, a washer, or a dryer. The symbol for a *resistance* (a device that changes electrical energy into heat energy) is also shown in Figure 1–4. As you will learn, different schematic symbols are used to represent different kinds of loads.

1–5 CURRENT

In working with electrical circuits, it is necessary to know the *rate of flow* of free electrons through the circuit, that it, how many free electrons pass a given point in the circuit in 1 second(s).

The rate of flow of free electrons is called *current*; it is designated by the letter symbol I, which indicates the *intensity* of electron flow. When a certain very large number of electrons (6.24×10^{18} to be exact) flow past a given point in 1 s, the rate of current flow is 1 *ampere*, (A). Hence, the *unit* of electric current is the ampere.

Prefixes

Very often, the basic unit of an electric quantity is either too small or too large for practical use. This difficulty is overcome by the use of *prefixes* that indicate either multiples or fractions of the basic units. The prefixes that an electrician encounters frequently are shown in Table 1–1.

TABLE 1–1. Common Prefixes

Prefix	*Meaning*	*Abbreviation*
Mega	One millionth	M
Kilo	One thousand	k
Milli	One thousandth	m
Micro	One millionth	μ

Measuring Current

Aside from the fact that electrons are invisible, it would be impossible to count them manually as they flowed by a given point in a circuit. Fortunately, an electrical instrument called either an *ammeter, milliammeter,* or *microammeter*, depending on its *range*, indicates directly the amount of current passing through a circuit.

The scale of a typical ammeter is calibrated (marked) as shown in Figure 1–5. The maximum current-carrying capability of this particular meter is 1 A, the highest value indicated on the scale. Each major calibration mark represents 0.1 A (100mA).

Usually, several sets of calibration marks are shown on the same scale, and the appropriate range is selected either by means of a *range switch* or by inserting plugs into appropriately marked *jacks*. Always read the manufacturer's literature that comes with any instrument before you try to use it.

Since an ammeter measures the current passing *through* a circuit, it is connected end to end with other components, as shown in Figure 1–6. Notice that the schematic symbol for an ammeter is a circle containing the letter symbol A. For a milliammeter the correct symbol is mA encircled, and for a microammeter, μA encircled.

When an ammeter is used to measure direct current, *it is necessary to observe polarity*. This means that the negative terminal of the meter must always be connected to a point of negative potential *with respect to* the positive terminal of the meter. If the meter is connected improperly, it can be ruined quickly.

When an ammeter is used to measure alternating current, it is *not* necessary to observe polarity.

Finally, whenever you use a current-measuring instrument, always make sure to use a range that is large enough to measure the circuit current. If the probable value of a circuit current is unknown, use the highest range available for the initial measurement. Then, if the current is small enough, you can always switch to a lower range.

FIGURE 1–5. Scale of a typical ammeter.

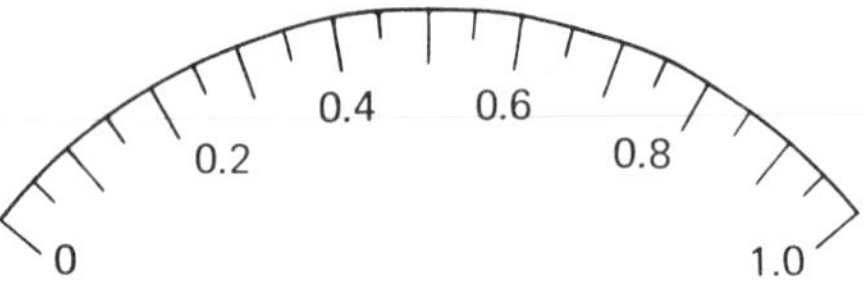

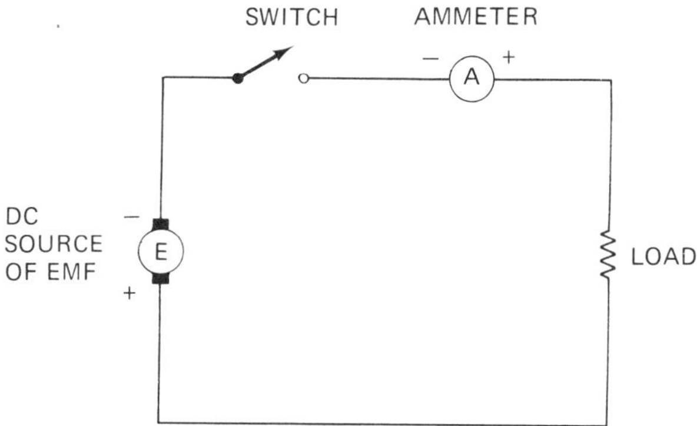

FIGURE 1–6. Correct method of connecting an ammeter in a dc circuit.

1–6 ELECTRIC PRESSURE

Whenever a source of electric energy is connected across the terminals of a *complete* electric circuit, it creates a *surplus* of free electrons at one terminal and a *deficiency* of free electrons at the other terminal. The terminal having the surplus of free electrons has a *net negative charge* and is indicated by a minus (−) sign. The terminal having the deficiency of free electrons has a *net positive charge* and is indicated by a plus (+) sign.

At the terminal charged positively, the free electrons are spaced farther apart than normal, and the forces of repulsion that act between them ("like charges repel") are reduced. This force of repulsion is a form of *potential energy*; it is also called the *energy of position*. Let us see what the term potential energy means.

Energy is often defined as the ability (capacity) to do work. To illustrate, when a hammer is raised above a nail, the hammer, as a result of its position, has the potential ability to do work; that is, it possesses *potential* energy. As the hammer drops, it gains *kinetic* energy, the energy associated with movement. Finally, when the hammer strikes the nail, its energy is expended (dissipated) and work is accomplished; the nail is driven into the surface on which its point had rested.

The electrons in a conductor possess potential energy in much the same manner as the hammer, and work is done when the electrons repel (push) one another to a new position in the conductor.

From the above discussion, it is apparent that the potential energy of free electrons at the positive terminal of a circuit is *less* than the potential energy of those at the negative terminal. Thus, a *potential energy differ-*

ence, a term usually shortened to *potential difference*, exists across the terminals.

Whenever a potential difference exists between two points in a metallic conductor, the free electrons try to distribute themselves uniformly throughout the conductor; in so doing, they flow from the point of negative potential toward the point of positive potential, that is, from − to +. Thus a *potential difference is the pressure that creates current in an electric circuit.*

Because the source creates the potential difference that forces current through a circuit, it is called a source of *electromotive*, or electron moving *force*. The usual abbreviation is EMF.

Now that we know both what is meant by the expression "source of EMF," and that in a metallic conductor free electrons move from negative to positive, we can give a precise definition for current. *Current is the movement of free electrons through a circuit under the influence of an applied EMF.*

In some cases, the terminals of the source always maintain the same polarity, and current flows in the same direction at all times through the circuit external to the source. Sources of this type are called *direct-current sources*, and the current that they create is *direct current*, abbreviated dc. In other cases, the terminals of the source reverse polarity periodically. Sources of this type are called *alternating-current sources*, and the current that they create is *alternating current*, abbreviated ac. For the remainder of this chapter, however, our only concern is with direct current. Chapter 2 considers alternating current in detail.

In Figure 1–7, we see that a dc source of EMF, a switch (control device), and a lamp (load) are connected by conductors. Since the source maintains a potential difference, it may be thought of as a *potential rise*. Moreover, since the electric energy provided by the source is used up in

FIGURE 1–7. Simple electric circuit.

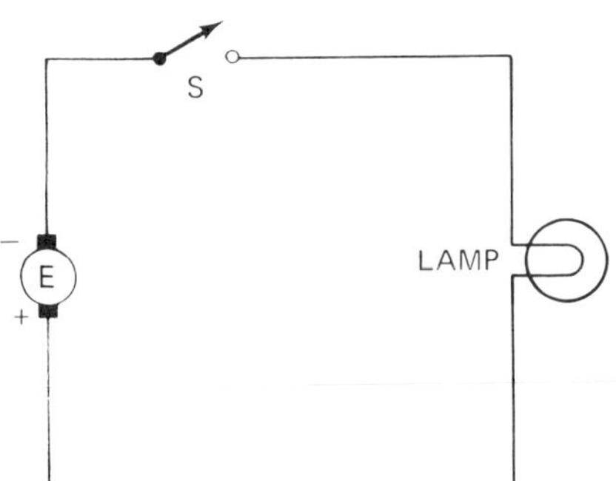

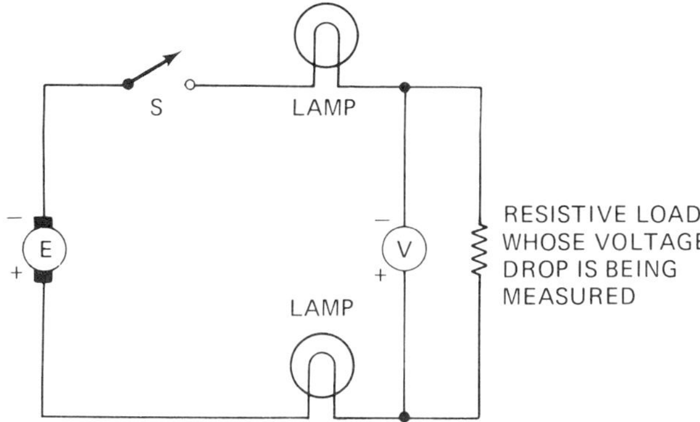

FIGURE 1–8. Correct method of connecting a voltmeter into a dc circuit.

doing work, such as lighting the lamp in Figure 1–7, the load represents a *potential drop*.

In Sections 1–12 and 1–13 you will see that the potential drops around an electric circuit are always equal to the potential rise; therefore, we use the same unit of measurement to represent either a potential rise or a potential drop. *The volt is the basic unit of potential difference*.

The letter symbol E is used to represent the potential rise occurring at the source; the letter symbol used to represent the potential drop occurring across a load is V. Both E and V are, of course, given in volts.

When the volt is either too small or too large for practical use, we can employ any of the prefixes introduced in Section 1–5.

Many electricians do not distinguish between potential rise and potential drop and refer to both as *voltage*. This interchangeable usage may be fine later on, but while you are using this text, please keep the distinction in mind.

Measuring Potential Differences

A *voltmeter* is used to measure either a potential voltage rise, E, or a voltage drop, V, across the terminals of a device. Thus a voltmeter is always connected *across* a device, as shown in Figure 1–8. In this case, the voltmeter is being used to measure the voltage drop across one of several devices making up the total load.

Since this is a dc circuit, polarity must be observed in connecting the meter, that is, − to − and + to +. If the voltmeter is connected in the wrong polarity, the meter pointer will move off scale and the meter may be damaged. It is also necessary, of course, to use a range suitable for the voltage rise or drop being measured. The correct symbol for the voltmeter is a circle enclosing the letter V.

1–7 RESISTANCE

As free electrons pick up velocity in their movement along the conductor, potential energy from the source is transferred to them in *kinetic* form; that is, the electrons acquire kinetic energy (the energy of motion). Before the electron has gone very far, however, it collides with a relatively large metal *ion*. An ion is simply an atom or group of atoms that by the loss or gain of one or more free electrons has acquired an electric charge. The ions take up fixed positions and give the metal conductor its shape. As a result of the collisions between free electrons and ions, the free electrons give up some of their kinetic energy to the ion in the form of *heat* energy. That is why a conductor feels warm to the touch when it is passing current. Since the heat energy released by the collisions into the surrounding air does not perform useful work, it is wasted energy. More important, if the amount of heat produced in a conductor is more than it can dissipate, the temperature of the conductor will rise and create a fire hazard. Many buildings have been destroyed as a result of overheated conductors.

In passing from one point to another in an electric circuit, a given free electron is involved in many collisions. Since current is the movement of free electrons, the collisions oppose current. Another word for oppose is *resist*; thus, we call the opposition to current *resistance*. Stated formally, *resistance is that property of an electric circuit that opposes current.*

The unit of resistance is the *ohm*, and the unit symbol is the Greek letter Ω (omega). The letter symbol R is used to indicate resistance in schematic diagrams.

When the ohm is too small a unit for practical use, we employ either the kilohm (kΩ), which is equal to 1000 ohms, or the megohm (MΩ), which is equal to 1,000,000 ohms.

All components used in electric circuits contain some resistance. Of particular interest to electricians is the resistance of the conductors.

Resistance of Metallic Conductors

Four factors affect the resistance of a metallic conductor: (1) *length*, (2) *cross-sectional area* (thickness), (3) *the type of material from which the conductor is made*, and (4) *temperature*.

The resistance of a metallic conductor is directly proportional to its length. That is, as the length of the conductor increases, its resistance increases. The truth of this statement is apparent when we consider that, the farther the free electrons move through the conductor, the more collisions there are with ions; hence, the greater the resistance.

The resistance of a metallic conductor is inversely proportional to the cross-sectional area (thickness). That is, as the thickness of the conductor increases, its resistance decreases, and as its thickness decreases, its resistance increases.

The resistance of a metallic conductor is dependent on the type of material used to form the conductor. The truth of this statement can be seen when we recall that the distinction between conductor and insulator materials is based on the number of free electrons that the material of the one or the other possesses per unit volume.

For most conductors, resistance increases when the temperature rises and decreases when the temperature drops. A good example of this is the incandescent lamp with a tungsten filament. The *cold* resistance of the filament is about 18 Ω, but its *white hot* resistance is about 240 Ω. As a result of its relatively low cold resistance, the tungsten filament experiences a heavy *surge* of current initially. As the filament heats and its resistance increases, current gradually declines to a normal value. In some devices, such as the heating elements of electric stoves, heavy surges cannot be tolerated and special metal alloys, such as Nichrome II(T), whose resistance values remain reasonably constant over a wide temperature range must be employed.

Measuring Resistance

The instrument most often used for the direct measurement of resistance is the *ohmmeter*. The ohmmeter contains its own source of EMF (electromotive force) in the form of a battery, so it is never used to make measurements in a *hot* (energized) circuit. *All power to the circuit under test must be turned off before any attempt is made to connect the ohmmeter into the circuit*. Failure to do so will result in destruction of the meter.

An ohmmeter is connected *across* the component or circuit whose resistance is to be measured.

Since ohmmeters with different scale and control arrangements are available commercially, no attempt at physical description will be made here. Again, be sure to consult the manufacturer's literature supplied with the instrument as a guide to its use.

Multimeters

To avoid carrying around separate ammeters, voltmeters, and an ohmmeter, most electricians use the *multimeter*, which incorporates all three meters into a single instrument. The quantity to be measured is selected by means of a function switch. A range switch provides maximum flexibility.

1–8 OHM'S LAW

In 1825 a German scientist, Georg Simon Ohm, performed a series of experiments that led to a statement of one of the most important laws that govern the behavior of all electric circuits. Working with a circuit similar to that shown in Figure 1–9, Ohm discovered that every time he closed the switch the circuit current had the same constant value. He also found that, *as long as the temperature of the conductors did not change*, the circuit current was *directly proportional* to the applied EMF. In other words, if the EMF was doubled, the current was also doubled. Conversely, if the EMF was reduced to one half its original value, the current also fell to one half its original value.

Carrying his experiment one step further, Ohm discovered that the value of current changed whenever he changed either the *type* or *thickness* or *length* of the conductor, or the *type* of bulb. Why did these changes occur? He found the answer, of course—every time the conductor or the bulb was changed, the resistance of the circuit changed. When the resistance of the circuit *increased*, the current *decreased*. When the resistance of the circuit *decreased*, the current *increased*.

On the basis of these discoveries, two very definite statements can be made about the behavior of electric circuits. (1) When the resistance and temperature of the circuit are constant, current is directly proportional to voltage. If the voltage increases, current increases, and if the voltage decreases, current decreases. (2) When the source of EMF provides a

FIGURE 1–9. Circuit similar to that used by Ohm.

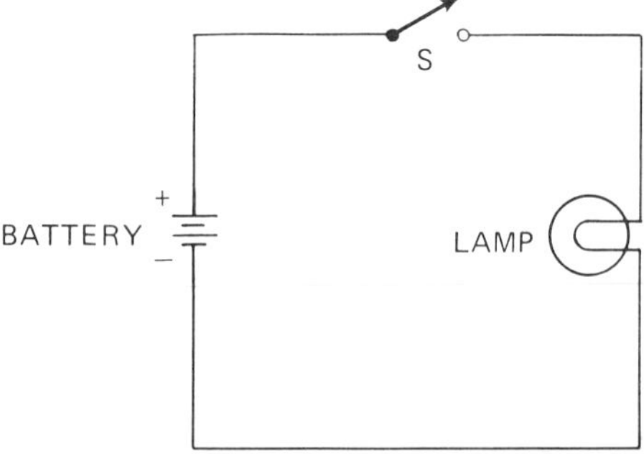

constant potential difference in the circuit and the temperature is constant, current is inversely proportional to resistance: when resistance goes up, current goes down; when resistance goes down, current goes up.

It was to honor the memory of George Ohm that the unit of resistance was called the ohm, and the relationship between voltage, current, and resistance is summed up in *Ohm's law*. The three forms of Ohm's law follow:

$$R = \frac{E}{I} \qquad I = \frac{E}{R} \qquad E = IR \tag{1–1}$$

where R is the resistance in *ohms*, E is the applied EMF in *volts*, and I is the current in *amperes*.

A few simple examples will serve to demonstrate the usefulness of Ohm's law. In each case, the circuit will be similar to that shown earlier in Figure 1–9.

Let $E = 30$ V and $I = 3$ A. What is the value of R?

$$R = \frac{E}{I} = \frac{30}{3} = 10\ \Omega$$

Let $R = 20\ \Omega$ and $E = 40$ V. What is the value of I?

$$I = \frac{E}{R} = \frac{40}{20} = 2\ \text{A}$$

Let $I = 3$ A and $R = 20\ \Omega$. What is the value of E?

$$E = IR = 3 \times 20 = 60\ \text{V}$$

1–9 POWER

In everyday life, it is important to know how much work is accomplished by a given process and how long it takes to do a given amount of work. The same situation holds true in dealing with electric circuits.

In electric circuits, the *rate of doing work* is called *power* and is indicated by the letter symbol P. To honor the memory of James Watt, inventor of the steam engine, the unit of power was named the *watt*, abbreviated W.

To calculate power in an electric circuit we use the relationship

$$P = EI \tag{1–2}$$

where P is the power in watts, E is the applied EMF in volts, and I is the circuit current in amperes. If several load devices are connected in the circuit, and you wish to determine the power dissipated by a single device, V (voltage drop) is used in place of E in Equation 1–2.

Suppose that a lamp is connected to a 120-V source and draws a current of 0.5 A. To determine the rate at which electric energy is supplied to the lamp, we use Equation 1–2.

$$P = EI = 120 \times 0.5 = 60 \text{ W}$$

Electric lamps, motors, and clothes dryers are just a few of the devices an electrician encounters that are rated in watts. The *wattage rating* of any device *indicates the rate at which the device converts energy from one form to another*, and often indicates the *safe operating limit* of the device.

Since power is dissipated by the resistance of any practical electric circuit, it is often convenient to express power in terms of R. By Ohm's law, $E = IR$. If we substitute IR for E in Equation 1–2, it becomes

$$P = EI = IRI = I^2R \qquad (1\text{–}3)$$

If, for example, the resistance of the lamp filament in the previous example is 240 Ω, the power dissipated by the lamp is

$$P = I^2R = 0.5^2 \times 240 = 60 \text{ W}$$

Since $P = 60$ W in both examples, it is apparent that Equations 1–2 and 1–3 can be used interchangeably.

The above example shows that the *power developed in a circuit of fixed* resistance is *directly proportional to the square of the current*. If the current is doubled, the power is increased by a factor of 4, since $2^2 = 4$. If the current is tripled, the power is increased by a factor of 9, since $3^2 = 9$. By similar reasoning, if the current is halved, the power is reduced to one fourth of its original value, and so on.

We can derive another useful relationship from Equation 1–2 by substituting E/R for I. This gives

$$P = EI = E \times \frac{E}{R} = \frac{E^2}{R} \qquad (1\text{–}4)$$

Now, if a 240-Ω lamp is connected to a 120-V source of EMF, the rate at which the lamp uses power is

$$P = \frac{E^2}{R} = \frac{120^2}{240} = 60 \text{ W}$$

The above example shows that Equations 1–2, 1–3, and 1–4 may be used interchangeably. Which equation you choose is determined by the known quantities. In practice, the value of any one quantity may be determined by measurement or, in most cases, by reference to the stamp or rating plate found on most electrical devices.

The last example also shows that *the power developed in a circuit of*

TABLE 1–2. Typical Power Consumption of Motors

Motor Size (hp)	*Typical Consumption (W)*
¼	300–400
½	450–600
Over 1/2, per hp	950–1100

1–11 THE WATTHOUR

To measure the theoretical electrical energy consumed by all devices connected to an electric circuit involves the determination of how much power is used and for what length of time. The *watthour* (Wh) may be used for this purpose and is simply the product of power in watts and time in hours. More often, however, use is made of the *kilowatthour* (kWh) in which kilowatts are used in place of watts.

Suppose, for example, that a 250-W lamp burns for 10 hours (h). The power consumed by the lamp is

$$250 \times 10 = 2500 \text{ Wh}$$

or, since 250 W = 0.25 kW,

$$0.25 \times 10 = 2.5 \text{ kWh}$$

It is on the basis of the kilowatthour that we pay our electric bill. To illustrate, suppose that six lamps, each rated at 100 W, operate 8 h per day for 30 days, and that the utility company charges \$0.05 per kilowatthour. The cost to operate the lamps is determined as follows:

$$\text{Total wattage} = 6 \times 100 = 600 \text{ W} = 0.6\text{kW}$$

$$\text{Daily kWh} = 8 \times 0.6 = 4.8\text{kWh}$$

$$\text{kWh for 30 days} = 30 \times 4.8 = 144\text{kWh}$$

$$\text{Cost} = \$0.05 \times 144 = \$7.20$$

A kilowatthour meter is installed in all residences and is read monthly by a representative of the local electric supplier. Some kilowatthour meters have a *cyclometer* dial similar to that shown in Figure 1–10(a), where the indicated number of kilowatthours is 2436. If the same meter is read one month later and shows 2981, for example, the electric consumer is charged for 545 kWh, or the difference between 2981 and 2436.

Other kilowatthour meters contain four dials that are marked as shown in Figure 1–10(b). From left to right, the *first* and *third* dials are read in a *counterclockwise* direction, while the *second* and *fourth* dials are read in the *clockwise* direction. The meter reading is determined by the *fixed resistance is directly proportional to the square of the applie voltage*. Thus, when the voltage is doubled, power increases by a fact of 4. Similarly, when the voltage is halved, power decreases to one fourt its original value.

In working with practical circuits, always keep in mind that power directly proportional to the square of *either voltage or current*. Thus, if circuit is modified in any way and the value of circuit resistance changed, *before* power is applied to the circuit, make sure that the wattag ratings of all devices connected thereto will not be exceeded.

Because the watt is a very small unit, it is often more convenient t express power in multiples of 1000 watts, or kilowatts (kW).

Measuring Power

Although we can calculate the power in a load by taking separate ammet and voltmeter readings, a direct-reading *wattmeter* is a very useful instru ment, particularly for ac circuits.

A wattmeter containing an *electrodynamometer movement* can b used in either ac or dc circuits. Inside the instrument case there are tw coils, a *current* coil and a *voltage* coil. To assist the user in connecting wattmeter properly, it is customary in practice to identify one end of eac coil with ± marks. Standard connections require that we connect th identified terminal of the current coil to the generator and the identifie terminal of the voltage coil to the side of the line containing the curre coil.

More will be explained about the use of wattmeters in Chapter : Before you attempt to use any wattmeter, be sure to read the manufa turer's literature supplied with the instrument.

1–10 HORSEPOWER

Electric motors, except for a few miniature types, are rated in *horsepowe* (hp). When it is necessary to convert from horsepower to watts, or vic versa, use the relationship

$$1 \text{ hp} = 746 \text{ W}$$

Theoretically, then, the rate at which electric energy is supplied to a 1-h motor is 746 W. In actual practice, however, the motor will consume mor than 746 W from the source. The difference between the theoretical an actual power requirements is consumed in overcoming such things a bearing friction and the air resistance of moving parts. Typical moto consumptions in watts are shown in Table 1–2.

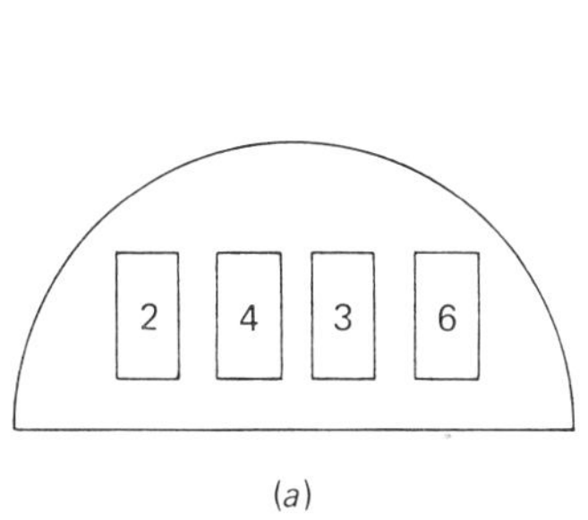

(a)

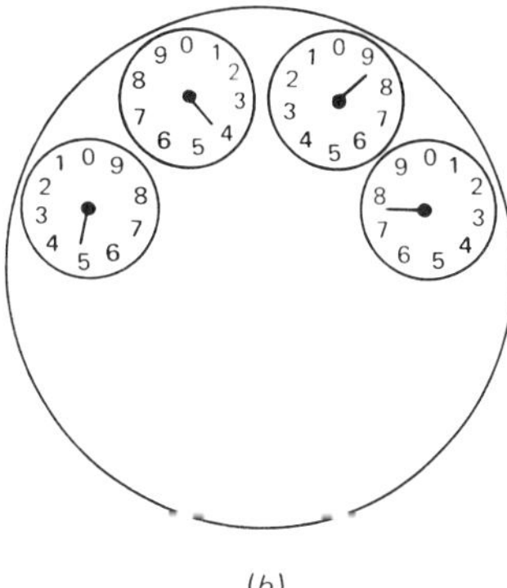

(b)

FIGURE 1–10. Kilowatthour meters with (a) cyclometer dial and (b) four dials.

last number the pointer has passed on each dial. On the first dial, the last number passed is 4. On the second dial the pointer rests directly on 4; however, until the pointer on the third dial passes zero the proper reading for the second dial is 3. The third dial reads 8 and the fourth dial 7. Thus the total indicated is 4387 kWh.

1–12 SERIES CIRCUITS

To work intelligently, an electrician must be able to make certain tests in any type of circuit. On the basis of these tests he must then decide if the circuit is operating properly or if a *fault* exists at some point in the circuit. If a fault exists, he must be able to find and correct the fault quickly. Fortunately, the distribution of voltage, current, and resistance in electric circuits is not random but follows definite patterns.

Distribution of *E*, *I*, and *R*

Let us first see how *E*, *I*, and *R* are distributed in a properly operating *series* circuit. An electric circuit in which all the component parts are connected *end to end*, as in Figure 1–11, is called a series circuit. Later we shall see what happens when a circuit fault upsets this distribution.

- In a complete series circuit, that is, one is which all switches and other control devices are closed, current has but *one* path to follow: from the negative terminal of the source, through each load device in turn, and back to the positive terminal of the source. Thus, *the same current exists in all parts of a series circuit;* accordingly, an ammeter connected to any point in the circuit will give the same reading.

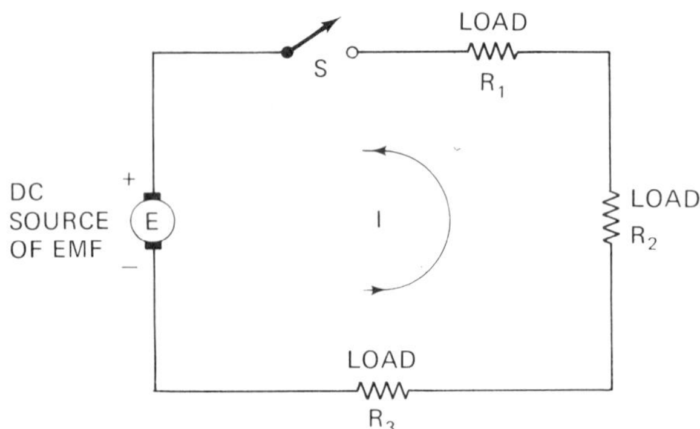

FIGURE 1–11. Series circuit.

- The component parts of a series circuit are connected end to end, and the same current passes through each part in turn. Since, with a given source voltage, the circuit current is determined solely by the resistance, it should be apparent that *the total resistance of the circuit is equal to the sum of the individual resistances*. Suppose, for example, that three load devices are connected in series with a source of EMF, and the resistances of these devices are 75, 200, and 165 Ω, respectively. Then the total resistance of the circuit is 75 + 200 + 165 = 440 Ω.
- In a series circuit, *the sum of the voltage drops across the individual load resistances must equal the applied EMF*. If, for example, 75-, 200-, and 165-Ω-load devices are connected in series with a 220-V source of EMF, the circuit current, by Ohm's law, will be $I = E/R = 220/440 = 0.5$ A. Then, since the same current passes through each load device in turn, the voltage drop across the devices ($V = IR$) will be 37.5, 100, and 82.5 V, respectively. The sum of these voltage drops is 220 V, the value of the applied EMF.
- If the value of any resistance is changed while the applied EMF is held constant, the values of the total resistance and the circuit current must also change. Thus *the change in resistance value of a single component in a series circuit results in a complete redistribution of voltage in the circuit*. To illustrate, we shall again use the 75-, 200-, and 165-Ω-load devices. Let us suppose, however, that for some unknown reason the resistance of the 200-Ω device changes to 100 Ω. Then the total resistance in the circuit will be 75 + 100 + 165 = 340 Ω, and the circuit current will be 220/340 = 0.647 A. The voltage drops across the individual devices become 48.5, 65.0, and 106.5 V, respectively. Notice that the sum of the drops is still equal to the applied EMF even though a redistribution of voltage has taken place in the circuit.

Series-Circuit Faults

Now let us look at common circuit faults and their effect on circuit operation.

Incorrect EMF. If the applied EMF is too *low*, the circuit current will also be abnormally low. Since $P = I^2R$, load devices will not develop the power required for proper operation. If the applied EMF is too *high*, the circuit current will be abnormally high. Thus the power-dissipation capability of one or more devices connected in the circuit may be exceeded, causing the device to *burn out*.

Wrong replacement part. If the resistance of a replacement part is lower than that of the original component, circuit current will increase. Since $P = I^2R$, one or more of the devices connected in the circuit may be burned out.

If the resistance of a replacement part is considerably higher than that of the original component, current will decrease. Again, since $P = I^2R$, the devices connected to the circuit may not be able to develop the power required for proper operation.

Finally, even if the resistance of the replacement part is correct, its power-dissipation capability must be *equal to* or *greater than* that of the original part.

To provide a *safety margin*, the wattage rating of any device should be higher than its maximum anticipated dissipation. To be on the safe side, always use exact replacement parts if the circuit operated properly before a fault occurred.

Open circuit. An open circuit, usually referred to as an *open*, is simply a break in the conducting path. The resistance of an open circuit is infinite for all practical purposes, since air replaces the conducting material. Air, you will recall, is an excellent insulator.

When an open occurs in a series circuit, there is no complete path through which current can pass. The current in all parts of the circuit is therefore zero. For example, if the circuit contains a string of series-connected lamps and the filament of one lamp burns out, all the lights are extinguished, since the circuit current is zero.

Short circuit. A short circuit, usually referred to as a *short*, occurs when a portion of the resistance normally encountered in an electric circuit is bypassed or short-circuited.

Consider what happens when two bare conductors come into contact

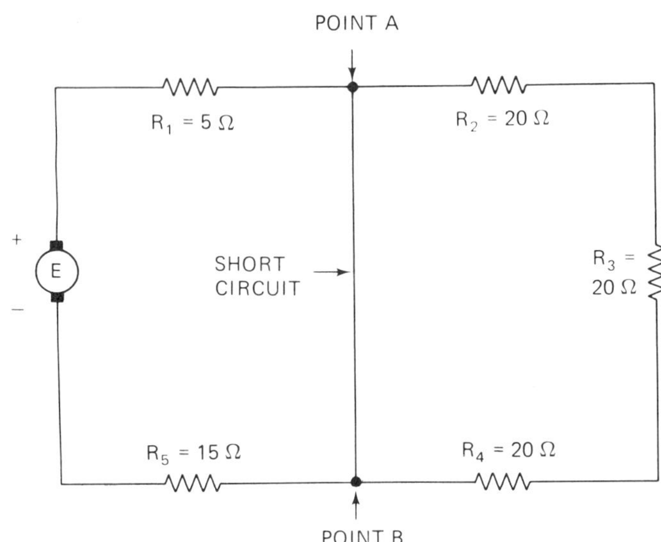

FIGURE 1–12. Short circuit.

with one another at points A and B, thereby forming a short. See Figure 1–12. The only resistance between points A and B, when this occurs, is the very small resistance of the wire. As a result, essentially the total *line* (circuit) current passes from point A to point B; the current through resistances R_2, R_3, and R_4 drops to zero—for all practical purposes. (Notice the use of numerical subscripts to distinguish one resistive load device from another.) The effective removal of R_2, R_3, and R_4 from the circuit causes a sharp increase in current. To illustrate, the normal circuit current is 1.5 A (E divided by the total resistance). When the short occurs, total resistance drops to 20 Ω ($R_1 + R_5$), and the current becomes 6 A.

Before the short occurs, the power dissipated by R_1 *and* R_5 is 11.25 and 33.75 W, respectively ($P = I^2R$). Under short-circuit conditions, R_1 dissipates 180 W and R_5 dissipates 540 W. Unless R_1 and R_5 are capable of handling such excessive amounts of power—which is extremely unlikely—these elements will be destroyed by the excessive heat. Once the devices represented by resistances R_1 and R_5 are burned out, an open circuit exists.

To summarize, in a *series* circuit, a short usually results in the destruction of those devices connected between the source of EMF and the point at which the short occurs. It should be understood, of course, that the source of EMF itself may also be either damaged or destroyed, depending on the nature of the source.

In view of the dangerous conditions that can result from a short circuit, extensive use is made of *protective devices* that open the main line automatically and disconnect the source of EMF from the circuit in which the short occurs. We shall examine these devices in Chapter 5.

1–13 PARALLEL CIRCUITS

A *parallel* circuit, such as that shown in Figure 1–13, is one in which the components are connected in a *side-by-side* manner and *more than one* path exists for the passage of current. Each current pathway is referred to as a *branch*.

Distribution of *E*, *I*, and *R*

Let us first see how E, I, and R are distributed in a properly operating parallel circuit, and then what happens when this distribution is upset by a circuit fault.

- Referring again to Figure 1–13, we notice that each of the three resistive load devices, R_1, R_2, and R_3, is connected directly across the terminals of the source of EMF. Thus *the same voltage* (that of the source of EMF) *exists across all branches of a parallel circuit.*
- Also notice in Figure 1–13 that a *separate* current passes through each branch of the parallel circuit. The value of each individual branch current is calculated from Ohm's law by using the familiar relationship $I = E/R$, where E is the source of EMF and R is the resistance of the branch.
- Finally, notice that the total current must pass from the negative terminal of the source of EMF to point A. At point A, the current divides, with a portion passing through branch R_1 and the remainder going to point B. At point B, the current again divides, with a portion passing through branch R_2 and the remainder through branch R_3. At point C, the

FIGURE 1–13. Parallel circuit.

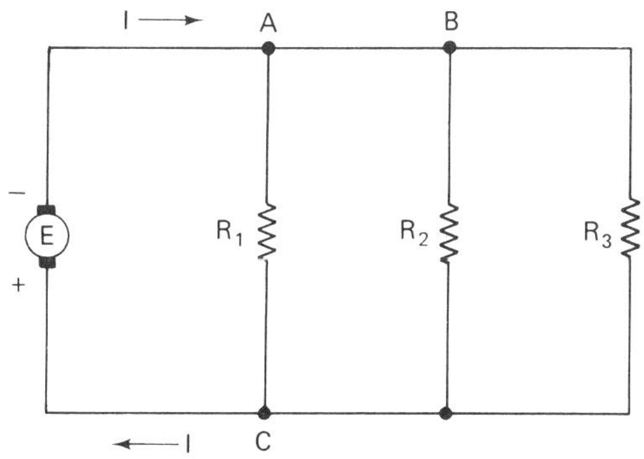

currents passing through all three branches recombine and return to the positive terminal of the source of EMF. From this we conclude that *the total current in a parallel circuit is equal to the sum of the individual branch currents.*

- In a *series* circuit, the equivalent resistance (that is, the total effective resistance) was found to be equal to the sum of the individual resistances. In a *parallel* circuit, a much different condition holds true. *Whenever two or more branches are connected in parallel, the total equivalent resistance of the circuit is always less than the value of resistance in the branch having the least resistance.*

EXAMPLE: If three branches are connected in parallel across a given source of EMF and the resistances of the branches are, say, 25, 50, and 75 Ω, the total effective resistance of the circuit, as far as the source is concerned, is *less* than 25 Ω.

With a little thought, this seemingly strange state of affairs comes as no great surprise. Because the full applied EMF is across each branch, more current is required from the source, and the equivalent resistance of the circuit must decrease to meet this demand for increased current.

Since the practical electrician is seldom, if ever, called upon to compute the total effective resistance of a parallel circuit, we shall not spend time in studying the equations needed for such computations. The effect of adding parallel branches should, however, be kept in mind when working with circuits of this type.

Parallel Circuit Faults

In a parallel circuit, the effects of an incorrect EMF or a wrong replacement part are essentially the same as in a series circuit.

Open circuit. In a parallel circuit, the *location* of the open determines what effect it has on circuit operation. In Figure 1–14, where six lamps are connected in three parallel branches, an open in the main line at point *A* has the same effect as in a series circuit, because the total current passes through point *A*. Thus an open at point *A* would cause *all* the lights to be extinguished. On the other hand, if the filament of any *one* lamp burned out and created an opening in *one* of the branches, both of the lamps located in that branch would be extinguished because the current in that branch would be zero. All the remaining lights would continue to glow, however, because each of the branches in which they are located would still be connected across the source of EMF and would continue to form a complete path for the passage of current. *An opening in any one branch*

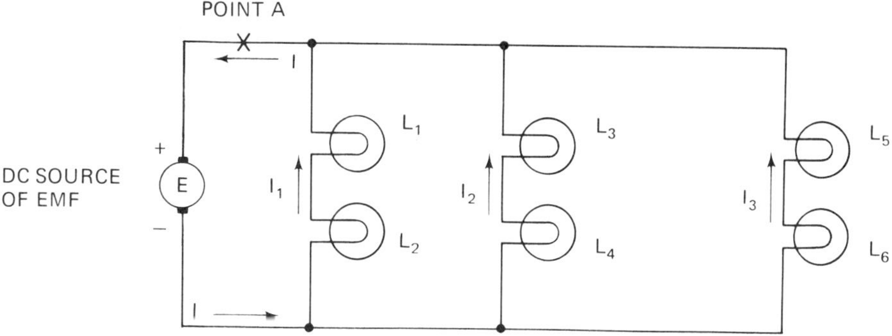

FIGURE 1–14. In a parallel circuit, the effect of an open depends upon its location.

has no effect on any other branch. The current in each remaining branch stays the same, because $I = E/R$ and neither E nor R changes. The total circuit current decreases, however, by an amount equal to the current that normally passes through the branch in which the open occurs.

Short circuit. In the parallel circuit of Figure 1–15, a short is assumed to exist between points A and B. Because the resistance of the short-circuit path is practically zero, essentially the entire circuit current passes through the source of EMF, through the conductors connecting points A and B to the source, and through the short circuit itself. Because the effective circuit resistance is practically zero, the current reaches a very high value of, perhaps, several hundred amperes, and it produces so much heat that the source, the conductors, or both may ignite and burn.

FIGURE 1–15. Short-circuited parallel circuit.

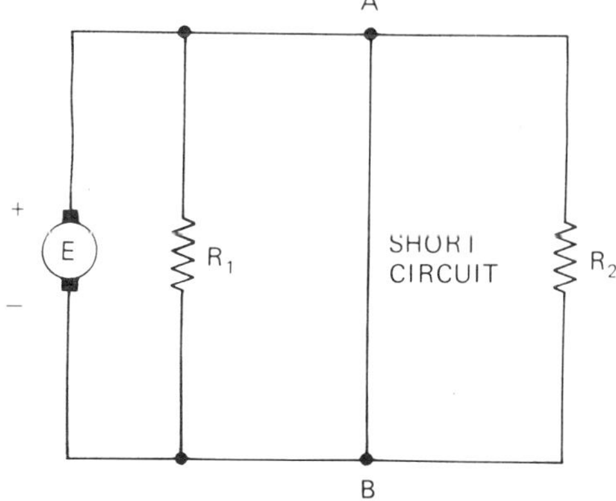

The devices represented by resistances R_1 and R_2 in Figure 1–15 are made inoperative as a result of the short circuit, since the current through these branches drops to essentially zero; however, the devices are not damaged and may be reused once the circuit fault is repaired.

To sum up, a short in a parallel circuit causes a dangerously high line current that may either damage or destroy the source. In any event, rapid and severe overheating of the source and conductors between the source and the short will occur, and a fire is very possible. The devices connected in the normal branches are not affected by the short, but they are made inoperative until the circuit fault is corrected.

1–14 SERIES–PARALLEL CIRCUITS

Series–parallel circuits are simply a combination of the series and parallel arrangements. A typical series–parallel circuit is shown in Figure 1–16. Since the total circuit current passes through R_1, the device represented by this resistance is in *series* with the source of EMF. At the junction of R_2 and R_5, the total current splits, with part of it passing through branch A and the remainder through branch B. Notice, however, that branch A contains three series-connected resistances, while branch B contains four series-connected resistances. When the two branch currents recombine, they flow through R_9. Thus R_9 is a *series*-connected conponent.

FIGURE 1–16. Series–parallel circuit.

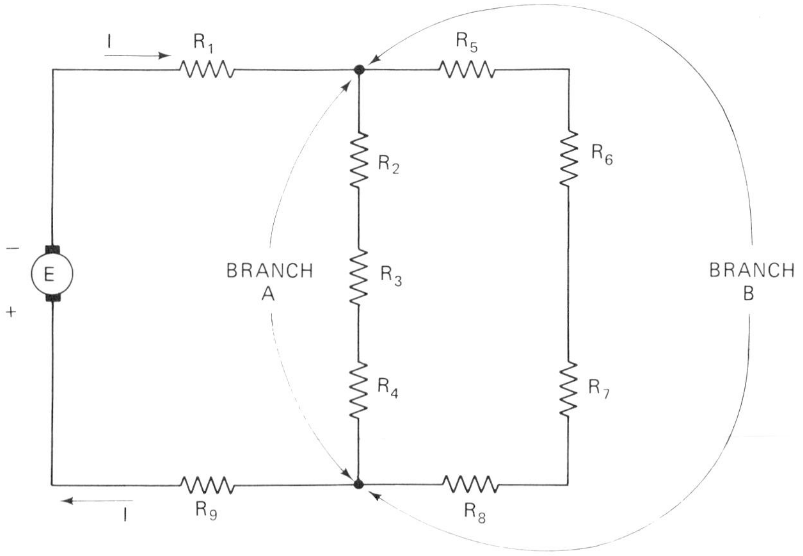

The same voltage is impressed across both the parallel branches. The value of this voltage is E minus the voltage drop occurring across resistances R_1 and R_9. If, for example, V_1 is 20 V, V_9 is 20 V, and E is 120 V, the voltage appearing across both branches is $120 - 40 = 80$ V.

Although, series–parallel circuits may appear to be somewhat involved, just remember that all the characteristics of series circuits apply for all series-connected components. Similarly, all the characteristics of parallel circuits apply when the components are in parallel.

1–15 POWER IN SERIES AND PARALLEL CIRCUITS

Whenever current passes through a resistance, a certain amount of power is dissipated. This power must come from the source; the circuit arrangement cannot change this requirement in any way. Thus, regardless of whether resistances are connected in series, parallel, or some combination thereof, the total power taken from the source is always equal to the sum of the individual powers dissipated by the circuit components. Since the total power requirement is supplied by the source by means of its total line current, the total power requirement is always equal to the product of E and I (total).

1–16 MAGNETISM

The ancients knew that special properties characterized *lodestones*, which were magnetic iron ores. Two of those magnetic forces that are utilized by modern technology follow: (1) When a piece of lodestone is suspended so that it can swing freely in a horizontal plane, it invariably assumes a definite position relative to the earth's north and south directions; (2) Lodestone specimens attract small fragments of iron.

The attractive force that a bar magnet exerts on iron filings is concentrated near its ends. These ends are called the *magnetic* poles of the bar magnet. When the bar is pivoted horizontally, the end pointing approximately toward the geographic north of the earth is the north-seeking pole or simply the *north pole* of the magnet and the other, south-seeking end, is the *south pole* of the magnet. Any fragment of a magnetic substance, no matter how small, has both a north and south pole. A single magnetic pole cannot exist.

If two compass needles are brought close to each other, the following law of magnetic poles is *seen: like poles repel; unlike poles attract*.

Only a few materials exhibit marked magnetic properties. Iron and steel and their alloys are strongly magnetic; on the other hand, cobalt and

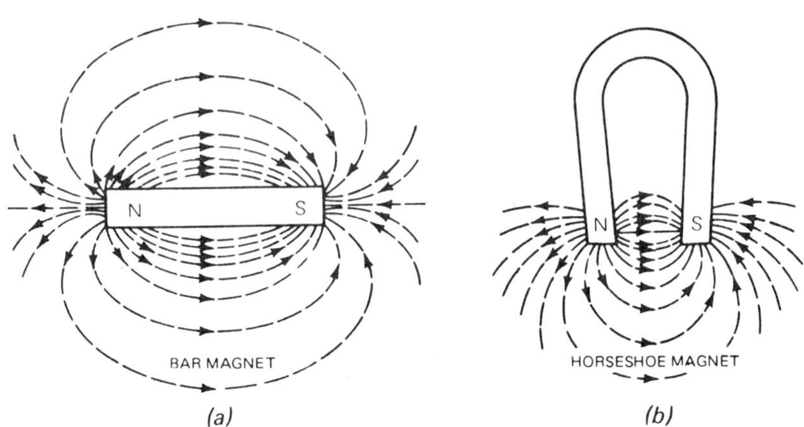

FIGURE 1–17. Magnetic fields of (*a*) bar and (*b*) horseshoe magnets.

nickel have very negligible magnetic properties in comparison with iron. Soft iron is more readily magnetized than steel, but steel retains its magnetism to a greater extent than soft iron. Consequently, magnets made of steel or steel alloys are good *permanent* magnets. Soft-iron magnets are good *temporary* magnets, but they either lose or reverse their polarity quickly. When a compass needle is brought near a magnet, a force acts on the needle and the needle assumes a definite direction with respect to the magnet. The region about a magnet in which this force is present is called a *magnetic field*; it is made up of *magnetic lines of force*, also called *flux*.

Since the compass needle assumes a particular position when placed near a magnet, a magnetic field has *direction*. The direction of a magnetic field at a certain point is defined as the direction indicated by the north pole of a compass needle when placed at this point.

It is convenient to represent the direction and intensity of a magnetic field by lines of force. The *intensity* of a field is represented by the density of the lines and the *direction* by arrowheads on the lines. See Figure 1–17. All lines of force are *closed* paths directed from the north pole to the south pole in the region *outside* the magnet and returning *inside* the magnet from south to north. The field existing between two like poles (which repel) is shown in Figure 1–18.

1–17 MAGNETIC FIELDS AROUND CONDUCTORS

If a *straight* conductor carrying a steady current is brought near a compass needle, the needle is deflected. There is no deflection if current does not flow. Therefore, a magnetic field exists about the conductor only when a

FIGURE 1–18. Field existing between two like poles.

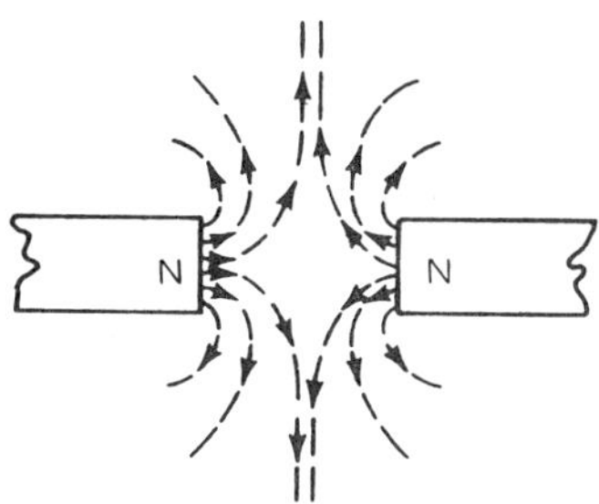

current flows through it. By moving the compass needle around the conductor and noting the direction in which the north pole points, the lines of force of the field are observed to form concentric circuits about the conductor. See Figure 1–19(a). If the direction of the current is reversed, the lines of force reverse, too. See Figure 1–19(b).

The following *left-hand rule* relates the direction of the lines of force to the direction of current. *If the conductor is grasped in the left hand with the thumb pointing in the direction of electron flow, the fingers point in the direction of the magnetic field around the conductor.*

The intensity of the field is greatest close to the conductor and rapidly decreases with distance from the conductor. The intensity of the field is increased if the current strength is increased.

The intensity of the magnetic field about a straight conductor is very small, even with large currents. When, however, the conductor is bent in the form of a loop, as in Figure 1–20, a much stronger field is produced inside the loop, since the lines of force established in each small section of the conductor combine as they flow in the same direction in the center of the loop.

If the conductor is coiled into a number of turns as in Figure 1–21,

FIGURE 1–19. Magnetic field existing around a straight conductor carrying a steady current: (*a*) current moving left to right; (*b*) current moving right to left.

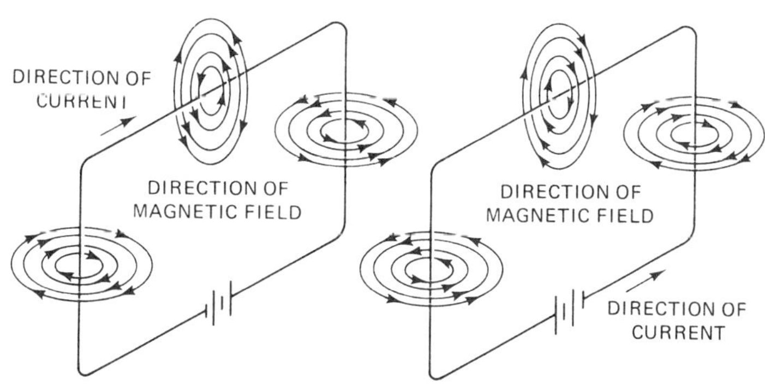

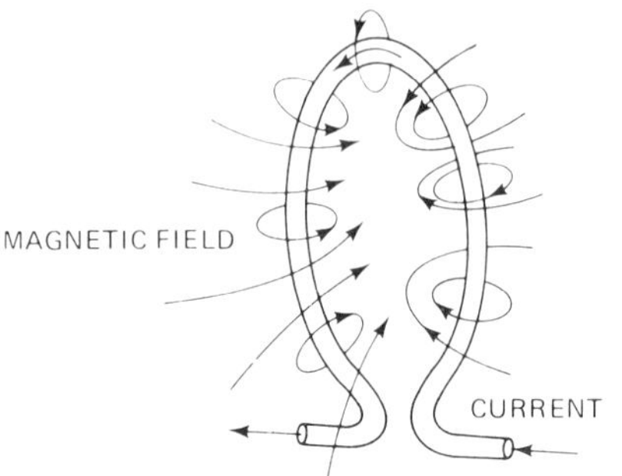

FIGURE 1–20. Magnetic field around a conductor bent in a loop.

the lines of force within each loop combine to produce a still larger field. Such a coil or helix, with its length great as compared to its diameter, is called a *solenoid*. The diagram of Figure 1–21 shows that the field about a solenoid is like the field about a bar magnet. The end of the coil from which the lines emerge acts as a north pole, and the end into which they enter acts as a south pole. The following rule relates current direction in the solenoid to the polarity of its ends. *If the conductor is grasped in the left hand with the fingers pointing in the direction of electron flow, the outstretched thumb will point to the north pole of the magnetic field around the coil.*

If a solenoid is supplied with a core of magnetic material, its magnetic field is strengthened. The combination of a coil and its magnetic core is called an *electromagnet*. The fact that the electromagnet is a temporary magnet is a distinct advantage. That is, it is surrounded by a magnetic field only as long as current flows through its turns. This permits *on–off* control of the magnetic field merely by the flick of a switch.

If a solenoid is provided with a movable soft-iron core, called a *plunger*, when a current is made to flow through the turns of the coil, the magnetic field tends to pull the plunger into the center of the coil. The coil together with its movable core is usually called a *solenoid* (although, strictly speaking, only the coil is the solenoid). A solenoid can also be

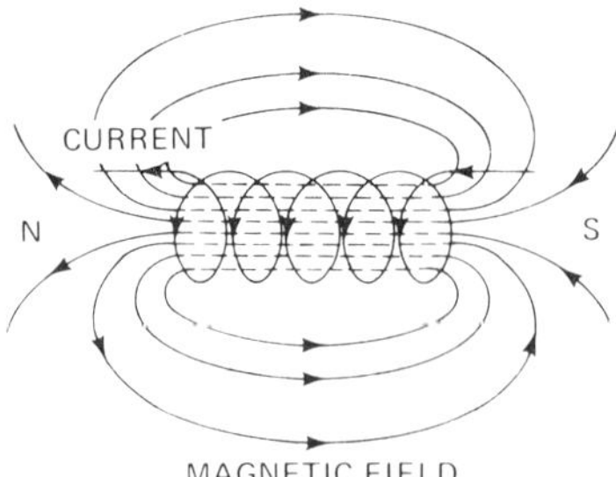

FIGURE 1–21. Magnetic field around a solenoid.

arranged so that its plunger is pushed out of the center of the coil when current through the coil is discontinued.

1–18 ELECTROMAGNETIC INDUCTION

Since a magnetic field is always associated with an electric current, it seems logical that a current is always associated with a magnetic field. The following experiments show that this is true, provided certain conditions are fulfilled.

A coil of wire is connected to a galvanometer, which is a very sensitive current-measuring instrument with zero at center scale. Then a magnet is moved into and out of the coil as indicated in Figure 1–22. The following data can be observed: (1) as the magnet moves into and out of the coil, a current flows, as indicated by the galvanometer's recordings; (2) when the magnet is at rest there is no current; (3) the more rapidly the magnet is moved, the greater the strength of the current measured by the galvanometer; (4) if an unmagnetized bar of iron is moved in and out of the coil, there is no current.

The current flow that is observed must be due to an EMF. This EMF is produced in the coil *only* when the *moving* magnetic field of the magnet is cutting across the turns of the coil. Furthermore, the strength of this EMF depends on the *speed* of the moving field. An EMF that is set up in a conductor when it is cut by a moving field is called an *induced* EMF, and the process is known as *electromagnetic induction*. If the magnetic field is held in a fixed position and the coil moved up and down over the magnet, the same observations can be made.

The results of the experiments described above are summed up in the two laws of induction known as Faraday's laws: (1) *When the magnetic flux (lines of force) linked with a circuit changes, an EMF is induced in the circuit.* (2) *The strength of the induced EMF is proportional to the rate of change of flux linkage.*

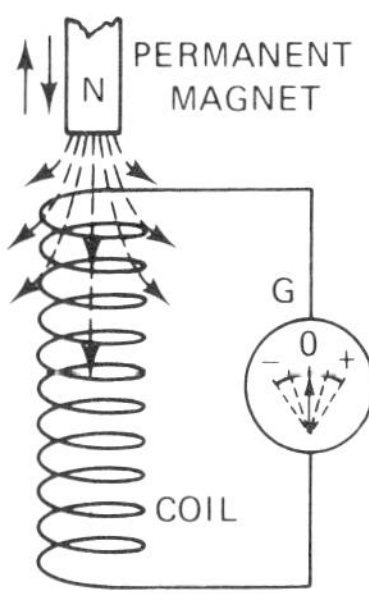

FIGURE 1–22. Electromagnetic induction.

In dealing with electromagnetic induction caused by a changing current rather than by mechanical motion, Faraday's second law may be restated as follows: *The magnitude (strength) of the induced EMF is directly proportional to the time rate of change of current.*

In Figure 1–23, coil A is placed within coil B with no electrical connection between the two coils. A galvanometer is connected in series with coil B. When switch S is closed, current I_A flows through coil A in a counterclockwise direction. At the same time, the galvanometer indicates a current, I_B, through coil B. This induced current lasts only a very short time and dies out quickly, even though current I_A is maintained steady by the battery.

When switch S is opened, the galvanometer again shows an induced current, but now the direction of the current is reversed; that is, I_B now flows in the same direction as I_A. As before, current I_B dies out rapidly.

Notice that the induced current, I_B, exists when, and only when, there is a change in current I_A. This phenomenon can be further verified experimentally by inserting an appropriate ammeter in series with coil A so that it can record the fact that I_A builds up gradually to a final steady value, and that I_B flows only during this buildup of I_A. Similarly, when switch S is opened, I_A does not die out immediately, but gradually, and I_B flows only during this time.

Also note that when switch S is closed, the direction of I_B is *opposite* to the direction of I_A. When the switch is open, the direction of I_B is the *same* as that of I_A.

In both cases the induced currents *oppose the change* that takes place in I_A. First, current I_B opposes the setting up of I_A, and then it opposes the stopping of I_A. As soon as the change in I_A is over, I_B dies out.

The behavior of the induced current can be best understood in terms of the magnetic fields set up by the currents flowing in the coils. A little reflection shows that, when I_A begins, it sets up a north magnetic pole at the top of coil A. Current I_B opposes this action by setting up a south

FIGURE 1–23. Demonstrating Lenz's law.

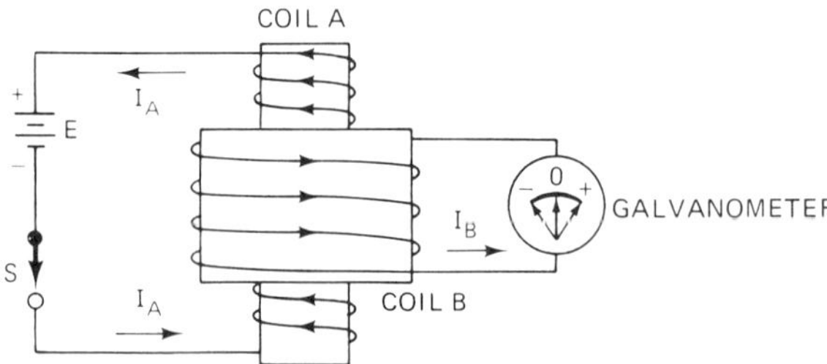

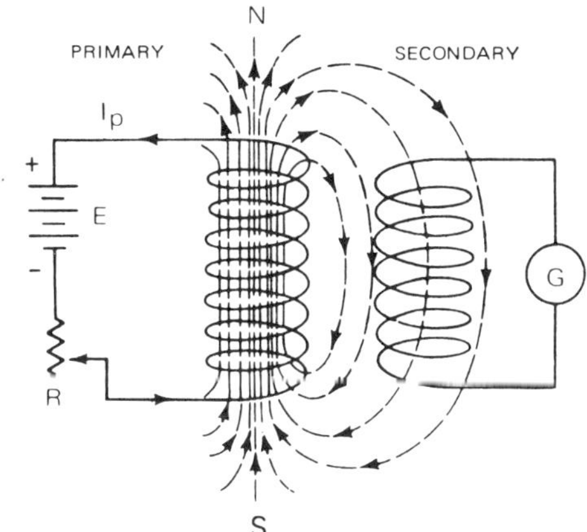

FIGURE 1–24. Mutual induction.

magnetic pole at the top of coil B, tending to neutralize the field set up by I_B. Once the field is set up by I_A, current I_B ceases to flow. When switch S is opened, the field set up by I_A begins to decay, and I_B once again produces a field that opposes this decay. In each case, I_B sets up a field that opposes the change taking place within the cores of the two coils as a result of the change in current I_A.

This phenomenon has come to be known as Lenz's law, which may be generalized as follows: *An induced current always flows in such a direction that the magnetic field it produces opposes any change in the existing field produced by the inducing current.*

The important concept in this generalization is that of *change*. The field of the induced current always opposes any change taking place in an existing field.

1–19 MUTUAL INDUCTION AND MUTUAL INDUCTANCE

In Figure 1–24, two coils are placed near each other. The *primary* coil is connected in series with a dc source of EMF and a *rheostat*. In electric circuits, resistance is used to control (limit) the amount of current. Although we could use high-resistance conductors to place the necessary resistance in the circuit, this would not be a very practical solution. Instead, the extra resistance is lumped into a single component called a *resistor*. Remember that resistance is a *property* of a circuit, while a resistor is a *component* used in a circuit. Since resistance is used to control current, heat is generated when current passes through a resistor; therefore, resistors are made in various sizes and shapes so that they can safely dissipate a certain amount of heat. The correct size must be selected for a given application.

A *rheostat* is simply a special form of resistor. One connection is made to an end of the rheostat, while the other connection is made to a movable contact located inside the rheostat casing. As the arm (wiper) moves across the resistance material, the value of the resistance is varied. Thus a rheostat is simply a *variable resistor*.

If connections are made to both ends and to the wiper of a variable resistor, the device is called a *potentiometer*.

Returning to Figure 1–24, the *secondary* coil has a galvanometer connected across its terminals. If the primary current is held constant, the magnetic field about the primary coil is constant. The magnetic lines about the primary coil link the secondary coil. Since the secondary coil is not being cut by a moving field, however, there is no EMF induced, and there is no secondary current.

If the primary current is varied by means of the rheostat, the field of the primary changes. As the primary current increases, the field increases; the magnetic lines spread outward across the secondary coil, and an EMF is induced in the secondary circuit. As the primary current decreases, the field decreases; the magnetic lines collapse inward across the secondary coil, and an EMF is induced in the secondary circuit. If the rate of change of the primary current is made greater, more lines cut across the secondary coil per second and the induced EMF is greater. These observations confirm Faraday's first and second laws.

The process by which an EMF is induced in a secondary winding by a changing current in a primary winding is called mutual induction, and the coils are said to possess mutual inductance.

Before we discuss the factors that determine the mutual inductance of two coils, let us first define *permeability*. By definition, *permeability is a figure that indicates the ability of a material to permit the setting-up of magnetic lines of force*.

The permeability of a *vacuum*—and, for practical purposes, air and all other nonmagnetic substances—is rated as 1. Thus permeability is the ratio between the magnetic flux produced in a magnetic material and the flux that would be produced if air were used instead, with the magnetizing force remaining constant.

Not all magnetic substances have the same permeability. For example, the permeability of soft iron is much greater than that of steel, and there are magnetic alloys that have a much greater permeability than soft iron.

The mutual inductance of two coils depends on the following interacting factors: (1) *construction*—mutual inductance increases with each increase in the number of turns per unit length (by the inch, for example) in each coil and also with any increase in the cross-sectional area or thickness of each coil; (2) *permeability*—an increase in the permeability of the cores or of the medium about the coils increases mutual inductance; (3) *relative position* of the two coils—the greater the distance between the coils, the less the mutual inductance; conversely, lesser distance means improved mutual inductance. Also, mutual inductance is greatest when the two coils are coaxial, and least when their axes are at right angles to each other.

1–20 SELF-INDUCTION AND SELF-INDUCTANCE

When a changing current passes through a conductor, a magnetic field grows and collapses through and about the conductor. Since the conductor is cut by its own changing field, we know from Faraday's law that an EMF must be induced in the conductor. This process is called *self-induction*.

According to the further developments noted in Lenz's law, the induced EMF must act in a direction that opposes an applied EMF. Hence the induced EMF is called a *counter EMF* (CEMF).

When the CEMF opposes the applied EMF, it must also oppose any change in current. *The property of an electric circuit that opposes any change in current in that circuit is called self-inductance or, simply, inductance.*

It must be realized that any conductor, whether in the form of a straight wire or a coil, possesses inductance. It must also be realized that the opposition offered to any change in current by inductance is a temporary condition. Thus inductance must never be confused with resistance. When the current eventually reaches a steady value, the lines of force around a conductor are no longer expanding or collapsing, and a CEMF is no longer present. Thus inductance may be thought of as a *temporary* opposition to establishment of the maximum value of current.

The inductance of a coil is increased substantially by increasing the number of turns on the coil and also by increasing the core's permeability.

1–21 PRECAUTIONS IN DC INDUCTIVE CIRCUITS

In the circuit diagrammed in Figure 1–25(a), a dc source of EMF, a switch, an iron-core coil (notice the symbol), and a resistance are connected in series. The resistance is that of the coil itself. When the switch is closed,

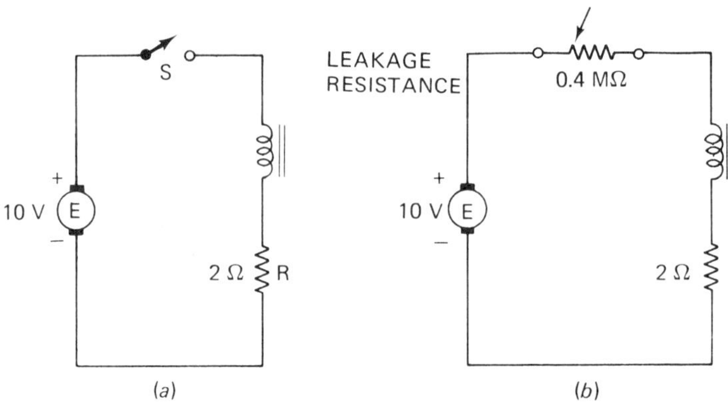

FIGURE 1–25. A dc inductive circuit: (*a*) switch open; (*b*) leakage resistance of open switch.

$$I = \frac{E}{R} = \frac{10}{2} = 5 \text{ A}$$

Next we see in Figure 1–25(b) that an open switch represents a very high value of resistance. We call this resistance *leakage resistance*. Let us assume that the leakage resistance in this particular case is 0.4 MΩ. When the switch is opened, the 5-A current that is still in the coil must pass through the leakage resistance of the switch. The voltage drop across the leakage resistance and, therefore, across the switch contacts is

$$\text{V} = IR = 5 \times 400{,}000 = 2{,}000{,}000 \text{ V}$$

This *extremely high* voltage, which is equal to many times the value of the EMF that is applied when the switch is closed, *is matched by the CEMF generated in the inductor by the collapsing magnetic field*. Obviously, voltages of this magnitude are dangerous. Furthermore, voltages of this magnitude will cause a breakdown in the insulation of the coil and a severe arcing across the switch.

To avoid the creation of such high voltages when switch contacts are opened in dc circuits that contain a coil, a *discharge device* (such as the *Thyrite* resistor) is connected across the coil. See Figure 1–26.

When the switch in Figure 1–26 is *closed*, a current of 1 A passes through the 10-Ω discharge resistor and the steady current through the branch containing the iron-core coil is 5 A. When the switch is *opened*, the 5-A current through the coil must pass through the discharge resistor and create a voltage drop of 60 V. (The total resistance of the circuit with the switch open is 2 + 10 = 12 Ω, and I = 5 A.)

The direction of current produced by the CEMF is the same as the original current through the coil. Thus the upper end of the discharge resistor becomes positive *with respect to* its lower end, and the voltage drop across the discharge path and the 10-V source of EMF are in *series* across the switch contacts. The total voltage drop across the switch contacts is 60 + 10 = 70 V. This is quite a reduction from the voltage appearing across the switch contacts when a discharge resistor was not used. The insulation of the coil is no longer subjected to abnormally high voltages, and there is no arcing across the open switch.

EXERCISES

1–1. Describe a typical application in which energy is converted from one form to another.

1–2. What happens when two *like* charges are brought close to each other?

1–3. What happens when two *unlike* charges are brought close to each other?

1–4. What do we mean when we say *polarity*?

1–5. What is the *net* electrical charge of a normal atom?

1–6. What is a *free* electron?

1–7. What is a *conductor*? An *insulator*?

1–8. Name the essential parts of any practical electric circuit.

1–9. What is *current*?

1–10. Why is a source of electricity sometimes referred to as a *potential rise*?

1–11. What is the difference between the letter symbols *E* and V?

1–12. How is a current-measuring instrument connected into a circuit?

1–13. How is a voltmeter connected into a circuit?

FIGURE 1–26. Use of a discharge resistor in a dc inductive circuit.

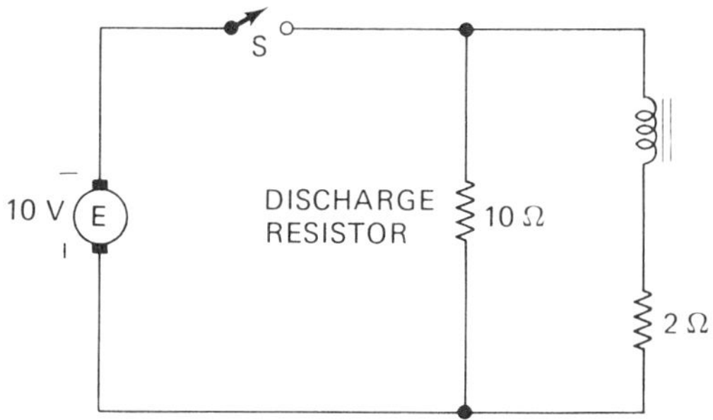

1–14. Define *resistance*.

1–15. What are the three basic forms of Ohm's law?

1–16. What factors determine the *resistance* of a metallic conductor?

1–17. Can a resistor of a given physical size be used in any electric circuit where its resistance value is correct? Explain your answer.

1–18. Define electric *power*.

1–19. Give three equations for computing power.

1–20. How is an *ohmmeter* connected into a circuit?

1–21. What is a *series* circuit? A *parallel* circuit?

1–22. What are the important characterisitcs of a *series* circuit?

1–23. What are the important characteristics of a *parallel* circuit?

1–24. How is a series circuit affected by an *open* circuit? A *short* circuit?

1–25. Define the terms north pole and south pole as applied to a magnet. State the law of magnetic poles and describe an experiment to demonstrate it.

1–26. Describe the magnetic field about a horseshoe magnet.

1–27. Describe the magnetic field between two like poles.

1–28. Describe the magnetic field around a straight conductor.

1–29. What effect is produced when current is passed through a solenoid?

1–30. The current through a coil is in a counterclockwise direction, as viewed from one end. What is the polarity of that end, and how is it determined?

1–31. What factors determine the intensity of the magnetic field surrounding a coil?

1–32. What is *permeability*?

1–33. What is an *electromagnet*? What are its advantages compared to a permanent magnet?

1–34. State Faraday's laws and explain their significance briefly.

1–35. State Lenz's law and explain its significance briefly.

1–36. Define *mutual* inductance.

1–37. Define *self-inductance*.

1–38. Distinguish between mutual induction and self-induction.

1–39. Upon what factors does mutual inductance depend? Self-inductance?

1–40. What precautions should be taken in a dc circuit containing one or more coils?

2

Alternating Current

2–1 INTRODUCTION

Recall from Section 1–6 that a current passing through a metallic conductor is the movement of free electrons under the influence of an applied EMF. Regardless of the polarity of the source of EMF, those free electrons always flow from negative to positive. Thus, if the source maintains a *fixed* polarity, the current always moves in the same direction and is called *direct current* (dc). If the source *reverses* its polarity periodically, however, the current also reverses direction at the corresponding periods, because it must maintain the negative-to-positive convention. A current of this type is called *alternating current* (ac).

For a variety of reasons, the power distribution systems of most, but not all, electric utility companies in the United States use alternating current. Let us take just a few moments to see why.

In the transmission of electric power from the generating station to the consumer, some waste (in the form of heat) occurs since even large-diameter conductors offer resistance to the passage of current. Remember that $P = I^2R$. To reduce the power loss, the utility companies can either lower the resistance of the transmission lines by using thicker conductors, or they can reduce the current passing through the conductors. If the resistance is lowered by half, the power loss is halved; but if the current is halved, the power loss is reduced by a factor of 4. Thus a current reduction is the best way to reduce transmission-line losses. At the consumer end, however, large amounts of current are needed. Fortunately, both the low transmission-line current and high consumer-end current can be realized by using a device known as a *transformer*. Transformers are examined in detail in Section 2–5. For the moment, let it suffice that you understand that these devices permit either the *stepping up* or *stepping down* of ac voltages and currents. Since the amount of power involved remains fairly constant (assuming minimum losses) and since $P = EI$, when the voltage is stepped up, the current is stepped down, and vice versa.

In a typical distribution system, power is generated initially at 13,800 V. By means of transformers, this voltage is then stepped up to anywhere between 23,000 and 765,000 V, depending on various factors, including the amount of power involved and the distance over which it must be transmitted. As noted earlier, when voltage is stepped up, current is stepped down. Thus current is reduced in the transmission line, and losses are minimized by the systematic use of transformers at various points throughout the transmission system.

At an area substation, another transformer is used to step down the extremely high transmission voltage to some lower value, such as 13,800 V. With the voltage back to its initial value, so also is the current—at least, for our purposes.

At the same substation additional transformers are often used for another step-down, either to 2300 or 4000 V. In this way, the current is increased beyond its initial value at the generating station.

Finally, in the immediate area of the residential power consumer, the voltage is further stepped down to 120/240 V, at which value it is used. The method of obtaining those two voltages, 120 and 240 V, from a single source is explained in detail in Section 2–5.

By means of transformers the utility company generates both the high voltage and relatively low current needed to reduce transmission-line losses and the relatively low voltage and high current needed by the residential consumer.

2–2 ALTERNATING CURRENT

Alternating-current power sources are called either *ac generators* or *alternators*, and combine physical motion and magnetism to produce ac voltage. In Section 1–18 you learned that when a conductor passes through a magnetic field and cuts the lines of force (flux) an EMF is induced in the conductor. This is the basic prinicple on which an alternator operates.

The simplest form of alternator is shown in Figure 2–1. To make it, a single loop of wire is placed between the poles of a permanent magnet and then made to rotate by some suitable mechanical device. As this loop rotates, it cuts the magnetic lines of force and an EMF is induced in the loop. The EMF produced exists between the two ends of the loop. *Slip rings* (one attached to each end of the loop) and *brushes*, in physical contact with the slip rings, apply the generated EMF to an external circuit.

Refer to Figure 2–2. When the loop is parallel to the magnetic lines of force, no flux is cut and no EMF is induced in the loop. As the loop begins to rotate, however, it cuts the magnetic flux at an increasing rate; therefore, the induced EMF becomes steadily larger. When the loop has rotated one-quarter turn (90°) from its starting position, the maximum EMF is induced in the loop, because it is now at right angles to the magnetic field. That is, it is cutting the maximum number of lines of force.

As rotation continues past the one-quarter point, the induced EMF is still in the *same* direction, but it is decreasing in value. At the half-revolution point (180°), the loop is again parallel to the magnetic field, and the induced EMF is zero. The loop has now made a one-half turn, during which the induced EMF gradually increased to its maximum and then gradually fell to zero. *Since the sides of the loop have now reversed position, the induced EMF reverses polarity.* The EMF, however, again increases to a maximum at the three-quarter point (270°) of rotation; then it declines to zero when the rotation is completed (360°). The action described is repeated during each rotation.

The *wave form* of the induced EMF is often referred to either as a *sine wave* or a *sinusoidal wave*. An alternating current has the same

FIGURE 2–1. The simplest form of alternator.

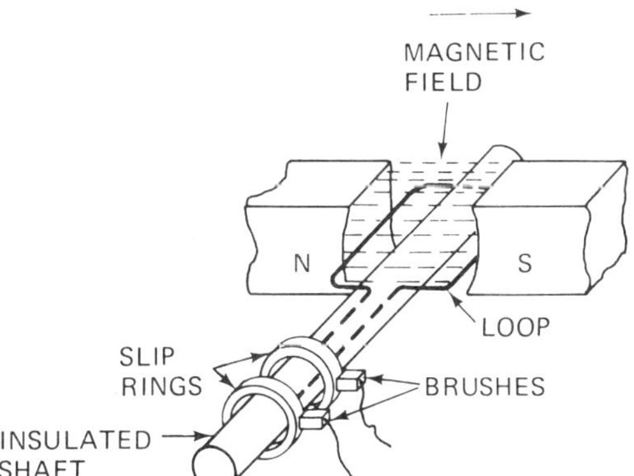

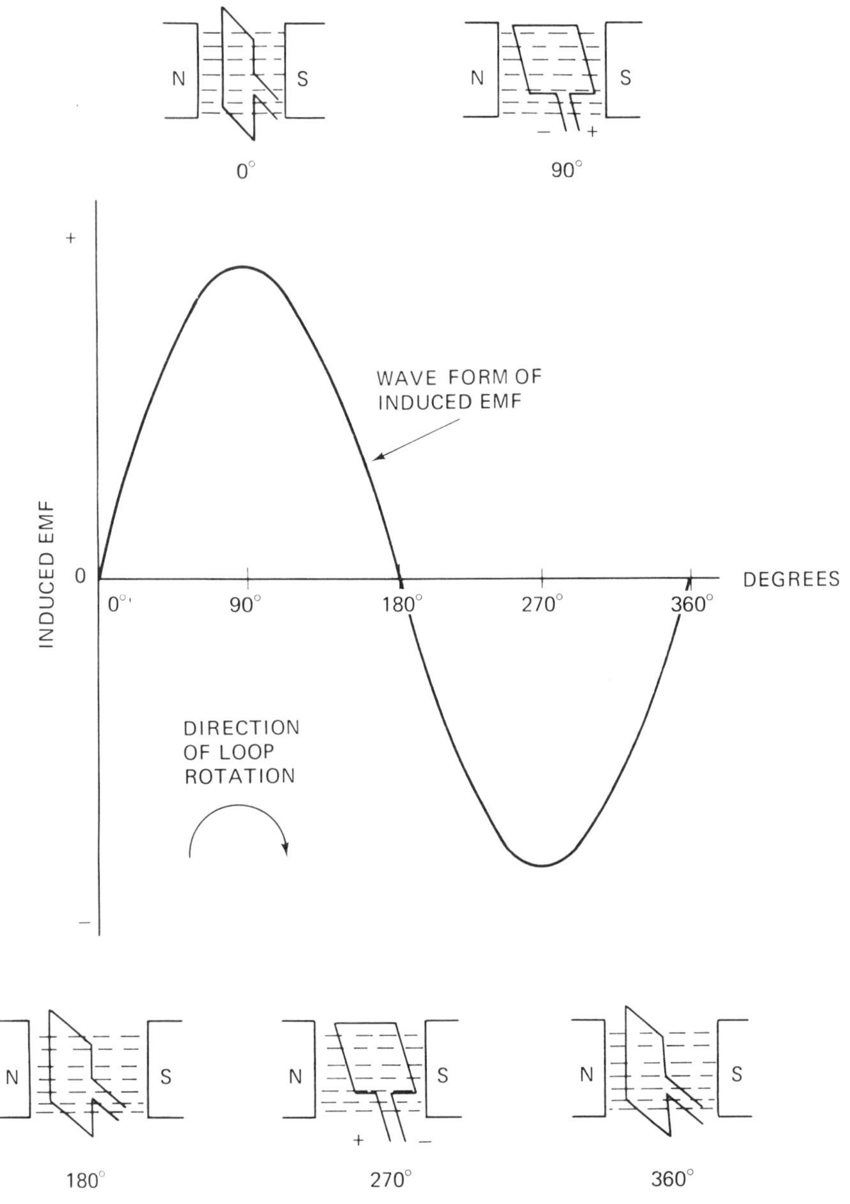

FIGURE 2–2. Generation of a sine-wave voltage.

general wave form. Notice that the portion of the wave form *above* the horizontal axis is considered to be *positive,* while that lying *below* the horizontal axis is considered to be *negative.*

Each complete revolution of the loop is one *cycle.* The number of cycles completed in one second (c/s), now ordinarily designated as hertz

(Hz), identifies the *frequency* of the voltage or current. The term honors the memory of Heinrich Hertz, an early pioneer in the study of electrical energy.

Most electric power generated in the United States has a frequency of 60 Hz, but there is still a limited use of 50 Hz. All appliances, both small and heavy, are marked for use at 50 to 60 Hz. Remember that operation at frequencies outside this range may either damage or destroy an appliance or make it operate improperly.

Although the wave form in Figure 2–2 is plotted on a horizontal axis marked in *degrees*, you will also find many wave-form representations where the horizontal axis is marked in terms of *elapsed time in seconds*.

If two simple alternators start at precisely the same instant and rotate in the same direction at exactly the same speed, wave forms representing the EMFs induced in their loops would start and end at the same points on a common graph, such as that shown in Figure 2–3. In addition, the maximum positive, the maximum negative, and the zero points would all correspond in time. The voltages or currents that the wave shapes represent are then said to be *in phase*. Thus the term *phase* indicates the *time* relationship between alternating voltages and/or currents. Notice, in particular, that the relative amplitudes or sizes of the wave forms are of no consequence in determining phase.

If the same two alternators start their rotation at different times or in different directions, or if their speeds of rotation are not identical, the wave shapes of the induced EMFs will be *out of phase*. The *phase difference* between the two wave forms is expressed in *degrees*. Figures 2–4(a), (b), and (c) show wave forms having phase differences of 90, 180, and 270°, respectively.

The terms *lead* and *lag* are used to describe the relative positions in time of voltages or currents that are out of phase. The quantity that is ahead in time leads; the one behind lags.

FIGURE 2–3. Two in-phase wave shapes.

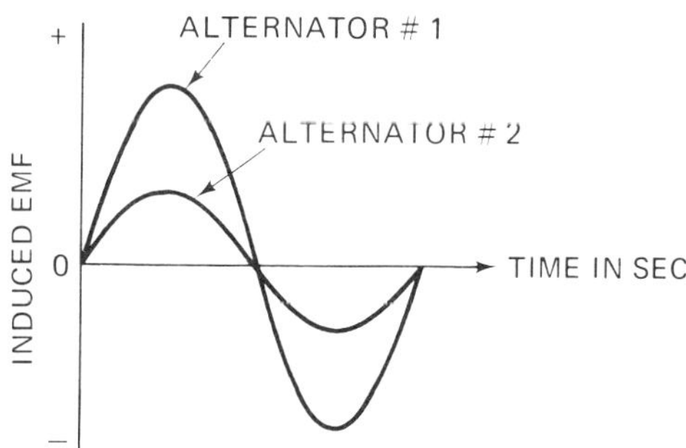

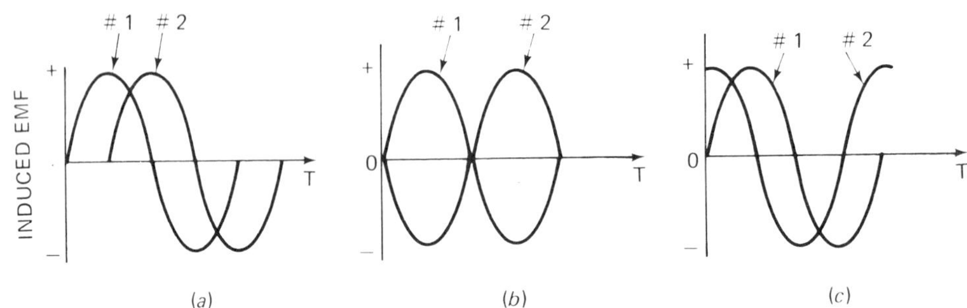

FIGURE 2–4. Wave forms with phase differences of (*a*) 90°, (*b*) 180°, and (*c*) 270°.

2–3 THE EFFECTIVE VALUE OF AC

As you know, direct current remains at a constant level at all times. Thus, when a direct current of, say, 1 A passes through a fixed resistance, it produces a definite amount of heat ($P = I^2R$). An alternating current having a *maximum (peak)* value of 1 A in passing through the same resistance would not produce the same amount of heat, however, since its value would be 1 A for only a very small part of the cycle and less than 1 A during the rest of the cycle. To compare direct and alternating current in terms of *heating effect,* alternating current is expressed in terms of its *effective* value. By definition, *the effective value of an alternating current or voltage is that value of the quantity which produces the same heating effect in a resistance as an equal value of direct current.* Mathematically, it can be shown that the effective value of an alternating current or voltage is equal to 0.707 or 70.7 percent of the peak (maximum positive or negative) value.

The 120/240 V ac found in residences is an effective value; that is, it will produce just as much heat in a resistance as 120/240 V dc. Moreover, the scale of instruments that you may use to measure ac voltage or current will have at least one scale calibrated in effective values.

Sometimes, the abbreviation rms, meaning *root mean square* and indicating the procedure for its mathematical derivation, is used in place of the term "effective." Just remember that rms and effective are one and the same thing.

2–4 POLYPHASE AC

The simple alternator that you considered in Section 2–2 contained a single loop that rotated in a magnetic field produced by a permanent

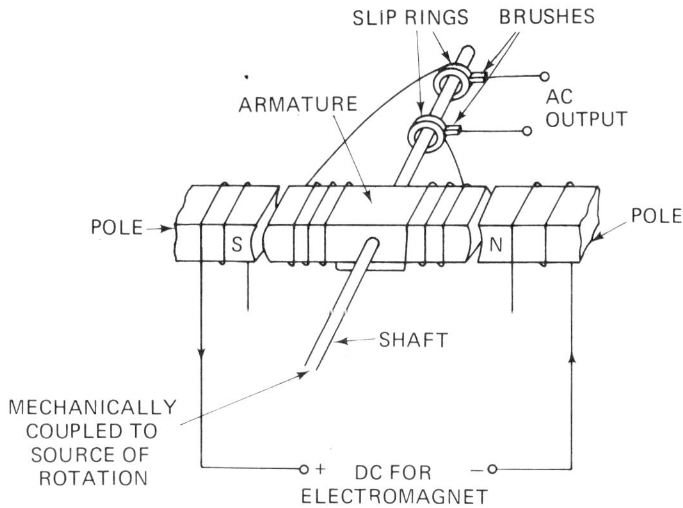

FIGURE 2–5. Simplified illustration of a practical alternator.

magnet. In practical alternators, that magnetic field is produced by an electromagnet. Since a steady magnetic field is required, direct current is sent through the electromagnet. In place of the single loop, many turns of insulated wire are wound on an iron core. The whole assembly is referred to as the *armature*. See Figure 2–5.

Suppose that two separate armature coils are wound upon the same core. If the coils are wound one over the other, are electrically separate, and each has its own set of slip rings and brushes, alternating current will be induced in each coil as the armature rotates. Because the coils are wound over each other and rotate together, their induced voltages will be in phase.

Now suppose that, instead of having the coils wound one over the other, they are wound at right angles to each other, as shown in Figure 2–6. At the instant armature coil *A* is cutting the maximum lines of force and has its maximum induced EMF, coil *B* is not cutting any lines and its induced voltage is zero. Because the two coils are at right angles to each

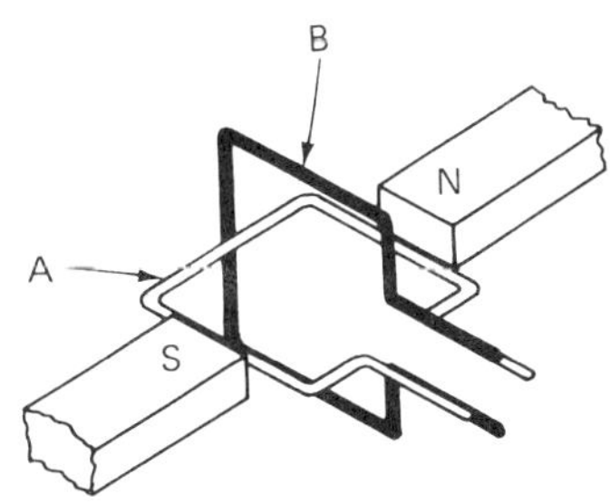

FIGURE 2–6. Two armature coils at right angles to each other.

other (90° apart), the phase difference between their induced EMFs is 90°.

Going one step further, if three coils, each spaced 120° from its neighbor, are wound on the same core, the phase difference between the three induced voltages is 120°. These relationships are shown graphically in Figure 2–7. Figure 2–7(a) shows the graph of the ac voltage produced by the alternator with a single armature coil. Such an alternator is called a *single-phase* alternator. Figure 2–7(b) shows the graph of the voltages produced by a *two-phase* alternator in which the two windings are spaced 90° apart. The EMF of phase *B* starts 90° behind that of phase *A* and this time relationship is maintained at all times. Figure 2–7(c) shows the graph

FIGURE 2–7. Alternators and the wave forms of their induced EMFs: (*a*) single phase; (*b*) two phase; (*c*) three phase.

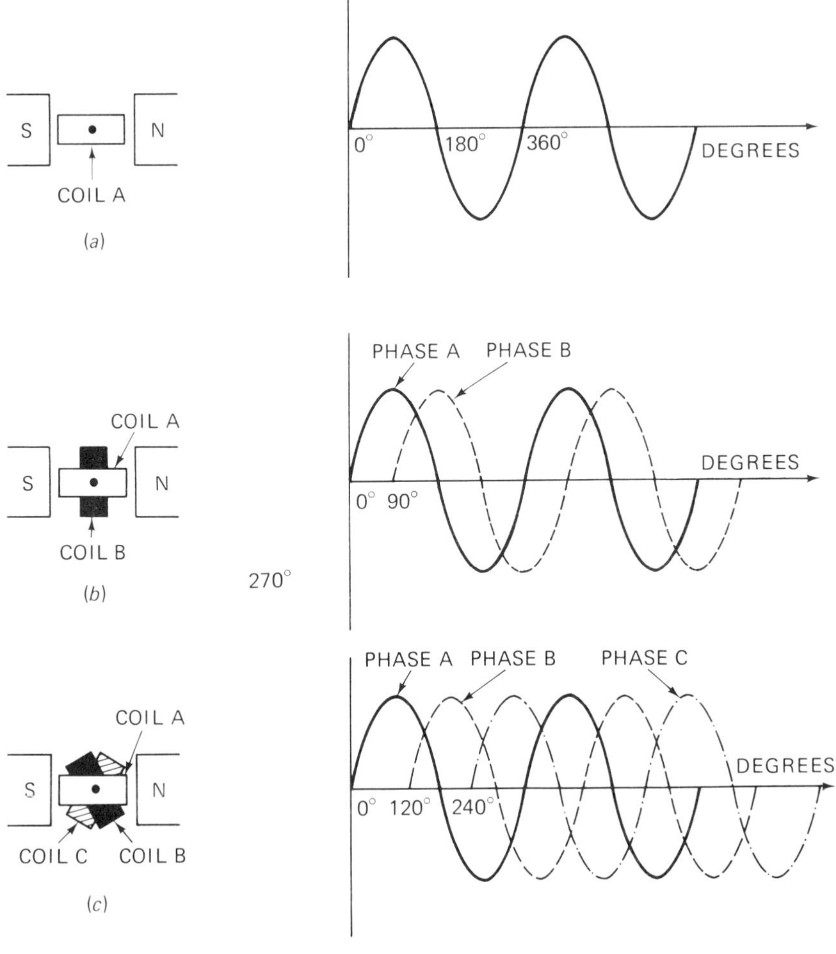

of the voltages produced by a *three-phase* alternator, that is, an alternator with three armature coils spaced 120° apart. The induced EMF of phase *B* lags behind that of phase *A* by 120°, and the induced EMF of phase *C* lags 120° behind that of phase *B*.

Alternators that have more than one set of armature coils are called *polyphase* alternators, and the alternating currents and voltages they produce are generally referred to as polyphase currents and voltages.

Most residential lighting systems and small appliances are constructed to operate on 120-V, 60-Hz, single-phase alternating current. (Actually, the electricity may be supplied at a voltage ranging anywhere between roughly 110 and 125 V, depending on many factors, but this is of no real consequence.) Electric ranges, water heaters, and some heavy appliances, such as large air-conditioning units, operate on 240-V, 60-Hz, single-phase alternating current. Both voltages are supplied to individual residences by means of the three-wire system explained in Section 2.6.

Two-phase systems are used in so few locations that they do not merit further consideration.

Three-phase alternating current is used in power-distribution systems and in commercial and industrial applications. Although three-phase alternating current is rarely encountered in residential wiring, some understanding of three-phase systems, such as how single-phase alternating current is derived from a three-phase system and how power is measured in a three-phase system, should prove beneficial. These subjects are examined, therefore, in Sections 2–6 and 2–7.

2–5 SINGLE-PHASE TRANSFORMERS

A simple single-phase transformer has two coils of wire that are *magnetically* coupled to, but *electrically* isolated from, each other. One coil, called the *primary,* is connected to an ac source of EMF; the other coil, the *secondary,* is connected to the load. When alternating current flows through the primary, the varying magnetic field of the primary cuts across the turns of the secondary and induces an EMF therein.

In an *ideal* or theoretically perfect transformer, all lines of force surround the primary link or cut across the turns of the secondary. Under these conditions, there is 100 percent magnetic *coupling*. In a practical device, however, some magnetic lines of force leak off into the air, and this is called *leakage flux*.

To reduce leakage flux, the primary and secondary windings are wound on a magnetic core that, because it has a much higher permeability than air, concentrates the lines of force and forms a closed magnetic circuit.

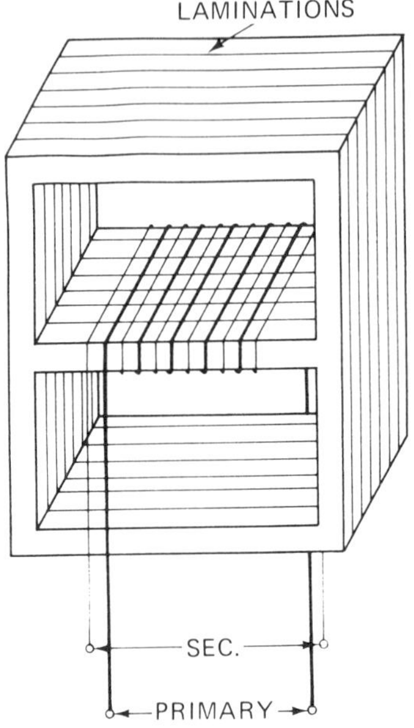

FIGURE 2–8. Shell-core transformer (the thickness of the laminations is grossly exaggerated).

In a *shell-core* transformer (Figure 2–8), the primary and secondary windings are placed one over the other on the center arm of the core. This type of transformer, which you will encounter often, is generally designed to produce a magnetic coupling that approaches 100 percent effectiveness.

In addition to losses resulting from leakage flux, a transformer has *copper* and *iron* losses. The copper loss is due to the resistance of the wire used in the windings. The iron losses result from *hysteresis* and *eddy currents*. Hysteresis loss is the loss of the energy required to make the core change the polarity of its poles in step with the frequency of the alternating current. This loss is reduced by the use of alloys with greater permeability than iron. Eddy currents are small circulating currents induced in the core by the changing magnetic field; they are reduced by making the core out of very thin strips called *laminations*. Each is coated with an insulation of varnish so that eddy currents cannot circulate through the core.

Figure 2–9 shows the schematic diagram of a transformer connected to a source and a load. Note the symbol for an ac source of EMF. If the secondary circuit is open at some point, only a small current flows through the primary, since the magnetic field, constantly cutting across the turns of the primary, induces in it a CEMF that is nearly equal to the applied EMF. The small amount of power consumed is used mainly to magnetize the core.

Notice, incidentally, that if direct current were applied to the primary, no CEMF would be developed, and current would be limited only by the low resistance of the winding. The resulting high current would burn out the primary. Thus a transformer is *never* used in a dc circuit.

If the secondary circuit is closed, current flows through the secondary winding and load. As we know from Lenz's law, the magnetic field around the secondary tends to neutralize a portion of the magnetic field around the primary. Thus the CEMF induced in the primary decreases, and the primary current increases.

As the load is increased, more current flows through the secondary circuit, the induced CEMF in the primary is decreased still more, and the primary draws more current from the source. As the load is decreased, however, less current flows through the secondary winding, the CEMF induced in the primary increases, and less current is drawn from the source. This shows us that, within limits, the transformer automatically adjusts itself to changes in the load. If the load is too great, however, the resulting high current through the primary circuit will burn out the primary winding.

To demonstrate another useful function of the transformer, assume an ideal transformer with 100 percent magnetic coupling and having a primary winding of 120 turns connected to a 120-V source of EMF. When the secondary circuit is open, it consumes no power. Since, in the ideal transformer, the powers consumed in the primary and secondary circuits are equal, no power is consumed in the primary circuit either.

The CEMF produced by the primary is equal to the voltage of the source. As the magnetic field cuts across the 120-turn primary, therefore, the EMF induced in each turn is 1 V. Since the same magnetic field cuts across the turns of the secondary, each turn of the secondary also has 1 V induced in it. If the secondary winding also has 120 turns, the induced voltage across the secondary is 120 V.

FIGURE 2–9. Schematic diagram of a transformer connected to a source and load.

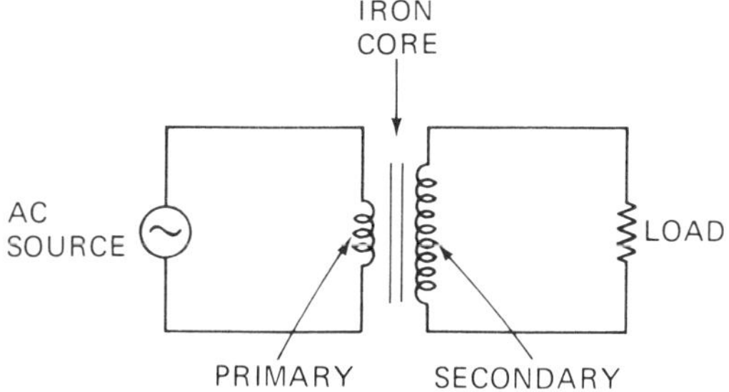

Suppose, however, that there are only 12 turns on the secondary. With 1 V induced in each turn, the secondary voltage is then 12 V. This arrangement is a *step-down* transformer. If the secondary contains 1200 turns, the secondary voltage is 1200 V, and this arrangement is a *step-up* transformer. Thus, as noted in Section 2–1, a transformer can be used to either step up or step down alternating voltage.

The above considerations show us that the ratio of the voltages across the transformer windings is equal to the turns ratio of the windings. This relationship is expressed by the formula

$$\frac{E_p}{E_s} = \frac{N_p}{N_s} \qquad (2\text{–}1)$$

where E_p is the voltage across the primary, E_s the voltage across the secondary, N_p the number of turns on the primary, and N_s the number of turns on the secondary.

If, for example, $N_p = 500$ turns, $N_s = 5000$ turns, and $E_p = 120$ V,

$$E_s = \frac{E_p N_s}{N_p} = \frac{120 \times 5000}{500} = 1200 \text{ V}$$

Again, in an ideal transformer, we assume no losses in our theoretical calculating. Thus the power of the secondary circuit ($E_s\ I_s$) is equal to the power of the primary circuit ($E_p\ I_p$). If, as in the above example, E_s is stepped up 10 times, I_s will be reduced to one tenth that of the primary. This relationship is given in the following formula:

$$\frac{I_p}{I_s} = \frac{N_s}{N_p} \qquad (2\text{–}2)$$

Using the same values for N_p and N_s as in the preceding example, and letting $I_p = 1$ A,

$$I_s = \frac{I_p N_p}{N_s} = \frac{1 \times 500}{5000} = \frac{1}{10} = 0.1 \text{ A}$$

So we see that a transformer can also be used to step up or step down alternating current. Of course, if the voltage is stepped up, the current is stepped down, and vice versa.

Transformers may be constructed with two or more secondary windings to obtain both step-up and step-down relationships.

A variation of the two-winding transformer is the *autotransformer* illustrated in Figure 2–10. This transformer has a single *tapped* coil. The turns between the tap and the two ends form the primary and secondary windings. A step-up arrangement is shown in Figure 2–10(a) and a step-down arrangement in Figure 2–10(b). Except in a few special applications,

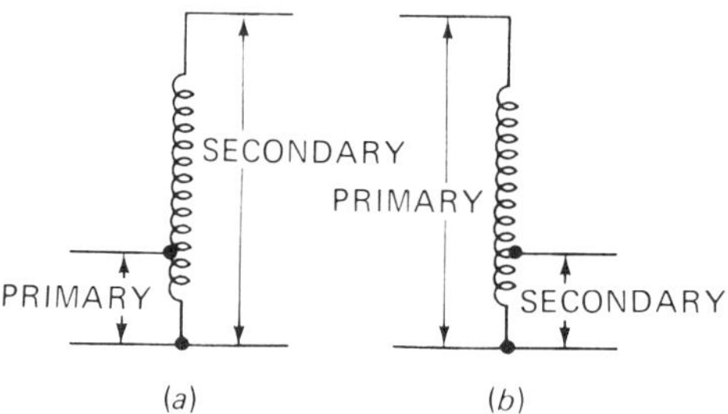

FIGURE 2–10. Autotransformer: (a) step up; (b) step down.

however, such as in connection with motor-starting devices, the use of autotransformers is prohibited by the National Electrical Code. [See NEC Secs. 210-9, 410-78, and 430-82(b).]

2–6 THREE-PHASE MACHINES AND TRANSFORMERS

Where large quantities of electric power are needed, three-phase alternating current, usually at 60 Hz, is supplied to the power mains. As noted in Section 2–4, such currents are generated by an alternator having three identical armature coils spaced 120° from each other. Since each armature coil has two ends, seemingly, six lines (two for each coil) are needed to transfer current to the load. Fortunately, it is possible to join one end of each coil at the alternator and transmit current over three lines, one for each phase. Two configurations, called the *delta* and *wye*, can be used to connect the armature coils.

Delta Configuration

In the delta configuration (Figure 2–11) the ends of each armature coil are joined to the ends of the other two coils. With this arrangement, the *net*

FIGURE 2–11. Delta configuration.

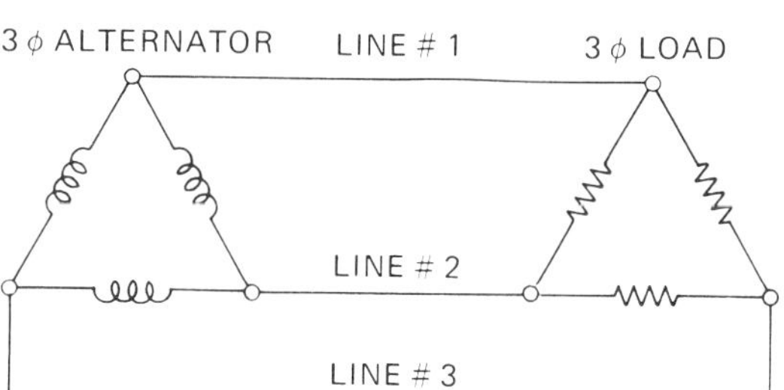

voltage around the triangle or delta (Δ) formed by the coils is zero. Of course, this means that current around the triangle is also zero.

The 3ϕ lines are connected to the 3ϕ load as shown in Figure 2–11. The voltage produced by an armature coil is called the *phase voltage,* and the voltage across the lines resulting from the phase voltage is the *line voltage. In a delta-connected circuit, the phase voltage is the same as the line voltage.*

The current through an armature coil, when it is connected to a load, is called the *phase current,* and the corresponding current flowing through the lines is the *line current. In the delta configuration, line current is equal to 1.73 times the phase current.* At any one instant, current flows from the alternator through one line to the load and returns via the other two lines; or conversely the current flows from the alternator through two lines and returns by way of the third.

In Figure 2–11 the load on the line is assumed to be *balanced;* that is, the load is such that all line currents are equal and have the same phase angle with respect to line voltages. Power companies always try to

FIGURE 2–12. Transformers: (*a*) three identical single-phase transformers with all primary and secondary windings delta connected; (*b*) the usual schematic representation of (*a*).

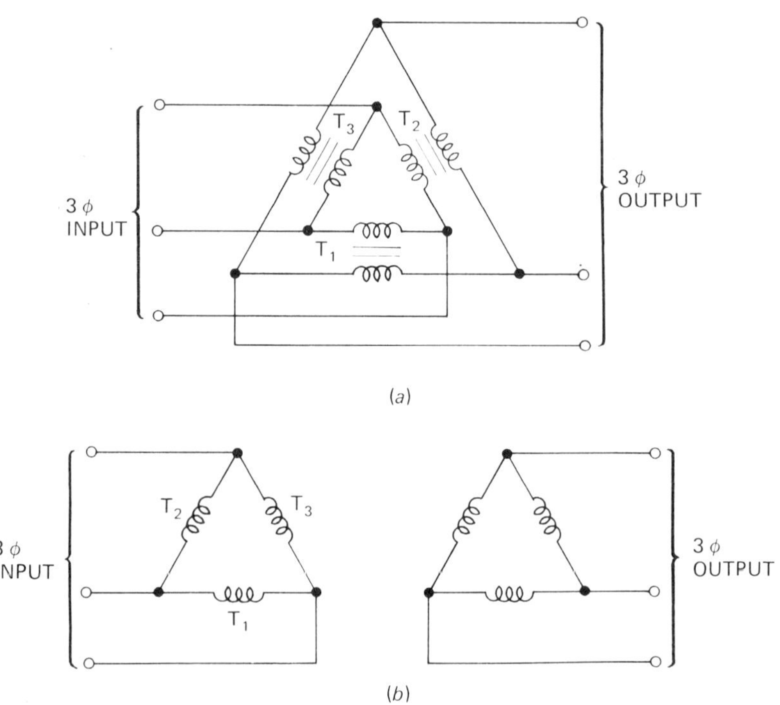

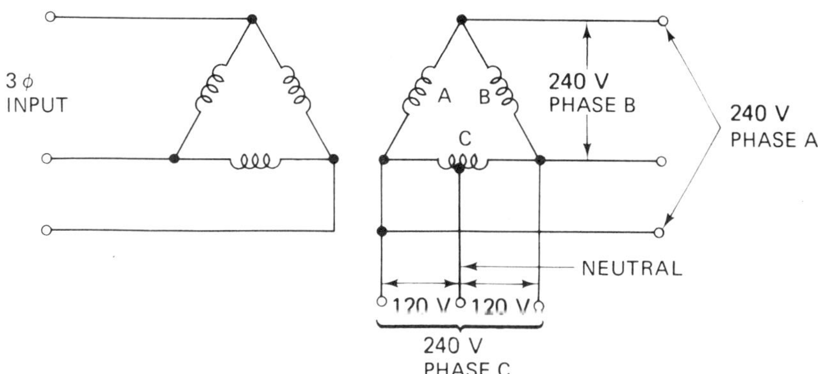

FIGURE 2–13. Obtaining 120/240-V single-phase alternating current from a delta-type transformer secondary.

maintain a balanced condition because it reduces power losses to a minimum. A balanced load is, therefore, assumed in all further discussion in this book.

To step up or step down 3ϕ alternating current, three identical single-phase transformers, one for each phase, can be used. See Figure 2–12(a). Notice that all primary and secondary windings are delta connected. Because the complexities of the diagram in Figure 2–12(a) tend to be confusing, the usual schematic representation of a 3ϕ transformer appears in the detail shown in Figure 2–12(b). In place of three single-phase transformers, a single three-phase transformer, with all windings on a single core, may be used. For either case, however, the schematic representation is the same.

To obtain 240-V single-phase alternating current from a delta-type secondary, connections are made across any *one* phase of the secondary. In Figure 2–13, phase *C* has been selected for this purpose. To obtain the 120 V needed for lighting circuits and small appliances, a *neutral* wire is connected to a center tap on phase *C*. In practically all modern installations the neutral wire is grounded. Assuming a grounded-neutral system, 240 V is applied to any device connected across the entire single-phase winding, while 120 V is delivered to any device connected between the neutral wire and either of the two outer (*hot*) wires. In actual wiring, the neutral line is *white*, and the hot wires are usually *black* (but *never* white or green).

Notice that 3ϕ alternating current at 240 V is also obtainable from the same secondary by connecting to all three wires. In Figure 2–13 the power is delivered at 240 V, and this is the usual arrangement for residences. For other applications, the power may be delivered at either 460 or 575 V. The voltage depends, of course, on the transformers selected for connection to the power company supply lines.

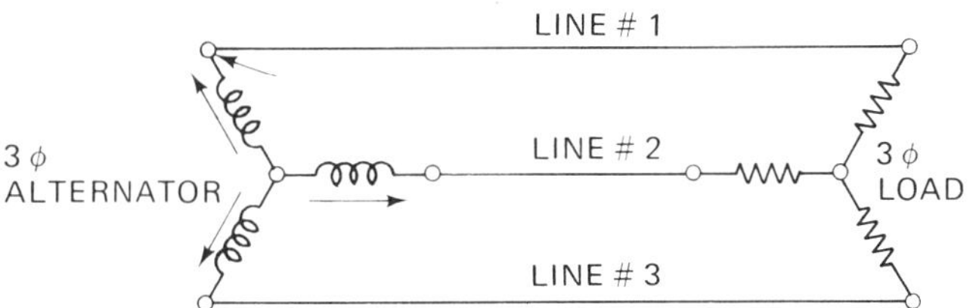

FIGURE 2–14. Wye configuration.

Wye Configuration

In this Y-shaped arrangement, one end of each armature coil is connected to one end of both the others. See Figure 2–14. The free ends of the coils are joined to the 3ϕ lines. The coils must be connected with correct polarities, as indicated. *In the wye configuration, the line current is the same as the phase current, but the line voltage is 1.73 times the phase voltage.* As in the delta configuration, single-phase alternating current is obtained across any two of the three lines.

A variation of the wye configuration is shown in Figure 2–15. Known as a three-phase, *four-wire* system, it has a grounded neutral connected to the junction of the transformer's secondary coils. This modified configuration offers certain advantages.

In a three-wire system, 3ϕ alternating current is obtained by connecting a load to all three wires. When single-phase alternating current is desired, the load is connected to any two of the wires. In the four-wire system, single-phase alternating current is obtained by connecting the single-phase load to the neutral and any one of the three *hot* wires.

Also in the four-wire system, if the single-phase current is obtained by connecting to the neutral and to one of the other three wires *and* if the

FIGURE 2–15. Three-phase, four-wire system.

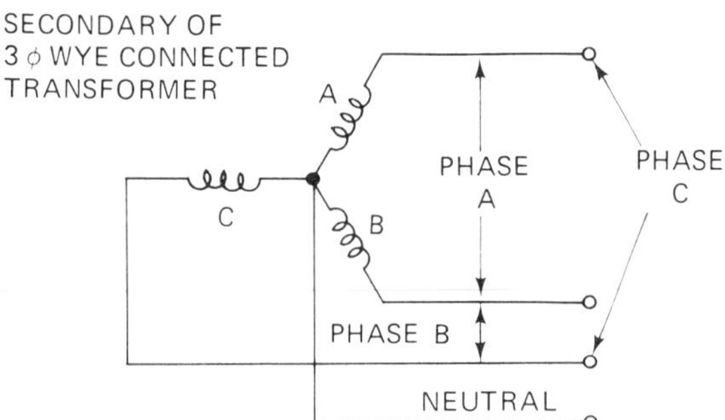

3ϕ voltage is 208 V (the usual case), the single-phase voltage will be 120 V. Thus by using four wires we can obtain either a 3ϕ current at 208 V for water heaters and similar large loads, or a 3ϕ current at 240 V for 3ϕ motors, or a single-phase current at 120 V for lighting and small appliances.

In conclusion, it is worth noting that for the same line voltage and current a three-wire, 3ϕ system will deliver 1.73 times as much power as a two-wire, single-phase system. Thus less copper is needed for the transmission of 3ϕ power than for an equivalent single-phase power.

2–7 POWER IN AC CIRCUITS

Reactive Power of an Inductance

Power supplied to an inductance to build up its magnetic field is stored as potential energy in the field itself. When the field collapses, this energy is *returned* to the circuit by induction. Since the inductance in an ac circuit alternately takes from and returns to the source equal amounts of energy, *no true power is dissipated by a pure inductor.*

In the case of a pure resistance, we are able to determine the true power by simply multiplying the effective root-mean-square values (rms) of voltage and current. In the circuit of Figure 2–16, even though the meters indicate the effective values of voltage and current for the pure inductance, the product of V and I does not represent true power since, as we have just seen, the true power is zero. The VI product is, however, directly proportional to the amount of energy stored and returned by the inductor each time that the current changes direction; therefore, the VI product in a pure inductance is called the *reactive power* of the inductance. The unit of reactive power is the *volt-ampere (reactive),* abbreviated var.

Reactive Power of a Capacitance

In working with electric motors, electricians often encounter a device called a *capacitor.* (Condenser is the obsolete name for capacitor.) *A*

FIGURE 2–16. Reactive power of an inductance.

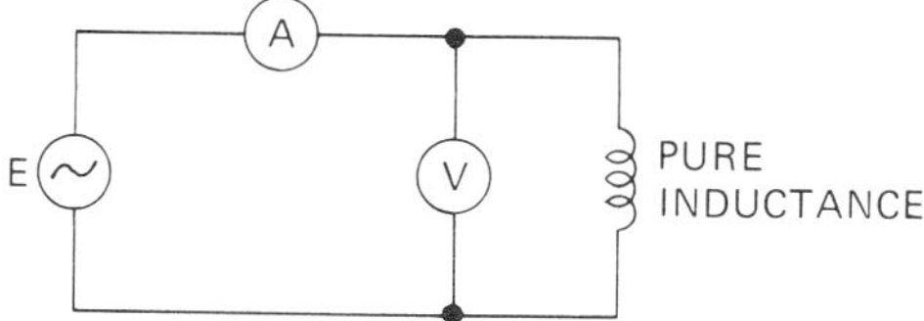

capacitor is simply a device that stores an electric charge under pressure. The ability of a capacitor to store an electric charge is called capacitance.

In physical form, a capacitor is simply two parallel conducting plates separated by an insulating material known as a *dielectric*. Small capacitors are often constructed as shown in Figure 2–17. Here two long strips of aluminum foil are rolled up with two strips of wax-paper dielectric. Leads are attached to each strip of foil for connection into a circuit, and the entire assembly is sealed in a cardboard casing. A capacitor of this type is referred to as a *lumped* capacitor since its total capacitance is lumped into a single component.

In electric circuits, *distributed* capacitance refers to that which is continuous along the whole length of a line, or is the set of capacitances between various parts of a coil. In the armature of an electric motor, for example, there are many parallel conductors separated by insulation. That is, the motor possesses distributed capacitance.

In the same way that a magnetic field is associated with an inductance, an *electric field* is associated with a capacitance—either lumped or distributed. *This field exists between the parallel conductors.*

Power supplied to a capacitor to build up its electric field is stored as electric energy in the field itself. If the direction of current reverses, this energy is *returned* to the circuit. Since the capacitances in an ac circuit alternately take from and return to the source equal amounts of energy, *no true power is dissipated by a capacitance.*

Again, the product of V and I in a pure capacitance does not represent true power; instead, VI represents the reactive power of a capacitance and is expressed in vars.

FIGURE 2–17. Typical construction of a small capacitor.

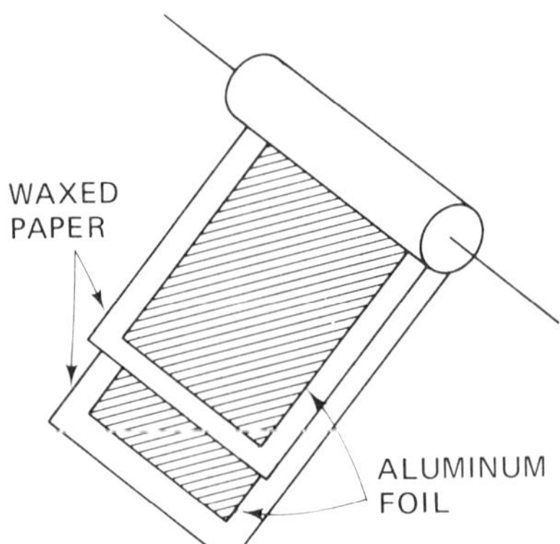

Apparent Power

If we use a voltmeter and an ammeter to measure the total voltage and total current in an ac circuit, their product represents neither the true power in watts nor the reactive power in vars; instead, the product is called the *apparent power*. Since apparent power is simply the product of volts and amperes, the unit of apparent power is the *volt-ampere* (VA).

Power Factor

The apparent power that a source must supply to a *reactive* load (one containing inductance and capacitance) is always greater than the true power that the load converts into some other form of energy.

The ratio between the true power (in watts) and the apparent power (in volt-amperes) of a load in an ac circuit is called the *power factor* of the load. From this definition,

$$\text{Power factor} = \frac{\text{watts}}{\text{volt-amperes}} \tag{2–3}$$

Power factor is often abbreviated *PF*. Since it is a simple ratio, power factor is a positive decimal fraction, which may be changed to a percentage simply by shifting the decimal point two places to the right.

To distinguish between inductive and capacitive loads, inductive loads are considered to have a *lagging* power factor and capacitive loads a *leading* power factor.

Although in residential wiring an electrician is not concerned with power factor, it is a very important consideration in installations such as those on a farm where a number of *induction* motors and transformers may be in operation. Since that type of load is inductive, the power factor will be lagging.

A voltmeter, ammeter, and wattmeter are used to determine power factor. Suppose, for example, that a single-phase induction motor (Section 12–9) is connected to a 240-V source and draws 80 A. The wattmeter reads 12.5 kW. The power factor is

$$PF = \frac{12{,}500}{240 \times 80} = \frac{12{,}500}{19{,}200} = 0.651 = 65.1\% \text{ lagging}$$

If a motor is operated from a 3ϕ circuit, multiply the volt-amperes by 1.73. If, for example, a 3ϕ motor connected to a 240 V source draws 14 A and the wattmeter indicates 4200 W,

$$PF = \frac{4200}{1.73(240 \times 14)} = \frac{4200}{5813} = 0.723 = 72.3\% \text{ lagging}$$

In most cases, power factor improves as the horsepower of the motor is increased. Typical power factors range from a low of about 50 percent for small fractional horsepower motors to a high of about 92 percent for 25-hp motors.

Importance of High-Power Factor

Since reactive power, which does no useful work, must be supplied by the power source, the source must be large enough to furnish both the reactive power and the true power that does useful work. Thus larger and heavier equipment and conductors are needed to provide a given amount of useful power to a load having a low-power factor than to one having a relatively high-power factor. Since it costs electric suppliers more to deliver such power, they charge the user a higher rate to compensate for their investment.

Conductors in the consumer's installation must also be larger than necessary to safely convey both the true power and reactive power. Heavier-than-necessary conductors can represent a substantial and needless investment, however.

Power-Factor Correction

The cheapest and most widely used method for correcting the lagging-power factor caused by inductive loads is to introduce a leading-power factor into the circuit. This is accomplished by connecting a *fixed* capacitor in *parallel* with the load.

Table 2–1 (see pp. 60–61) shows how to calculate the size capacitor needed to obtain an *improved* power factor. Suppose, for example, that we want to improve a lagging-power factor from 72 to 90 percent. Present power factors are listed in the left column and desired power factors across the top of Table 2–1. Entering the table at 72 percent and moving to the right, we find the factor 0.48 in the column headed 90 percent. When a true power is multiplied by 0.48, the product is the *kilovar* (kvar) of reactive power needed to raise the power factor to 90 percent. If the true power is, for example, 12,500 W, the needed reactive power is

$$12{,}500 \times 0.48 = 6000 \text{ var} = 6 \text{ kvar}$$

Capacitors for this purpose come in standard units whose nameplates list, among other factors, the rated voltage and frequency at which the unit is to be operated and the reactive power in kvar that it will introduce into the circuit. A standard unit having a kvar rating closest to the calculated value is connected in parallel with the load. If necessary, several standard capacitors may be hooked in parallel to obtain a value equal to the *sum* of the kvar ratings of the individual units.

Although a 100 percent power factor is the best theoretical solution, the cost of achieving it must be weighed against the practical cost of operating at a somewhat lower power factor. The capacitors use oil as a dielectric, are hermetically sealed in a steel can, and are expensive.

Since a capacitor stores an electric charge, there also is the danger of receiving a *severe* shock when one is handled. To minimize this danger, the National Electrical Code requires that a resistor or inductor must be included in the circuit to form a discharge path having a specified time constant. When a time equal to five time constants has elapsed, the capacitor may be removed from the line without danger of shock. See Secs. 460-28 and 460-6(b).

In 3ϕ circuits, each phase must be corrected. A three-phase capacitor, having three capacitors connected in either a delta or wye configuration, is contained in a single case. Separate connectors are provided on top of the case to connect one capacitor to each line.

Either of two methods can be used in connecting capacitors for power-factor correction. One method is to calculate the size of capacitor needed to correct the power factor for the complete installation and then to connect this capacitor to the main power bus near the *service entrance.* (The general term "service entrance" includes all wires and equipment from the outside of the building up to and including the meter; a switch to disconnect the entire installation from the power supplier's lines; overcurrent protection, such as circuit breakers or fuses, and the ground wire.) Since the capacitor corrects the power factor from the point of insertion *back* to the source, the correction is not applied on lines leading to individual appliances. This method reduces the penalty imposed by the electric supplier for a low-power factor.

A second method is to connect a capacitor at each appliance. Although this method also corrects the low-power factor on individual feeder lines inside the plant, it is usually more costly that the method described above. The choice of method is usually dictated by economics.

Measuring Power in Three-Phase Circuits

When two wattmeters are connected as shown in Figures 2–18(a) and (b), the total power in either a balanced or unbalanced load is the algebraic sum of the two meter readings.

TABLE 2–1. Factors for Calculating the Size Capacitor Needed to Obtain an Improved Power Factor with an Inductive Load

Westinghouse Electric Corp

Present Power Factor	Resultant or Desired Power Factor (%)						
	80	*81*	*82*	*83*	*84*	*85*	*86*
50	0.98	1.01	1.03	1.06	1.09	1.11	1.14
51	0.94	0.96	0.99	1.01	1.04	1.07	1.09
52	0.89	0.92	0.95	0.97	1.00	1.02	1.05
53	0.85	0.88	0.90	0.93	0.95	0.98	1.01
54	0.81	0.83	0.86	0.89	0.91	0.94	0.97
55	0.77	0.79	0.82	0.85	0.87	0.90	0.93
56	0.73	0.76	0.78	0.81	0.83	0.86	0.89
57	0.69	0.72	0.74	0.77	0.80	0.82	0.85
58	0.65	0.68	0.71	0.73	0.76	0.78	0.81
59	0.62	0.64	0.67	0.70	0.72	0.75	0.77
60	0.58	0.61	0.64	0.66	0.69	0.71	0.74
61	0.55	0.57	0.60	0.63	0.65	0.68	0.71
62	0.51	0.54	0.57	0.59	0.62	0.64	0.67
63	0.48	0.51	0.53	0.56	0.59	0.61	0.64
64	0.45	0.48	0.50	0.53	0.55	0.58	0.61
65	0.42	0.44	0.47	0.50	0.52	0.55	0.58
66	0.39	0.41	0.44	0.47	0.49	0.52	0.54
67	0.36	0.38	0.41	0.44	0.46	0.49	0.51
68	0.33	0.35	0.38	0.41	0.43	0.46	0.49
69	0.30	0.32	0.35	0.38	0.40	0.43	0.46
70	0.27	0.30	0.32	0.35	0.37	0.40	0.43
71	0.24	0.27	0.29	0.32	0.35	0.37	0.40
72	0.21	0.24	0.26	0.29	0.32	0.34	0.37
73	0.19	0.21	0.24	0.26	0.29	0.32	0.34
74	0.16	0.18	0.21	0.24	0.26	0.29	0.32
75	0.13	0.16	0.18	0.21	0.24	0.26	0.29
76	0.10	0.13	0.16	0.18	0.21	0.23	0.26
77	0.08	0.10	0.13	0.16	0.18	0.21	0.24
78	0.05	0.08	0.10	0.13	0.16	0.18	0.21
79	0.03	0.05	0.08	0.10	0.13	0.16	0.18
80	0.00	0.03	0.05	0.08	0.10	0.13	0.16
81		0.00	0.03	0.05	0.08	0.10	0.13
82			0.00	0.03	0.05	0.08	0.10
83				0.00	0.03	0.05	0.08
84					0.00	0.03	0.05
85						0.00	0.03
86							0.00
87							
88							
89							
90							
91							
92							
93							
94							
95							
96							
97							
98							
99							
100							

87	88	89	90	91	92	93	94	95	96	97	98	99	100	*Present Power Factor*
1.16	1.19	1.22	1.25	1.28	1.30	1.34	1.37	1.40	1.44	1.48	1.53	1.59	1.73	*50*
1.12	1.15	1.17	1.20	1.23	1.26	1.29	1.32	1.36	1.39	1.43	1.48	1.54	1.69	*51*
1.08	1.10	1.13	1.16	1.19	1.21	1.25	1.28	1.31	1.35	1.39	1.44	1.50	1.64	*52*
1.03	1.06	1.09	1.12	1.14	1.17	1.20	1.24	1.27	1.31	1.35	1.40	1.46	1.60	*53*
0.99	1.02	1.05	1.07	1.10	1.13	1.16	1.20	1.23	1.27	1.31	1.36	1.42	1.56	*54*
0.95	0.98	1.01	1.03	1.06	1.09	1.12	1.16	1.19	1.23	1.27	1.32	1.38	1.52	*55*
0.91	0.97	0.97	1.00	1.02	1.05	1.08	1.12	1.15	1.19	1.23	1.28	1.34	1.48	*56*
0.87	0.90	0.93	0.96	0.99	1.01	1.05	1.08	1.11	1.15	1.19	1.24	1.30	1.44	*57*
0.84	0.86	0.89	0.92	0.95	0.98	1.01	1.04	1.08	1.11	1.15	1.20	1.26	1.40	*58*
0.80	0.83	0.86	0.88	0.91	0.94	0.97	1.00	1.04	1.08	1.12	1.16	1.23	1.37	*59*
0.77	0.79	0.82	0.85	0.88	0.90	0.94	0.97	1.00	1.04	1.08	1.13	1.19	1.33	*60*
0.73	0.76	0.79	0.81	0.84	0.87	0.90	0.94	0.97	1.01	1.05	1.10	1.16	1.30	*61*
0.70	0.72	0.75	0.78	0.81	0.84	0.87	0.90	0.94	0.97	1.01	1.06	1.12	1.26	*62*
0.67	0.69	0.72	0.75	0.78	0.80	0.84	0.87	0.90	0.94	0.98	1.03	1.09	1.23	*63*
0.63	0.66	0.69	0.72	0.74	0.77	0.80	0.84	0.87	0.91	0.95	1.00	1.06	1.20	*64*
0.60	0.63	0.66	0.68	0.71	0.74	0.77	0.81	0.84	0.88	0.92	0.97	1.03	1.17	*65*
0.57	0.60	0.63	0.65	0.68	0.71	0.74	0.77	0.81	0.85	0.89	0.93	1.00	1.14	*66*
0.54	0.57	0.60	0.62	0.65	0.68	0.71	0.74	0.78	0.82	0.86	0.90	0.97	1.11	*67*
0.51	0.54	0.57	0.59	0.62	0.65	0.68	0.72	0.75	0.79	0.83	0.88	0.94	1.08	*68*
0.48	0.51	0.54	0.56	0.59	0.62	0.65	0.69	0.72	0.76	0.80	0.84	0.91	1.05	*69*
0.45	0.48	0.51	0.54	0.56	0.59	0.62	0.66	0.69	0.73	0.77	0.81	0.88	1.02	*70*
0.42	0.45	0.48	0.51	0.54	0.56	0.60	0.63	0.66	0.70	0.74	0.78	0.85	0.99	*71*
0.40	0.42	0.45	0.48	0.51	0.53	0.57	0.60	0.63	0.67	0.71	0.75	0.82	0.96	*72*
0.37	0.40	0.42	0.45	0.48	0.51	0.54	0.57	0.61	0.64	0.68	0.73	0.79	0.94	*73*
0.34	0.37	0.40	0.42	0.45	0.48	0.51	0.55	0.58	0.62	0.66	0.70	0.77	0.91	*74*
0.31	0.34	0.37	0.40	0.43	0.45	0.49	0.52	0.55	0.59	0.63	0.67	0.74	0.88	*75*
0.29	0.31	0.34	0.37	0.40	0.43	0.46	0.49	0.53	0.56	0.60	0.65	0.71	0.85	*76*
0.26	0.29	0.32	0.34	0.37	0.40	0.43	0.47	0.50	0.54	0.58	0.62	0.69	0.83	*77*
0.24	0.26	0.29	0.32	0.35	0.37	0.41	0.44	0.47	0.51	0.55	0.59	0.66	0.80	*78*
0.21	0.24	0.26	0.29	0.32	0.35	0.38	0.41	0.45	0.48	0.52	0.57	0.63	0.78	*79*
0.18	0.21	0.24	0.27	0.29	0.32	0.35	0.39	0.42	0.45	0.50	0.54	0.61	0.75	*80*
0.16	0.18	0.21	0.24	0.27	0.29	0.33	0.36	0.39	0.43	0.47	0.51	0.58	0.72	*81*
0.13	0.16	0.19	0.21	0.24	0.27	0.30	0.33	0.37	0.41	0.45	0.49	0.56	0.70	*82*
0.10	0.13	0.16	0.19	0.22	0.24	0.28	0.31	0.34	0.38	0.42	0.46	0.53	0.67	*83*
0.08	0.11	0.13	0.16	0.19	0.22	0.25	0.28	0.32	0.35	0.39	0.44	0.50	0.65	*84*
0.05	0.08	0.11	0.14	0.16	0.19	0.22	0.26	0.29	0.33	0.37	0.42	0.48	0.62	*85*
0.03	0.05	0.08	0.11	0.14	0.17	0.20	0.23	0.26	0.30	0.34	0.39	0.45	0.59	*86*
0.00	0.03	0.05	0.08	0.11	0.14	0.17	0.20	0.24	0.27	0.32	0.36	0.42	0.57	*87*
	0.00	0.03	0.06	0.08	0.11	0.14	0.18	0.21	0.25	0.29	0.34	0.40	0.54	*88*
		0.00	0.03	0.05	0.09	0.12	0.15	0.18	0.22	0.26	0.31	0.37	0.51	*89*
			0.00	0.03	0.06	0.09	0.12	0.15	0.19	0.23	0.28	0.34	0.48	*90*
				0.00	0.03	0.06	0.09	0.13	0.16	0.21	0.25	0.31	0.46	*91*
					0.00	0.03	0.06	0.10	0.13	0.18	0.22	0.28	0.43	*92*
						0.00	0.03	0.07	0.10	0.14	0.19	0.25	0.39	*93*
							0.00	0.03	0.07	0.11	0.16	0.22	0.36	*94*
								0.00	0.04	0.08	0.12	0.19	0.33	*95*
									0.00	0.04	0.09	0.15	0.29	*96*
										0.00	0.05	0.11	0.25	*97*
											0.00	0.06	0.20	*98*
												0.00	0.14	*99*
													0.00	*100*

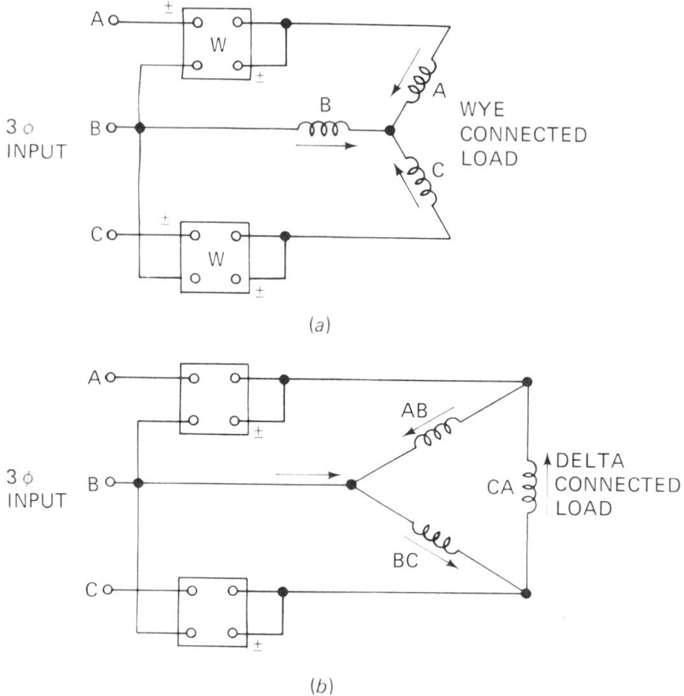

FIGURE 2–18. Measuring power in (*a*) 3ϕ wye and (*b*) 3ϕ delta systems.

EXERCISES

2–1. What is an *alternating* current?

2–2. Which of two possible methods is ordinarily used to reduce power losses on electric transmission lines? Why?

2–3. What is the basic principle on which an alternator works?

2–4 Describe the operation of a simple (single-loop) alternator.

2–5. In connection with alternating current, what is meant by the term hertz? 60 Hz?

2–6. When are two wave forms *in phase? Out of phase?*

2–7. What is meant by the *effective* (rms) value of an alternating voltage or current?

2–8. What is meant by the terms *polyphase voltage* and *polyphase current?*

2–9. How are the armature coils of an alternator arranged to produce 3ϕ alternating current?

2–10. Describe the physical arrangement of a single-phase transformer.

2–11. What do we mean by the expression *ideal transformer?*

2–12. What is *leakage flux?*

2–13. Describe the physical arrangement of a *shell-core* transformer.

2–14. In connection with transformer losses, what is meant by the terms *hysteresis* and *eddy currents?*

2–15. In practical transformers, how are hysteresis and eddy-current losses reduced?

2–16. If the secondary circuit of a two-winding transformer is open, is a large or small current passing through the primary circuit? Why?

2–17. Is it possible for a transformer to be used in a dc circuit? Explain in detail.

2–18. What happens to the primary current when the load on the secondary *increases?* Why?

2–19. What happens to the primary current when the load on the secondary *decreases?* Why?

2–20. What is a *step-up* transformer? A *step-down* transformer?

2–21. What relationship exists between the *voltage* and *turns ratios* of a transformer primary and secondary?

2–22. What relationship exists between the *current* and *turns ratios* of a transformer primary and secondary?

2–23. What is an *autotransformer?* Are autotransformers widely employed in electric circuits? Explain in detail.

2–24. In a *delta* configuration, what relationship exists between the *phase voltage* and the *line voltage?*

2–25. In a *delta* configuration, what relationship exists between *line current* and *phase current?*

2–26. How is 120/240 V single-phase alternating current obtained from a *delta*-type secondary?

2–27. In a wye configuration, what relationship exists between the *line current* and *phase current?* Between the *line voltage* and *phase voltage?*

2–28. What is a three-phase *four-wire* system?

2–29. How is single-phase alternating current obtained from a 3ϕ four-wire system?

2–30. Is any *true* power dissipated by an inductance? Explain in detail.

2–31. What is the *unit* of reactive power?

2–32. What is *capacitance?* A *capacitor?*

2–33. *Where* is the electric charge stored in a capacitor?

2–34. Does a capacitance dissipate any *true* power? Explain in detail.

2–35. What is meant by the term *apparent power?* What is the *unit* of apparent power?

2–36. What is meant by the term *power factor?*

2–37. Describe the procedure for determining the power factor of the load in a single-phase system.

2–38. Describe the procedure for determining the power factor of the load in a 3ϕ system.

2–39. In what two ways can the power factor of a load be expressed?

2–40. What types of loads exhibit a *lagging* power factor? A *leading* power factor?

2–41. Why is a *high* power factor important?

2–42. What is the cheapest and most widely used method for improving the *lagging* power factor caused by inductive loads?

2–43. Suppose that you want to improve a lagging power factor from 70 to 92 percent. By what *factor* would you multiply true power to determine the reactive power required?

2–44. Describe two methods for connecting capacitors to correct a lagging power factor.

2–45. How can the power in a three-phase circuit be measured?

3

Illumination

3–1 BASIC CONCEPTS

In this text our previous references to lighting loads have been only in terms of power. Too many variables affect lighting, however, to say with certainty that a given amount of power dissipated will always provide a desired amount of illumination. Some of the variables affecting lighting effectiveness are the type of bulb used—incandescent or fluorescent, whether the bulb is clear or frosted, how much surrounding areas reflect or diffuse the light, how efficiently the fixture is constructed for the concentration of illumination, and finally the distance from the lamp to the area where the level of lighting is measured.

Two interrelated units, the *footcandle* and the *lumen,* are important in any consideration of illumination values. In Figure 3–1, a candle is located in the center of a sphere having a radius of 1 foot (ft). Every point on the sphere is said to be illuminated to a level of 1 footcandle (fc), since each point is located 1 ft from the light source. Notice that the footcandle is a measure of illumination at a *point*; no area is involved.

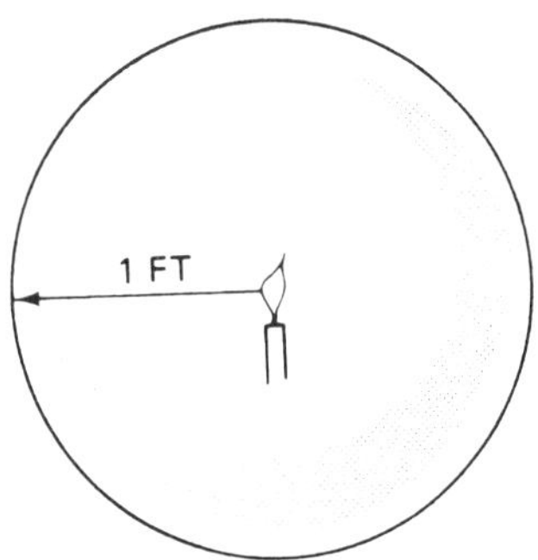

FIGURE 3–1. Defining the footcandle.

When area is involved, the quantity of light is measured in lumens (lm). One lumen is the amount of light falling on each square foot (ft^2) of the sphere diagrammed in Figure 3–1. One lumen of light is needed to produce a uniform illumination of 1 fc over an area of 1 ft^2. To produce a level of illumination of, say, 10 fc, we must, therefore, provide 10 lm for every square foot of area involved. Thus 1 fc = 1 lm/ft^2.

The intensity of illumination can be measured with a *footcandle meter* (commonly called a *light meter*), which is simply a *photoelectric cell* used in combination with a microammeter. When light falls on the cell, the cell generates an EMF that is proportional to the intensity of the light. Current produced by this EMF, in passing through the microammeter, deflects a pointer across a scale calibrated in footcandles so that readings can be made.

One other characteristic of light requires explanation at this point: *light varies inversely as the square of distance from the light source*. To illustrate this characteristic, suppose that a hole with an area of 1 ft^2 is cut in the sphere of Figure 3–1. With a light source of one candle in the center of the sphere, exactly 1 lm of light will escape through the opening. If rounded screens are placed at distances of 2 and 3 ft from the light source, as shown in Figure 3–2, the illuminated areas of the screens measure 4 ft^2 and 9 ft^2, respectively. Since the amount of light remains the same (1 lm), the inner screen receives ¼ lm/ft^2, while the outer screen receives only $^1/_9$ lm/ft^2.

To determine the relative illumination of the outer screen to the inner screen, divide the square of the shorter distance by the square of the longer:

$$^4/_9 = 44.4 \text{ percent}$$

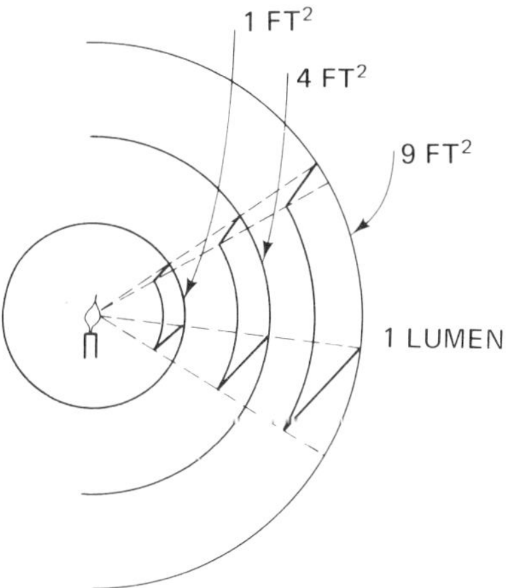

FIGURE 3–2. Light varies inversely as the square of distance from the light source.

3–2 INCANDESCENT LAMPS

Incandescence means *a glowing due to heat*. Because *tungsten* has a very high melting point, it is the principal light source in incandescent lamps. Incandescent lamps are reasonably inexpensive, have fair life expectancies, can be used in practically any environment, will continue to operate with wide variations in applied voltage, and are generally used with simple fixtures. Moreover, because their filaments are purely resistive, these lamps have a unity (1) power factor.

The incandescent lamps most often used in residential lighting range in size from about 25 W up to about 500 W and have standard *screw-shell* bases. The *mogul* is the largest of the screw-shell bases and is used mostly on lamps from 300 W upward. The wattage rating, initial output in lumens, average output in lumens, and life expectancy in hours for typical incandescent lamps are shown in Table 3–1.

As a lamp ages, its inner surface becomes blackened from minute tungsten deposits and the lumen output drops correspondingly as the tungsten filament burns. There is also a proportionately small decrease in power consumption. In areas where lighting levels are critical, replacement at regular intervals is recommended.

The lowest overall cost of illumination is achieved by operating lamps at the voltage for which they are designed. Although operation at a lower voltage prolongs the life of a lamp bulb, the actual watts dissipated, the

TABLE 3–1. Characteristics of Incandescent Lamps

Rating (W)	Initial Output (lm)	Average Output (lm)	Normal Life Expectancy (h)
25	265	230	1000
40	470	415	1000
60	840	795	1000
100	1640	1540	750
150	2700	2420	750
200	3800	3400	750
300	5750	5150	1000
500	9900	8750	1000

total output in lumens, and the output of lumens per watt show sharp declines. At higher than design voltages the opposite characteristics prevail. See Table 3–2.

"Long-life" lamps are a poor investment. They last longer simply because they are designed to operate at voltages that are 10 to 15 percent higher than the standard 120 V. Thus they deliver far fewer lumens per watt than ordinary incandescent lamps. And this means that they cost substantially more to operate.

In cases where lamp replacement is difficult and, therefore, costly, you might consider the use of *extended-life* lamps that are designed for about 2500 h of service. These lamps, too, deliver fewer lumens per watt than ordinary lamps, but their cost is appreciably less than that of the "long-life" variety.

Types R and PAR lamps are another alternative. These lamps are built with an aluminum or silver reflector deposited directly on the inside

TABLE 3–2 Voltage Characteristics in Percentages of Incandescent Lamps

Rated Voltage	Normal Life Expectancy	Actual Watts Dissipated	Total Output in Lumens	Output Lumens per Watt
85	800	78	58	72
90	400	85	68	80
95	200	93	83	90
100	100	100	100	100
105	58	108	118	109
110	37	116	140	120
115	18	124	162	131

of the bulb. Because the enclosed reflector cannot get dusty or dirty, lamp efficiency is unaffected by dirt. Lamps of this type are often a good choice for such applications as merchandise display or in residential applications where a high concentration of light is desired (bathroom mirrors, game rooms, entryways, and the like).

Type R lamps are available in wattages ranging from 15 to 1000 W. Type PAR lamps have ratings ranging from 75 to 500 W. Also, 1000-W type PAR lamps are available in a "Quartzline" construction.

3–3 FLUORESCENT LAMPS

Fluorescent lamps offer several advantages over the incandescent variety: (1) high efficiency, with two to three times more lumens per watt (in color, up to about 100 times more); (2) less heat (an important consideration in air-conditioned spaces); (3) light with less glare or shadow, which reduces eyestrain since it is distributed over a larger lamp-surface area; (4) under ordinary operating conditions, five to ten times the life expectancy.

The disadvantages of fluorescent lamps as compared to their incandescent counterparts include (1) temperature sensitivity (lamps intended for residential use operate at temperatures approaching 50°F; special lamps are available, however, for operation at lower temperatures); (2) high humidity may necessitate the use of a protective enclosure; (3) life expectancy is affected to a great extent by the number of times the lamp is turned of and off (published life expectancies assume that lamps operate continuously for 3 h each time that they are turned on); (4) a power factor of less than unity; and (5) a higher initial cost.

The fluorescent lamp is illustrated in Figure 3–3. A tungsten filament is sealed into each end of a long glass tube whose inner surface is coated with a suitable *phosphor*. Essentially all air remaining in the tube is then pumped out, and very small volumes of *argon gas* and *mercury* are introduced.

In operation, the passage of current through the filaments produces sufficient heat to vaporize the mercury so that the tube is filled with mercury vapor. Then a relatively high voltage is impressed across the tube, using the filaments as electrodes. The high voltage *ionizes** the argon and mercury vapor and so produces ultraviolet rays. When these rays strike the phosphor coating, the coating *fluoresces* (glows) and produces visible light. Once the tube starts to glow, the filament circuit is opened,

*Recall from Section 1–7 that when valence electrons break away from their parent atoms the atoms become ions. This process is called ionization.

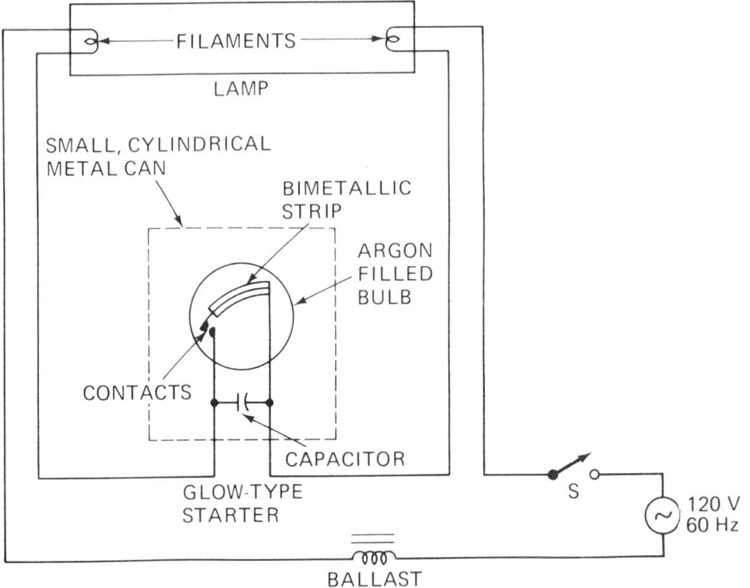

FIGURE 3–3. Typical fluorescent fixture.

but the heat produced by ionization of the gases keeps the mercury vaporized so that the lamp continues to emit light.

As shown in Figure 3–3, the filament circuit is controlled by a *glow-type starter*. Its movable and fixed contact points are enclosed in a small glass bulb containing argon gas. The movable contact point is attached to a *bimetallic strip*.* (Normally the fixed point is not; that is, the two points are separated.) But when the main switch is closed, the 120-V current ionizes the gas in the bulb so that a small current flows between the movable and fixed contact points in the bulb. This current produces heat, which in turn causes the bimetallic strip to bend until the two contact points touch each other. When that occurs, the flow of current through the gas ceases; it now flows instead through the closed contact points and then through the filaments of the lamps.

When current stops flowing through the gas, the bimetallic strip cools off and the contact points separate to open the filament circuit. Enough residual heat remains, however, to keep the contacts closed long enough for the filaments to vaporize the mercury in the lamp.

When ionization occurs within the tube, the voltage across the tube

*A bimetallic strip consists of two strips of dissimilar metal that are welded or riveted together along their length. One end of the strip is rigidly fixed and the other carries a contact. Heat causes the metal strips to expand. One metal expands more than the other so that the arm bends.

drops to a very low value. Since the starter is connected in parallel with the tube, the voltage across the starter is at the same low value and is insufficient to cause ionization of the argon gas in the starter bulb. As a result, the contact points remain separated and the filament circuit of the lamp remains open.

A device called a *ballast* is connected in series with the lamp and starter; it consists of many turns of fine wire on an iron core. Because of self-induction, a high CEMF (counterelectromotive force) is generated by the ballast at the instant that the contact points of the starter separate and break the circuit. This high-voltage surge is applied across the filament electrodes of the lamp, ionizing the mercury vapor. Once this ionization starts, current continues to flow through the lamp even though the voltage across the electrodes drops to a low value.

After ionization starts, the ballast performs a second function. Once started, ionization tends to grow and, unless limited, the resulting excessive current would destroy the lamp. The ballast, however, provides the needed safety action. After it has set up the high-voltage surge that starts ionization, the ballast acts as a *reactor* that keeps current through the lamp within safe limits.

The capacitor across the starter contact points reduces sparking across these points as they are separated. Without the capacitor, the electrical pressure would cause current to arc across the gap formed by the separating contacts. This arcing would ionize the gap, and current would continue to flow even after the contact points were a considerable distance apart. Since the electrical pressure is used to charge the capacitor rather than in forming an arc, sparking is reduced.

Although the fluorescent lamp is essentially an ac device, it can be operated on direct current. The high-voltage surge set up at the time the contact points open is produced by either type of current. The protective action of the ballast is absent, however, since no CEMF is generated by direct current. It is necessary, therefore, to connect a current-limiting resistor in series with the ballast. This resistor causes a power loss, of course, and there is some deterioration in lamp performance.

The schematic diagram of a typical two-lamp fluorescent *fixture* (often called a *luminaire* in engineering publications) is shown in Figure 3–4. Here an autotransformer steps up the supply voltage to the level needed for *starting*. The starters (notice symbol in the figure) permit current flow through the lamp filaments. After a few seconds, the heated filament emits electrons, and as the starter opens, a high voltage causes current to flow from the filament through the mercury vapor to the phosphor coating. The lamp current is then limited by the reactor characteristic of the ballast.

The capacitor in series with the filament of lamp 1 causes the current to be out of phase with that of lamp 2. As the alternating current goes

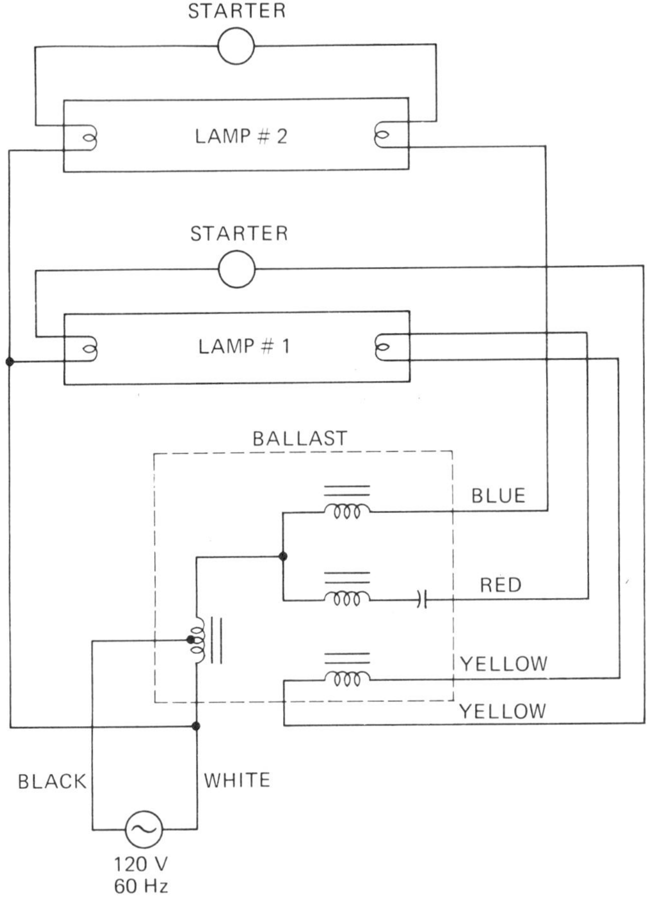

FIGURE 3–4. Typical two-lamp fluorescent fixture.

through zero when reversing, the lamps then go on and off at different times to avoid a *stroboscopic* effect on moving objects. The capacitor also improves the power factor of the unit to about 0.95. Without the capacitor, the high inductance of the ballasts produces a power factor ranging from 0.5 to 0.75. For large lighting loads, a low power factor is, of course, undesirable.

The type of starting described thus far is called *preheat starting*. Fluorescent lamps designed for preheat starting have bi-pin (two-pin) bases, and are available in sizes from 4 to 100 W.

An *instant-start* fluorescent lamp also has a filament at each end, but they are short-circuited inside the lamp. Thus current cannot flow through the filaments before the lamp lights. As we can see in Figure 3–5, no starter is needed, but the ballast provides an open-circuit voltage of from 450 to 600 V, which produces ionization immediately when the switch is closed. Most instant-start lamps are of the *Slimline*Ⓣ type and use a single-contact base. Sizes range from 40 to 75 W.

Another form of instant-start lamp is the *cold-cathode* type. Although the internal construction of this lamp is somewhat different than that of the conventional instant-start lamp, its operating principle is essentially the same. The voltages involved are considerably higher, however, being on the order of from 800 to 1000 V. For this reason, wire designed for a higher voltage than ordinary fixture wire must be used between the ballast and lamps. The maximum voltage at which ordinary fixture wire can be employed is 300 V.

Some newer fluorescent lamps are of the *rapid-start* type; that is, starting is delayed only a fraction of a second. No starter is employed and

FIGURE 3–5. Fluorescent fixture with instant-start lamps.

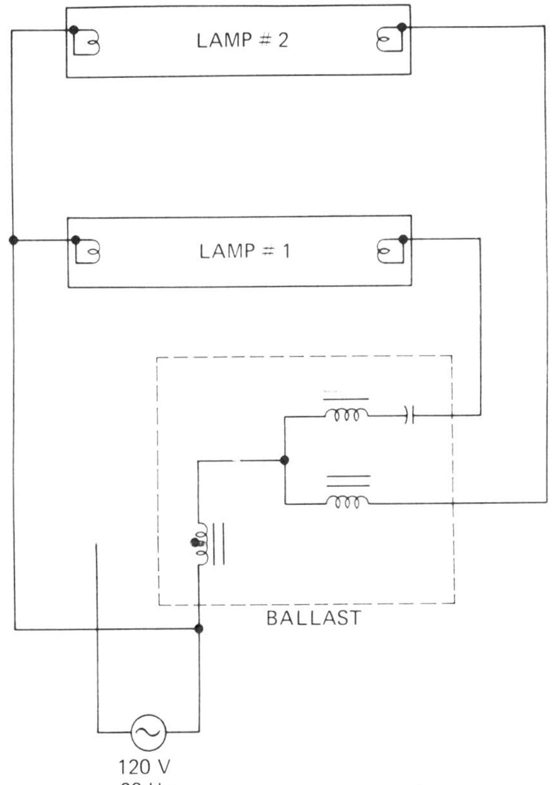

the ballast delivers an open-circuit voltage ranging from about 250 to 400 V. Nonshorted filaments, located at each end of the tube, are heated during operation of the lamp by special 3½-V windings on the ballast, plus the heat produced by current through the tube. Larger-sized rapid-start lamps use a recessed double-contact base; smaller sizes use the bi-pin base. Sizes run from 30 to 220 W.

A *preheat/rapid-start* lamp is also available. It may be used in a fixture designed for either preheat or rapid-start lamps. This is an exception to the general rule that lamps designed for one type of starting will not fit sockets designed for a different kind.

Fluorescent lamps are manufactured in sizes ranging from 6 in. to over 4 ft in length. In addition, these lamps are made as straight tubes or in circles and semicircles. Different lamp colors are made by coating the tube interiors with various phosphors. Consequently, fluorescent tubes lend themselves readily to various decorative designs.

Finally, be sure to note that fluorescent lamps, unlike the incandescent types, are *not* rated by voltage. The transformers and ballasts are designed for specific voltages—usually 120, 208, 240, and 265 V.

Performance varies, or course, with deviation from the rated voltages of transformers and ballasts. In Figure 3–6 the percent of rated lumen output and the changes in power input to ballasts, transformers, and lamps as voltage changes are detailed. Moderate deviations from the rated voltages have little effect on lamp life.

3–4 HOW TO OBTAIN GOOD LIGHTING

The preferred method for general lighting calculations is called the *zonal-cavity method*, from the technique of dividing a room into zones (also called cavities). This method is preferred not because it is *necessarily* more accurate, but because it is more flexible, can be applied to every type of interior space, and allows more of the actual lighting situation to be taken into account during the calculations. Numerical results are therefore generally more representative of the actual lighting situation than those arrived at by older calculation systems, especially when odd sizes and shapes of rooms are involved, or when surface-mounted, recessed, or luminous ceiling systems are used.

Calculating Illumination

Interior general lighting calculations are usually used to determine (1) how many luminaires (fixtures) are needed to provide an average given illumination level in an interior space, and (2) how the luminaires should be

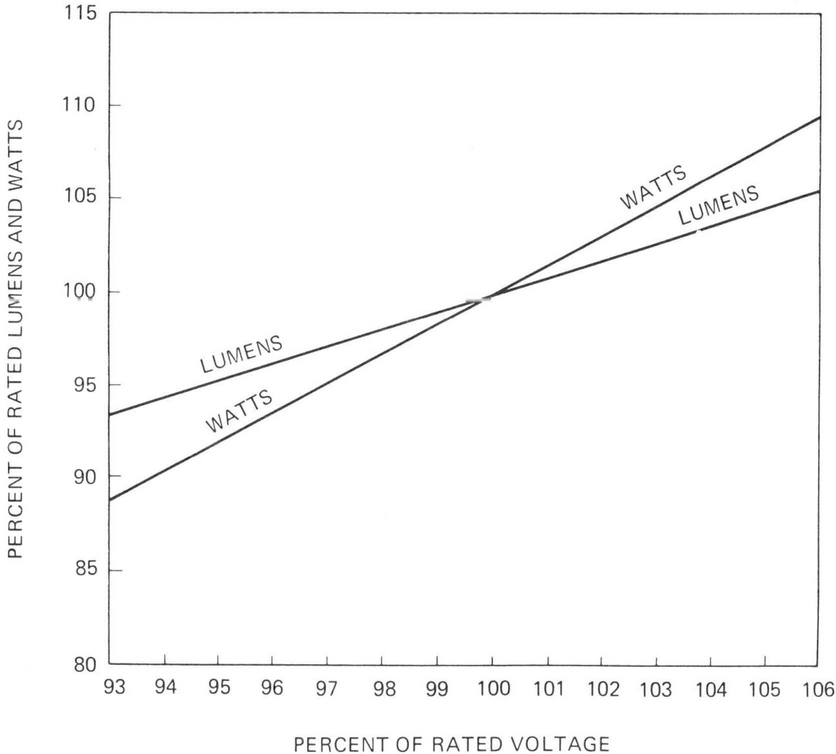

FIGURE 3–6. Operating characteristics of fluorescent units.

arranged to provide uniform illumination throughout the space. (A luminaire is a complete lighting unit and consists of a light source, together with a globe, reflector, refractor, socket, housing, and other parts integral to the housing.)

Conversely, if the quantity and type of equipment is known for a given space, the illumination level may then be calculated. A practical equation for illumination calculations must take into account the varying degrees of absorption of walls, ceiling, and floor surfaces, the reflection of light, the efficiency of distribution of the luminaire, the shape of the room, and the point at which the illumination is needed (for example, floor, desk top, or workbench).

A quantity that takes into account all the influences mentioned above is the *coefficient of utilization* (CU). To account for the depreciation of illumination over time due to dirt on room and luminaire surfaces, and again of the light sources, two factors must be included in the equation: *lamp lumen depreciation* (LLD) and *luminaire dirt depreciation* (LDD).

When all factors are assembled, we have:

$$\text{fc} = \frac{\text{lm}}{A} \times \text{CU} \times \text{LLD} \times \text{LDD} \tag{3–1}$$

where fc = average maintained illumination in foot candles
lm = lumens generated by light sources
A = lighted area in square feet (length × width)
CU = coefficient of utilization
LLD = lamp lumen depreciation
LDD = luminaire dirt depreciation

To determine the number of luminaires needed to provide a given average illumination level, we simply rearrange Equation 3–1 in this form:

$$\text{Total luminaires} = \frac{\text{fc} \times A}{\text{CU} \times \dfrac{\text{lm}}{\text{lamp}} \times \dfrac{\text{lamps}}{\text{luminaire}} \times \text{LLD} \times \text{LDD}} \tag{3–2}$$

An alternative equation that provides luminaire spacing information is

$$\frac{A}{\text{luminaire}} = \frac{\dfrac{\text{lm}}{\text{lamp}} \times \dfrac{\text{lamps}}{\text{luminaire}} \times \text{CU} \times \text{LDD} \times \text{LLD}}{\text{fc}} \tag{3–3}$$

Then the total number of luminaires is

$$\text{Total luminaires} = \frac{\text{total room area}}{A/\text{luminaire}} \tag{3–4}$$

A Practical Example

The first floor of an old home has been made into a single large room that is to be used for work requiring generally excellent illumination. The room dimensions are 45-ft length, 30-ft width, and 10.5-ft height. The luminaire mounting height is 9 ft and the work-plane height is 3 ft. Lights are to be replaced at burnout. What type and how many luminaires are needed, and what is the most satisfactory arrangement of luminaires?

Seven tables are used in the solution that follows. For convenience, they are located in the Appendix at the rear of this book and are numbered 1 through 7. All the tables have been reproduced with permission from the *Illuminating Engineering Society* (IES) *Handbook*. If you become involved in sufficient illumination work, you should purchase a copy of this handbook for ease of reference.

Step 1. Decide on the average maintained illumination required. If you need help in making this decision, contact the engineering department of your local electric utility company. Currently, you may find considerable differences of opinion on this subject, since government agencies concerned with energy conservation are recommending levels of illumination that are considerably lower than those deemed appropriate by many illumination engineers. For example, most illumination engineers would recommend about 75 fc of illumination at an office desk top. But government agencies now recommend 50 fc for the same situation.

In making your decision, give due weight to such factors as operating economy, safety, efficiency of workers (or occupants), and the physical characteristics of the individuals involved (for example, average age, vision, physical handicaps, and so forth).

In this particular example, we shall arbitrarily assume an average maintained illumination of 150 fc, since the data given specify the need for generally excellent illumination.

Step 2. Determine the *reflectances* (percentages of light reflected by ceiling, wall, and floor surfaces). For best results, ceilings and walls should be a color that reflects as much light as possible. But the finish should be dull or flat to avoid the mirror effect of glare-producing spots.

Paint and wood manufacturers can generally provide reflectance values for their products. Typical values for white, pale pink, pale yellow, or ivory approximate 80 percent; buff, light gray, or light blue, about 60 percent; most dark colors, around 10 percent; and black, 2 percent. Natural-finish woods may reflect anywhere from 5 to about 45 percent.

To avoid unnecessary complication, we shall assume the following reflectances: ceiling 80 percent, walls 50 percent, and floor 30 percent.

Step 3. Select the luminaire. Use Table 1 (or similar manufacturer's literature) to make your selection. Let us assume that we elect to use a two-lamp, prismatic lens, bottom unit with open top. Notice that the LDD maintenance category for this luminaire is VI, and that to obtain the typical distribution pattern shown the ratio of maximum spacing between luminaire centers to the mounting (or ceiling) height above the work plane (Max. S/MH_{wp}) is 1.2.

Step 4. Select the lamp type. Use Table 2 (or similar manufacturer's literature) for this purpose. For this problem we pick the fluorescent type F48T12/CW/HO, rated at 60 W. (Although it is not shown in Table 2, a lamp of this type provides approximately 80 lm/W, or a total of 4800 lm.) Since the lamps are to be replaced on burnout, we shall use the LLD factor of 79 percent.

Step 5. Determine the room ratios by using Table 3. With suspended luminaires and a work plane not on the floor, there are three cavities: (1)

ceiling (space from ceiling to luminaire), (2) room (space from luminaire to work plane), and (3) floor (space from work plane to floor). In this case, the ratios turn out to be 0.4, 1.7, and 0.8 for ceiling, room, and floor.

Step 6. Use Table 4 to determine the percent of *effective* ceiling- and floor-cavity reflectances for various reflectance combinations. These values are shown as 74 percent (ceiling) and 27 percent (floor).

Step 7. Refer back to Table 1 to determine the coefficient of utility. In the table, notice that the percent of effective ceiling-cavity reflectance is abbreviated p_{cc}, the percent of wall reflectance, pw, and the room-cavity ratio, RCR. The actual value of p_{cc} in this problem is 74 percent. But the closest value shown on the table is 70 percent, and this will suffice. Similarly, the actual room-cavity ratio is 1.7, but the nearest value shown in Table 2 (and the value we shall use) is 2. By using p_{cc} = 70 percent and RCR = 2, the value of CU turns out to be 0.58. But notice that the CU values given in the table are for 20 percent effective floor-cavity reflectance (p_{FC}). Since our value for p_{FC} is 30 percent, we use Table 5 to find the correction factor that is a multiplier of 1.06. Using this factor, CU = 0.61.

Step 8. Calculate the work-plane area. $45 \times 30 = 1350$ ft^2.

Step 9. Determine the LDD. The LDD depends on the degree of dirt and the maintenance category of the luminaire (already determined to be VI). Table 6 provides guidance in selecting the degree of dirt. For our facility, we shall use the classification of very clean.

Now we go to Table 7 to find the LDD factor. Since the room is newly finished and will probably be cleaned often, we shall use an LDD of 90 percent.

The preceding steps give us all the information that we need to use Equation 3–2. For convenience, that equation is repeated here.

$$\text{Total luminaires} = \frac{\text{fc} \times A}{\text{CU} \times \dfrac{\text{lm}}{\text{lam}} \times \dfrac{\text{lamps}}{\text{luminaire}} \times \text{LLD} \times \text{LDD}}$$

By substituting the values arrived at in our example, we obtain

$$\text{Total luminaires} = \frac{150 \times 1350}{0.61 \times 4800 \times 2 \times 0.79 \times 0.90}$$

$$= \frac{212{,}500}{4164}$$

$$= 51$$

Space limitations between semidirect luminaires are related to their mounting height above the work plane. From Table 1 we know that the S/MH_{wp} ratio for the luminaires being used is 1.2. So our spacing between rows should not exceed $1.2 \times 6 = 7.2$ ft. And the end-to-end spacing cannot exceed the room-cavity height minus 4 ft, or $6 - 4 = 2$ ft.

With the 4-ft fluorescents being used, the best arrangement would be to use three rows of eight luminaires and three rows of nine luminaires. The end-to-end spacing between the luminaires in the shorter rows would be 1½ ft, and the end luminaires would be 1¼ ft from the walls. The end-to-end spacing in the longer rows would be 1 ft, and the end luminaires would be 6 in. from the walls. Notice that the luminaire spacings are substantially less than the maximum permissible allowances. This practice is recommended for all lighting, and is particularly important for direct and semidirect equipment.

3–5 CURRENT AND POWER REQUIREMENTS

For incandescent lighting systems, the power factor is unity (1), and the total power is simply the sum of the wattage ratings of the individual lamps. The current is simply $I = P/E$, where P is the total power and E is the applied EMF.

The NEC does not require that the power rating of fluorescent units be given. It does require, however, that all such units be plainly marked with their operating voltage, current—including ballasts and transformers, and frequency.

Without correction, the power factor of fluorescent units most often ranges from 0.5 to 0.6. When power-factor correction is provided, by means of a capacitor built into the ballast, the unit is identified in technical data as HPF (high power factor), and the power factor is about 0.9.

To illustrate the wisdom of using HPF units, assume the use of a fixture that has two 40-W lamps operating at 120 V, 60 Hz.

The total *lamp* power is 80 W. To allow for losses by auxiliary equipment, add 20 percent. In this case, $0.20 \times 80 = 16$ W, and the total *fixture* power is $80 + 16 = 96$ W.

With a HPF unit,

$$I = \frac{P}{E(PF)} = \frac{96}{120(0.9)} = 0.888 \text{ A}$$

If a noncorrected unit is used,

$$I = \frac{P}{E(PF)} = \frac{96}{120(0.5)} = 1.60 \text{ A}$$

If the lamps were to be installed in a circuit carrying a maximum of 15 A, we could use 16 of the HPF units, but only 9 of the noncorrected units.

Although incandescent lighting fixtures may be less expensive to install, they produce less lumens per watt than fluorescent units. Thus, for a given level of light, operating costs are higher with incandescent units. In air-conditioned spaces, the use of incandescent fixtures also increases the air-conditioning load.

EXERCISES

3–1. Name five variables that affect lighting.

3–2. Define the term *footcandle*.

3–3. Define the term *lumen*.

3–4. What relationship exists between footcandles and lumens?

3–5. What is a *light meter*, and how does it work?

3–6. Give an example of how light varies with distance.

3–7. What advantages are associated with incandescent lamps?

3–8. What blackens the inner surface of an incandescent lamp as the lamp ages?

3–9. How is an incandescent lamp affected when it is operated at higher-than-design voltages?

3–10. How is an incandescent lamp affected when it is operated at lower-than-design voltages?

3–11. What are the advantages of fluorescent lamps compared to those of the incandescent variety?

3–12. What are the disadvantages of fluorescent lamps compared to the incandescent variety?

3–13. Explain how a fluorescent tube using *preheat* starting operates.

3–14. Describe how a *glow-type* starter works.

3–15. How does a *bimetallic strip* operate?

3–16. What two functions are performed by a *ballast*?

3–17. Why is a *capacitor* connected across the starter contact points?

3–18. Explain how a fluorescent lamp can be operated from a dc source of EMF.

3–19. Why is a capacitor often placed in series with the filament of one lamp in a multilamp fluorescent fixture?

3–20. Name *three* starting methods for fluorescent lamps and describe each briefly.

3–21. How are fluorescent lamps rated? How does performance vary with deviations from rated voltages?

3–22. What is actually meant by the term *fixture*? What other name is used in some literature in place of fixture?

3–23. What is the *coefficient of utilization*?

3–24. What determines lamp lumen depreciation?

3–25. Explain the significance of the LLD.

3–26. What is the power factor in an incandescent lighting system? Why?

3–27. What is the power factor in fluorescent lighting systems? Why?

3–28. Explain how the power factor can be improved in a fluorescent lighting system.

3–29. What practical advantage is associated with the use of HPF fluorescent fixtures?

3–30. Over a span of time, which type of lighting system, incandescent or fluorescent, would prove the more economical? Why?

II

ELECTRIC CIRCUIT COMPONENTS

4

Electrical Conductors

4–1 THE NATIONAL ELECTRICAL CODE

The National Electrical Code (NEC) first adopted in 1897, is sponsored by the National Fire Protection Association (NFPA) under the auspices of the American National Standards Institute (ANSI). To assure the standardization and proper installation of electrical equipment, the code establishes *minimum* standards for electrical installations within buildings and other premises.

The code is divided into *articles*, each of which contains as many *sections* (subdivisions) as is necessary for clarification and comprehensiveness. Each article or group of articles is written by a panel of experts. Updates and changes are made from time to time so that a revised code is published about every 3 years.

The adoption and enforcement of the NEC is usually left to the discretion of *local* authorities rather than to state or national authorities. Local wiring inspectors may therefore differ in their interpretation of some provisions of the code, since NEC standards are minimum requirements. In the absence of any local authority, the NEC prevails, however—usually under the authority of the electric utility company.

The NEC is *not* an instruction manual, but it does set an official minimum standard for the determination of many electrical design problems.

So that you will be able to use the code intelligently, take a few moments now to scan its table of contents and to read Article 90, the Introduction, in its entirety. Read also Article 100, Definitions, and Article 110, Requirements for Electrical Installations. By so doing, you will acquire a "feel" for the language of the code.

Notice that a two-number method of identification is used in referring to a particular portion of the code. The first number identifies the article and the second number identifies a section of that particular article. For example, reference 90-1 indicates Article 90, Section 1.

A similar identification method is used to refer to sections of this text, with the first number designating the chapter and the second number designating the section. Since the lowest-numbered article in the code is 90, while the highest-numbered chapter in this book is 13, no confusion should result.

Product Listing and Labeling

Before starting the production of electrical products, most U.S. manufacturers submit samples of their product to a nationally recognized testing laboratory, inspection agency, or other organization concerned with product evaluation. Products that meet nationally recognized standards are then *listed* by these organizations. Listing simply means that the items have been found suitable for use in a specified manner.

Some listing organizations also require *labeling*; that is, the equipment must carry a label, symbol, or other identifying mark that indicates compliance with nationally recognized standards or tests to determine suitable usage in a specified manner.

It is always wise to check with a local inspector to see if you should buy listed and/or labeled electrical equipment.

Local Codes, Permits, and Licenses

Frequently, local ordinances limit or qualify the NEC standards. You must also know, therefore, the ordinances that apply in your region.

Always check with your local governmental agency to determine whether or not a permit is required before a wiring job is started. If one is required, a permit must be shown to the local inspector. Power companies may refuse to supply power in the absence of an inspection certificate.

If electrical wiring is your *business*, state laws usually dictate that you obtain a license. On the other hand, if you are doing wiring in your own home and do not engage in this activity as a business, no license is generally required. To avoid possible legal difficulties, however, verify the no-license requirement before you start work.

4–2 THE CIRCULAR MIL AND WIRE SIZES

As we saw in Section 1–9, when current passes through the resistance of any practical electric circuit, the *rate* at which electrical energy is converted to heat energy is called *power*, designated P. Mathematically,

$$P = I^2R \tag{4–1}$$

Equation 4–1 tells us that the amount of heat generated by a current-carrying conductor of fixed resistance is *directly proportional* to the *square* of the current passing through the conductor. Thus, if the current is doubled, the amount of heat that the conductor can dissipate to the ambient (surrounding environment) is quadrupled. If the current is tripled, the amount of heat that the conductor must dissipate is multiplied by a factor of 9. Obviously, with a conductor of given size, if the current is steadily increased, a point is eventually reached where the conductor cannot dissipate the amount of heat generated and a definite fire hazard exists.

As will be shown in Section 4–7, the NEC specifies the maximum amperage a given conductor can handle safely under varying conditions. At this point, however, let us turn our attention to the problem of how wire sizes are specified.

Conductors used for residential wiring are usually round, and the unit used for measuring their cross-sectional area is the circular mil (cmil). A mil is 0.001 in., and a conductor having a cross-sectional area of 1 cmil has a diameter of 1 mil.

The cross-sectional area of any round conductor is, therefore, equal to the square of the diameter in mils; that is,

$$A = D^2 \tag{4–2}$$

where A is the area in cmil and D is the diameter in mils.

If, for example, a round conductor has a diameter of 3 mils (0.003 in.), its cross-sectional area is 3^2, or 9 cmil. To take another example, if a round conductor has a diameter of 9 mils (0.009 in.), its cross-sectional area is 9^2, or 81 cmil.

TABLE 4–1. Properties of Conductors

AWG Size	Area (Cmil)	No. of Wires	DC Resistance at 25°C, 77°F (Ω/1000 ft)	
			Copper	Aluminum
18	1,620	Solid	6.51	10.7
16	2,580	Solid	4.10	6.72
14	4,110	Solid	2.57	4.22
12	6,530	Solid	1.62	2.66
10	10,380	Solid	1.018	1.67
8	16,510	Stranded	0.6404	1.05
6	26,240	Stranded	0.410	0.674
4	41,740	Stranded	0.259	0.424
3	52,620	Stranded	0.205	0.336
2	66,360	Stranded	0.162	0.266
1	83,690	Stranded	0.129	0.211
0	105,600	Stranded	0.102	0.168
00	133,100	Stranded	0.0811	0.133
000	167,800	Stranded	0.0642	0.105
0000	211,600	Stranded	0.0509	0.0836
	250,000	Stranded	0.0431	0.0708
	300,000	Stranded	0.0360	0.0590
	350,000	Stranded	0.0308	0.0505
	400,000	Stranded	0.0270	0.0442
	500,000	Stranded	0.0216	0.0354
	600,000	Stranded	0.0180	0.0295
	700,000	Stranded	0.0154	0.0253
	750,000	Stranded	0.0144	0.0236
	800,000	Stranded	0.0135	0.0221
	900,000	Stranded	0.0120	0.0197
	1,000,000	Stranded	0.0108	0.0177

The American Wire Gauge

To assure the manufacture of conductors in sizes suitable for all applications, the American wire gauge was developed to assign a number to a particular size of wire. The AWG numbers start at 40, the *smallest*, having a diameter of 3.145 mils, which is about the diameter of a human hair. The gauge numbers then *descend* in order to 0000, the *largest*, with a diameter of 460 mils. The size of any conductor larger than 0000 is expressed directly in its circular-mil area, beginning with 250,000 cmil. The largest recognized standard size is 2,000,000 cmil.

In some literature the reader will find circular mil abbreviated CM and the prefix M used to indicate thousand. Thus 250,000 cmil, for example, may also be written as 250 MCM. It is also common practice to designate size 00 as 2/0, size 000 as 3/0, and size 0000 as 4/0.

Table 4–1 lists the properties of standard-sized conductors from no. 18 to 1,000,000 cmil. Except as provided by NEC Sec. 310-3, all conductors no. 8 and larger are stranded.

The last two columns of Table 4–1 give the dc resistance in ohms per 1000 ft (Ω/1000 ft) of copper and aluminum conductors at room temperature, 25°C or 77°F. The practical application for information of this type is shown in Section 4–9.

A *wire gauge*, which is a hand-held, circular metal disk with slots around its edge, is handy for determining the exact size of a wire. Each slot is marked with its corresponding AWG number. The slot into which the wire fits, *not* the opening behind the slot, indicates the wire size.

Skin Effect

When alternating current flows in a conductor, it has a tendency to flow toward the outside of the conductor rather than through the cross section uniformly. This *skin effect* increases as the frequency of the alternating current is increased. Its effect is slight in small conductors, but it must be carefully considered in large conductors. To compensate for skin effect at a frequency of 60 Hz ac, multiply the dc resistance shown in Table 4–1 by the appropriate factor listed in Table 4–2.

TABLE 4–2. Multiplying Factors for Converting DC Resistance

AWG Size or Cmil	*For Nonmetallic-Sheathed Cables or in Air*		*For Metallic-Sheathed Cables or in Metallic Raceways*	
	Copper	*Aluminum*	*Copper*	*Aluminum*
Up to 3	1	1	1	1
2	1	1	1.01	1
1	1	1	1.01	1
0	1.001	1	1.02	1
00	1.001	1.001	1.03	1
000	1.002	1.001	1.04	1.01
0000	1.004	1.002	1.05	1.01
250,000	1.005	1.002	1.06	1.02
300,000	1.006	1.003	1.07	1.02
350,000	1.009	1.004	1.08	1.03
400,000	1.011	1.005	1.10	1.04
500,000	1.018	1.007	1.13	1.06
600,000	1.025	1.010	1.16	1.08
700,000	1.034	1.013	1.19	1.11
750,000	1.039	1.017	1.21	1.12
800,000	1.044	1.017	1.22	1.14
1,000,000	1.067	1.026	1.30	1.19

Table 4–2 shows that for sizes smaller than 2/0, skin effect requires little consideration; but for large sizes it is important, especially when conductors are enclosed in metal *raceways*, that is, any channel for holding wires, cables or bus bars that is designed expressly for and used solely for this purpose. Notice also that the correction factor for copper is appreciably *higher* than for aluminum.

Wire-Gauge Approximations

In setting up the AWG wire table, the cross-sectional dimensions are chosen so that each *decrease* of one gauge number represents a 25 percent *increase* in cross-sectional area. Thus a decrease of three gauge numbers represents an increase in cross-sectional area of $1.25 \times 1.25 \times 1.25$, or approximately a 2:1 increase. Similarly, a change of 10 wire-gauge numbers represents a change of 10:1 in cross-sectional area. Moreover, since doubling the cross-sectional area cuts the resistance in half, a decrease of three wire-gauge numbers cuts the resistance of a conductor of a given length in half. If we memorize the dimensions for any one wire-gauge number, therefore, we can learn the approximate dimensions for any other wire-gauge number. For this purpose, AWG no. 10 wire is quite convenient. For practical purposes, the diameter is taken as 100 mil, the area as 10,000 cmil, and the resistance for copper as 1 Ω/1000 ft at room temperature.

To illustrate the usefulness of the above approximations, let us determine the resistance of 200 ft of AWG no. 14 copper wire at room temperature. Since the cross-sectional area of AWG no. 14 wire is $1/(1.25)^4$ that of AWG no. 10 wire,

$$R = 2.5 \times \frac{200}{1000} \times 1\ \Omega = 0.5\ \Omega$$

4–3 NATIONAL ELECTRICAL CODE COLOR SCHEME

Any special exceptions will be discussed as the need arises, otherwise the general color scheme established by the NEC is explained below.

Wires

- Only *white* may be used for the grounded neutral wire.
- The grounding wire from a normally non-current-carrying component, such as the frame of a motor, may be *green, green and yellow stripe*, or *bare* (uninsulated). The green and yellow stripe is usually found on

appliances intended for export to European markets, as well as those for domestic sale.

- The color of the *hot* wire (or wires) used in conjunction with the white neutral wire depends on the number of wires in the circuit and is as follows:

  Two-wire circuit: black
  Three-wire circuit: black, red
  Four-wire circuit: black, red, blue
  Five-wire circuit: black, red, blue, yellow

Terminals

- On *grounding* receptacles, that is, two parallel slots plus a third round or U-shaped opening for a corresponding prong on the plug, the round or U-shaped opening leads to a special *green* terminal. When installing the receptacle, a wire (green, green and yellow stripe, or bare and uninsulated) must be connected from the green terminal to ground.
- On switch lugs, receptacles, and the like, the *hot* terminal is *natural copper* or *brass* and the *ground* terminal is a *whitish* color, such as *nickel, tin* or *zinc plate*.

4–4 INSULATION

The term *wire* generally refers to any *round* conductor up to AWG no. 4/0. A conductor larger than size 4/0 is a *cable*. Both wires and cables are distinguished from flat bars, called *bus bars*, which are often used in place of cables of 1,000,000 cmil and larger. Due to its increased rectangular surface area, a bus bar provides better cooling than round cable and is usually less cumbersome to install.

The *lightest*-sized wire permitted by the NEC for ordinary residential wiring is size AWG no. 14. Sizes greater than 4/0 are seldom encountered in such applications.

Locations

Wires of a given type are often restricted to use in locations referred to as *damp*, *dry*, and *wet*. Article 100 of the code defines these three types of locations.

- *Dry location*—A location not normally subject to dampness or wetness, but which may be temporarily subject to dampness or wetness, as is the case of a building under construction.

- *Damp location*—Partially protected locations under canopies, marquees, roofed open porches, and like locations, and interior locations subject to moderate degrees of moisture, such as some basements, some barns, and some cold-storage warehouses.
- *Wet location*—Installations underground or in concrete slabs or masonry in direct contact with the earth, and locations subject to saturation with water or other liquids such as vehicle-washing areas and locations exposed to weather and unprotected.

Insulation

For many years, *natural rubber* was the conductor insulating material most widely used. But, such insulation dried out, cracked, and became brittle from age and heat. So rubber has been largely replaced with *thermoplastic*. Other improved materials are also being developed constantly. *Cross-linked synthetic polymer*, for example, has insulating characteristics far superior to either rubber or thermoplastic. It may very well replace the other two almost entirely in the years ahead.

The amount and type of insulation to be utilized is determined by the potential difference (voltage) between conductors and by the location (damp, dry, or wet) in which the conductor is to be used.

To identify the numerous types of insulation, the NEC has established a letter-code system that describes their characteristics. Some of the most common types still being manufactured are shown in Table 4–3. The letters R, T, and X refer to the material—rubber, thermoplastic, or cross-linked synthetic polymer. H and HH will endure progressively higher temperatures. W indicates that a given insulation is moisture resistant; N indicates a final extruded jacket of nylon or the equivalent. The insulations listed in Table 4–3 are suitable for ordinary residential and farm wiring at 600 V or less. A complete listing of conductor insulations, maximum operating temperatures, application provisions, thickness of insulation, and type of outer covering is given in NEC Table 310-13.

4–5 CABLES

In the preceding section, any conductor larger than AWG no. 0000 was classified as a cable. The word *cable* is also generally taken to mean two or more insulated conductors enclosed by the manufacturer in a common external jacket, which may be either metallic or nonmetallic. In referring to such cables it is common to use a designation such as 12–4, for example; the 12 indicates wire size, the 4 the number of wires in the cable.

TABLE 4–3. The NEC Letter System for Insulation Identification

Trade Name	Type Letter	Maximum Operating Temperature(°F)	Application
Heat resistant	RHH	194	Dry locations
Moisture and heat-resistant rubber	RHW	167	Dry and wet locations
Thermoplastic	T	140	Dry locations
Moisture-resistant thermoplastic	TW	140	Dry and wet locations
Moisture- and heat-resistant thermoplastic	THW	167	Dry and wet locations
Moisture- and heat-resistant thermoplastic	THWN	167	Dry and wet locations
Heat-resistant thermoplastic	THHN	194	Dry locations
Moisture- and heat-resistant cross-linked synthetic polymer	XHHW	194 167	Dry locations Wet locations

Armored Cable (Type AC)

Type AC cable, commonly known as BX cable, and frequently used in residential wiring, is a fabricated assembly of insulated conductors in a flexible metallic enclosure. (See Article 333).

The covering of type AC cable is a flexible metal tape. At all points of termination, special fittings are used to protect the wires from abrasion (unless the box or fitting affords equivalent protection). An approved insulating bushing or its equivalent is placed between the conductors and the metal tape.

The conductors in type AC cables are rubber covered. When the conductors are thermoplastic covered, the cable is designated type ACT. Type ACL cable contains lead-covered conductors and is used in damp or wet locations.

Metal-Clad Cable (Type MC)

Type MC cable is a factory assembly of one or more conductors, each individually insulated and enclosed in a metallic sheath of interlocking tape, or a smooth or corrugated tube.

Cables of the MC series may be used in systems where the voltage exceeds 600 V. It is unlikely that you will encounter cable of this type in residential wiring.

For detailed information about type MC cables, see Article 334.

Nonmetallic-Sheathed Cable

A nonmetallic-sheathed cable is a factory assembly of two or more insulated conductors having an outer sheath of moisture-resistant, flame-retardant, nonmetallic material (Sec. 336-1).

Two types of nonmetallic-sheathed cable are permitted: NM and NMC. The use of type NM, often called *Romex*, is restricted to normally dry locations. For damp or wet locations, type NMC must be used.

Both types are available with either two or three conductors, and most also contain a bare, uninsulated wire for grounding. Conductor sizes range from nos. 14 to 2 AWG for copper, and from nos. 12 to 2 AWG for aluminum and/or copper-clad aluminum.

Service-Entrance Cable

Types SE and USE only are permitted. As the name implies, cables of this type are used primarily for service entrances, but they may also be used as specified for certain other purposes (Sec. 338-3). The prefix U indicates underground. These cables will be thoroughly described in Chapter 6.

Miscellaneous Cable Types

Type MI or mineral-insulated metal-sheathed cable (Article 330), type SNM or shielded nonmetallic-sheathed cable (Article 337), and type UF underground feeder and branch-circuit cable (Article 339) are all cables not ordinarily installed in normal residential wiring, but unusual circumstances may dictate their use.

4–6 WIRE MISCELLANY

Fixture Wire

Fixture wire is a special type used only for the internal wiring of lighting fixtures; it is never used in a circuit leading up to a fixture. The minimum size for fixture wire is no. 18 AWG.

Numerous types of fixture wire are available. The type you select for a particular application is determined primarily by the circuit amperage and ambient temperature.

To avoid a rather lengthy discussion of limited value in this text, the reader is referred to Article 402 of the code.

Low-Voltage Wire

When a potential of 30 V or less is applied to a circuit, *bell* wire (also called *annunciator* wire) is used. The conductors are no. 18 for residential wiring, and the insulation is paraffin impregnated cotton. Common applications include doorbells and thermostats.

4–7 AMPACITY

The maximum current in amperes that a conductor can carry safely and continuously is called its *ampacity*. In addition to its cross-sectional area, several factors affect the ampacity of a given conductor. If conductors are in free air where they can readily dissipate heat, their ampacity is much greater than if enclosed in a raceway—the outer sheath of cables is considered a raceway in this context. Moreover, when more than three conductors are in the same raceway, their ampacity is systematically reduced so that the heat of the ambient, plus that produced by the passage of current through the conductors, does not exceed the heat rating of the conductor insulation.

NEC Tables 310-16 through 310-19, and their associated notes, provide detailed information on the maximum, continuous ampacities of copper and aluminum conductors. Tables 4–4 through 4–6 are excerpts from those code tables. The following examples will show you some practical applications of these tables.

Let us determine the ampacity of size AWG no. 8 copper conductors with THW insulation, three in a cable. The installation is in a location where the ambient temperature is expected to reach a maximum of 104°F (40°C).

In Table 4–5 we see that the temperature rating of the conductor is 167°F (75°C), and the ampacity, based on an ambient temperature of 86°F (30°C), is 45 A. In Table 4–6, the derating factor at the elevated temperature is shown as 0.88. Thus the ampacity is

$$45 \times 0.88 = 39.6 \text{ A}$$

As another example, suppose that for a load rated at 30 A you want to know what size of copper conductors (type THW insulation) to use in

TABLE 4–4. Allowable Ampacities in Amperes of Single, Insulated Conductors in Free Air

Size AWG	*Copper*		*Aluminum or Copper-Clad Aluminum*	
	140°F (60°C) RUW, T, TW, UF	*167°F (75°C) RH, RUW, RUH, THW, THWN, XHHW, USE, ZW*	*140°F (60°C) RUW, T, TW, UF*	*167°F (75°C) RH, RHW, RUH, THW, THWN, XHHW, USE, ZW*
14	20	20		
12	25	25	20	20
10	40	40	30	30
8	55	65	45	55
6	80	95	60	75
4	105	125	80	100
3	120	145	95	115
2	140	170	110	135
1	165	195	130	155
0	195	230	150	180
00	225	265	175	210
000	260	310	200	240
0000	300	360	230	280

TABLE 4–5. Allowable Ampacities of Insulated Conductors, Maximum of Three in Raceway or Cable or Earth (Directly Buried)*

Size AWG	*Copper*		*Aluminum or Copper-Clad Aluminum*	
	140°F (60°C) RUW, T, TW, UF	*167°F (75°C) FEPW, RH, RHW, RUH, THW, THWN, XHHW, USE, ZW*	*140°F (60°C) RUW, T, TW, UF*	*167°F (75°C) RH, RHW, RUH, THW, THWN, XHHW, USE*
14	15	15	15	15
12	20	20	25	25
10	30	30	30	40
8	40	45	40	50
6	55	65	55	65
4	70	85	65	75
3	80	100	75	90
2	95	115	85	100
1	110	130	100	120
0	125	150	115	135
00	145	175	130	155
000	165	200	155	180
0000	195	230		

*Based on Ambient Temperature of 86°F (30°C) and Expressed in Amperes.

TABLE 4–6. Correction Factors: Ambient Temperature Higher than 86°F (30°C) for Conductors Approved for Indicated Temperatures

Ambient Temperature		*Correction Factor: Copper, Aluminum, or Copper-Clad Aluminum*	
°F	*°C*	*140°F (60°C)*	*167°F (75°C)*
86–104	31–40	0.82	0.88
105–122	41–50	0.58	0.75
123–141	51–60		0.58
142–158	61–70		0.35
159–176	71–80		

Reprinted with permission from the 1978 National Electrical Code, Copyright 1977, National Fire Protection Association, Boston, Mass.

a three-wire cable. Again, let us presume an ambient temperature reaching a maximum of 104°F (40°C).

From Table 4–6 we learn that the derating factor for 104°F is 0.88 when using copper conductors with type THW insulation. Thus the ampacity must be increased to

$$\frac{30}{0.88} = 33.1 \text{ A}$$

By checking Table 4–5, we can then see that an ampacity of 33.1 A requires three size AWG no. 8 conductors in a cable.

Where the number of conductors in a raceway or cable exceeds three, additional derating (reduction) of ampacity is necessary. Setting aside the exceptions that the code identifies in the notes for Tables 310-16 through 310-19, the required reductions are listed in Table 4–7.

Any exceptions to the above procedures for determining ampacity are discussed as the need arises.

TABLE 4–7. Reduction Factor Applied to Ampacities When the Number of Conductors in a Raceway or Cable Exceeds Three

Number of Conductors	*Percent of Values in NEC Tables 310–16 and 310–18*
4–6	80
7–24	70
25–42	60
43 and above	50

4–8 TEMPERATURE CONVERSIONS

At times, it may be very convenient to know how to convert Fahrenheit degrees to Celsius, or vice versa. To do this, we use the relationship

$$°\text{C} = \frac{5}{9}(\text{temp. in } °\text{F} - 32) \tag{4–3}$$

For example, we convert 140°F to the Celsius equivalent as follows:

$$°\text{C} = \frac{5}{9}(140 - 32)$$

$$= \frac{5}{9} \times 108$$

$$= 60$$

To convert degrees Celsius to Fahrenheit, use the relationship

$$°\text{F} = \frac{9}{5}(°\text{C}) + 32 \tag{4–4}$$

To express 60°C in Fahrenheit, for example, we have

$$°\text{F} = \frac{9}{5}(60) + 32$$

$$= 108 + 32$$

$$= 140$$

4–9 VOLTAGE DROPS

Two basic problems influence the selection of the *correct conductors* for a given application: *ampacity*, which we considered in Section 4–7, and *voltage drop*.

In Section 1–9, we learned that power is *directly proportional* to the *square* of the voltage drop appearing across the resistance of any practical conductor ($P = V^2/R$). Since power represents the rate at which electrical energy is converted to heat energy, and since the heat loss performs no useful function, it represents *waste*. As you well know from your monthly utility bill, electrical energy is too expensive to waste.

An equally important reason for minimizing such waste is that the quality performance of electrical devices is usually dependent on the voltage that they receive. Suppose, for example, that a lighting system, as a result of the voltage drop in the conductor connecting it to the source, operates at 95 percent of its rated voltage. Since

$$P = \frac{V^2}{R} = (0.95)^2 = 90 \text{ percent}$$

the lights will produce only 90 percent of their rated output.

From the above, it is apparent that we must select conductors of cross-sectional area sufficient to keep the voltage drop within satisfactory limits.The wire selected must also have sufficient ampacity. When voltage drop and ampacity are both considered, select the *largest* conductor.

Section 210-19(a) of the code reads as follows: "Conductors for branch circuits as defined in Article 100, sized to prevent a voltage drop exceeding 3 percent at the farthest outlet of power, heating, and lighting loads, or combinations of such loads and where the maximum total voltage drop on those feeders and branch circuits to the farthest outlet does not exceed 5 percent, will provide reasonable efficiency of operation."

In ordinary residential wiring, *branch circuits* start at the fuse (or circuit-breaker) panel and there are no *feeders*. Thus the voltage drop in the system should not exceed 3 percent of the supply voltage. On farms, feeders run from a pole-mounted meter to the point where they enter a building. (Complete definitions for the terms branch circuit and feeder are given in Section 6–1). Allowing a 2 percent drop on the feeders plus a 3 percent drop on branch circuits, the overall drop is 5 percent. Notice, however, that the recommended voltage drops are *maximums*. In practice, a 2 percent drop on both branch circuits and feeders is a wise and widely accepted choice when the cost of the heavier conductors is weighed against the cost of power loss in smaller-sized conductors over an extended period of time.

To explain the selection of satisfactory conductors, let us assume the use of copper conductors with type T insulation in a metallic raceway to supply an 85-A load. The load is located 150 ft from a single-phase, 120-V, 60-Hz source. The voltage drop is not to exceed 2 percent of the source voltage, and the ambient temperature will never rise over 86°F (30°C).

Given these specifications, the voltage drop must not exceed $0.02 \times 120 = 2.4$ V. Since the load is 150 ft from the source, the total length of wire required is $2 \times 150 = 300$ ft. The maximum allowable resistance of the conductor, by Ohm's law, is therefore

$$R = \frac{V}{I} = \frac{2.4}{85} = 0.0282\ \Omega \text{ for } 300 \text{ ft}$$

Table 4–1 showed the dc resistance in ohms per 1000 ft for copper conductors. Size AWG no. 0 has a dc resistance of 0.102 Ω/1000 ft. So the resistance of 300 ft of this conductor is

$$0.102 \times 0.3 = 0.0306\ \Omega$$

which is greater than the maximum allowable resistance to limit the voltage drop to 2.4 V.

The next larger sized conductor, AWG no. 2/0, has a dc resistance of 0.0811 Ω/1000 ft, and

$$0.0811 \times 0.3 = 0.0243\ \Omega/300\ \text{ft}$$

To compensate for skin effect, we refer to Table 4–2 and use the multiplying factor of 1.03. Then

$$\Omega/300\ \text{ft} = 0.0243 \times 1.03 = 0.025\ \Omega$$

which is less than the maximum allowable resistance. Thus the voltage drop will be somewhat less than 2 percent (85 × 0.025 = 2.125 V, or 1.77 percent of the supply).

Referring to Table 4–5, we find that AWG no. 2/0 wire has an ampacity of 145 A, which is more than adequate for this application.

Also notice in Table 4–5 that, *without considering the voltage drop*, the selection of size AWG no. 3 is indicated; and in Table 4–1, the dc resistance per 1000 ft of no. 3 is 0.205 Ω. Each 100 ft would, therefore, have a resistance of 0.0205 Ω, and the resistance for 300 ft would be 0.0615 Ω. Using these values, the voltage drop across the conductors would be

$$V = IR = 80 \times 0.0615 = 4.92\ \text{V}$$

which is more than twice the allowable voltage drop. This simply reinforces the earlier statement that, considering both the ampacity tables and the voltage drop, the largest conductor is the one to use.

Now let us take a look at another, more direct method for determining the wire size to use for a particular application. The equation to use is

$$A = \frac{ID\rho}{V} \tag{4–5}$$

where A is the cross-sectional area in circular mils, V is the allowable voltage drop across the conductors, I is the load current in amperes, D is the distance from the supply to the load in feet, and ρ (the Greek letter "rho") is the *resistivity* of the conductors.

By definition, resistivity is simply the resistance in ohms of a sample 1 ft long and 1 mil in diameter. The unit of resistivity is ohms-cmils per foot (Ω-cmil/ft).

At first glance, the introduction of resistivity seems to throw a very complicated factor into otherwise simple calculations. Fortunately, this is not so. For electrical work all you have to remember is that the resistivity of copper is 10.4 Ω-cmil/ft, and for aluminum, 17 Ω-cmil/ft.

Considering only the right-hand member of Equation 4–6, we have the units

$$\frac{A \cdot \text{ft} \cdot \dfrac{\Omega\text{-cmil}}{\text{ft}}}{\text{V}}$$

In the numerator, the unit feet cancels out leaving

$$A \cdot \frac{\Omega\text{-cmil}}{\text{V}}$$

Since V = IR, we can rewrite the above as

$$\frac{A \cdot \Omega\text{-cmil}}{A\Omega}$$

When A and Ω are canceled out, we are left with cmils, the unit required for area in Equation 4–6.

Before we can work with Equation 4–6, one other factor, distance D, requires special comment. For single-phase, two- or three-wire systems, distance D must be multiplied by *two*. The neutral conductor of a single-phase, three-wire system is *not* included, since it carries no current with a balanced load and, therefore, creates no voltage drop. A three-wire system is designed so that it is balanced when the total load is applied. If part of the load is not in use, there will be some current through and voltage drop across the neutral conductor; but for practical computations we set aside this drop and assume a balanced load.

In three-phase systems, distance D is multiplied by the factor 1.732, which is the square root of 3.

To illustrate the usefulness of Equation 4–6, let us use the same data as in the preceding example and see how the two answers compare. For convenience, the data are repeated here.

We want to use copper conductors with type T insulation in a metallic raceway to supply an 85-A load. The load is located 150 ft from a single-phase, 120-V, 60-Hz ac source, the voltage drop does not exceed 2 percent of the source voltage, and the maximum ambient temperature is 86°F (30°C).

The permissible voltage drop is 120 × 0.03 = 2.4 V, the resistivity of copper is 10.4 Ω-cmil/ft, I = 85 A, and D = 2 × 150 = 300 ft. Substituting these values in Equation 4–6, we obtain

$$A = \frac{85 \times 300 \times 10.4}{2.4}$$

$$= 110{,}500 \text{ cmils}$$

From Table 4–1, size AWG no. 2/0 has an area of 133,100 cmils and is the wire selected. This is in agreement with the more roundabout selection technique demonstrated earlier.

In practice, economic considerations may dictate the acceptance of a somewhat higher voltage drop. That would be the case, for example, if the load in the above example were used intermittently and for short periods of time. The cost of any power wasted over a good many years under such circumstances would probably be much less than the cost of the heavier wire needed for a 2 percent voltage drop. In other words, common sense plays a very important role in the specifications for any actual installation.

Where it is permissible, the use of 240 V instead of 120 V may also result in an appreciable materials savings and, at the same time, keep the voltage drop within acceptable limits.

To illustrate the validity of the above statement, recall that $P = EI$. For a given power, if the voltage is doubled, the current is halved. With the current halved, the voltage drop across the fixed resistance of a given length of a particular size of wire is also halved, since $V = IR$. If the voltage drop was 2 percent of the applied EMF with a 120-V source, it is reduced to 1 percent when we use a 240-V source. This means that we can use a smaller-sized wire and still limit the voltage drop to 2 percent of the applied EMF.

In the previous example, we used a 120-V source and found the required wire size to be 2/0. Since the voltage drop across the conductors was limited to 2.4 V, the voltage drop across the source was $120 - 2.4 = 117.6$ V. The power dissipated by the load is

$$P = EI = 117.6 \times 95 = 9996 \text{ W}$$

Using a 240-V source, we can now determine the size of the wire required to deliver the same 9996 W to the load. Again using Equation 4–6, but with I reduced from 85 to 42.5 A, and with an allowable voltage drop of $240 \times 0.02 = 4.8$ V,

$$A = \frac{42.5 \times 300 \times 10.4}{4.8}$$

$$= 27{,}625 \text{ cmils}$$

There is a big difference between the cost of size AWG no. 2/0 and size no. 4, which has an area of 41,740 cmils. See Table 4–1.

4–10 ALUMINUM WIRE

It was noted earlier that the resistivity of aluminum is 17 Ω-cmil/ft, whereas that of copper is 10.4 Ω-cmil/ft. This means that the conductivity of aluminum is considerably *lower* than that of copper, so a *larger*-sized aluminum conductor is needed to carry a given current. For ordinary wiring, aluminum conductors are usually two AWG sizes larger than would be necessary for copper. This specification is not *always* accurate, however. Always check the ampacities of aluminum conductors by referring to Tables 310-16 through 310-19 of the code.

Because copper is, at present, much more common than aluminum in residential wiring, this text is concerned mainly with copper.

Major drawbacks to the use of aluminum conductors in residential wiring include the relatively great difficulty of soldering it, its tendency to loosen at screw terminals, and the present limitations on the use of solderless connectors.

4–11 RACEWAYS AND WIREWAYS

As we noted earlier, a raceway is a channel for holding wires, cables, or bus bars. For cables, the raceway is an integral part of the complete assembly. Here, however, we are concerned with raceways that are completely separate from the conductors. Included in this group are *conduit, electrical metallic tubing* (EMT), *surface raceways*, and *wireways*.

Raceways provide mechanical protection for the conductors and are mechanically installed as a complete system with all necessary outlet boxes and fittings. Conductors are then pulled through the raceway.

Rigid Metal Conduit (Article 346)

In outward appearance, rigid metal electrical conduits look like the pipes used for water, gas, or steam. However, they are lined with a smooth enamel that makes it easy to pull wires through them. Manufactured in 10-ft lengths, such rigid metallic conduit is connected to junction, outlet, and switch boxes by means of locknuts and bushings. This sort of conduit is generally used where there is a strong chance of mechanical injury or where moisture presents a problem.

Flexible Metal Conduit (Article 350)

Generally called *greenfield*, flexible metal conduit is, in effect, the outer casing only of type AC armored cable. Its use is prohibited in wet locations unless the conductors are of the lead-covered type or other type approved for the specific conditions. Generally, the smallest size is the ½-in. electrical trade size. Use of greenfield is rather limited, but it may prove convenient when an occasional change in the location of the load is required. *(The term electrical trade size does not denote the actual physical dimensions*. See Table 4, Chapter 9 of the code.)

Electrical Metallic Tubing (Article 348)

The EMT-type raceway has a thin wall that does not permit threading. Compression rings or set screws are used to secure connectors and couplings. Tubing sizes range from a minimum of ½-in. electrical trade size to a maximum of 4-in. electrical trade size. The use of EMT is prohibited where it is subject to severe physical damage, in cinder concrete or in cinder fill where it is subject to permanent moisture unless protected on all sides by a layer of noncinder concrete at least 2 in. thick or unless the tubing is at least 18 in. under the fill. The use of dissimilar metals may cause galvanic action and is not recommended. EMT, like rigid conduit, is manufactured in 10-ft lengths.

Surface Raceways (Article 352)

Surface raceways may be either metal or nonmetallic troughs into which the conductors are installed. They are intended primarily for adding extensions to existing circuits rather than for wiring a complete building. The location must be dry and the raceway must be exposed. Elbows, couplings, and similar fittings are so designed that the sections can be mechanically coupled and, if made of metal, electrically coupled while protecting the wiring from abrasions. Holes inside the raceway are drilled so that the heads of screws or bolts are flush with the surface.

Wireways (Article 362)

Wireways are sheet-metal troughs with hinged or removable covers. Conductors are laid in place after the wireway has been installed as a complete unit. Wireways may be installed only for open work; those intended for outdoor use must be made rainproof.

4–12 TERMINAL CONNECTIONS AND OTHERS

Terminal Connections (Sec. 110-14)

First the end is prepared by removing the insulation from the wire that attaches to a terminal and by cutting the insulation at a 60° angle. This reduces the possibility of nicking and weakening the conductor. Then the insulation is pulled off the conductor and the conductor is cleaned thoroughly.

To attach the conductor to a screw-type terminal, the prepared wire end is bent into a loop, with the loop left wide enough for open insertion under the screw head. Next the loop is inserted under the screw head in such a way that it will close when the screw is tightened. Use long-nosed pliers to close the loop. Then tighten the screw. Correctly done, the insulation should nearly butt on the edge of the terminal.

Many modern devices do not have terminal screws. For a good permanent connection it is merely necessary to insert the bare conductor to a depth indicated by a marker on the device. Again, the bare portion of the conductor should extend no more than a fraction of an inch out of the terminal hole.

Solderless terminals with a clamping nut are also commonly used for size AWG no. 6 and heavier conductors. The stripped and cleaned conductor is simply inserted into the terminal and secured in place by tightening the nut.

Joints

Where joints (splices) are permitted, they must be mechanically and electrically as sound as a continuous conductor; and insulation equivalent to that removed must be used to cover the joint.

When you strip insulation in order to make a joint, taper the remaining insulation to resemble a sharpened pencil.

With solid or small-stranded wires, join the two stripped ends at about two thirds their length. Then wrap the free ends in opposite directions until the joint lies as flat as possible. Butt adjacent turns of the joint and make sure that the free ends terminate at a point close to the original insulation.

With larger-stranded wires, bend the individual strands at about their halfway point so that they resemble a cone. Next arrange the cones so that they intermesh. Then take one strand and wrap around the assembly in a clockwise direction. Finally, take an opposite strand and wrap it around the assembly in a counterclockwise direction. Make all wraps as tight as possible and with adjacent turns butting. Repeat the procedure with each set of strands until all are wrapped.

Taps

A tap is formed when insulation is removed from a section of continuous wire and a tap wire is joined to the exposed section. With solid or smaller-sized stranded wires, place the exposed end of the tap wire *under* the continuous wire. Next, twist the tap wire one full turn in a counterclockwise direction and pass it under itself. Then wrap the remaining length of tap wire around the continuous wire in a clockwise direction.

If you prefer, you can also place the tap wire *over* the continuous wire initially. Reverse the directions of the wraps from those previously described.

With larger-stranded wires, separate the strands of the continuous wire into two equal groups and insert the tap wire into the opening. Then divide the strands of the tap wire into two equal groups and wrap each group around the continuous wire in opposite directions.

Soldering

Once a joint or tap is completed, solder it carefully. Use rosin-core solder; never use acid-core solder. Before you use the soldering iron, make sure that its tip is clean and well *tinned*, that is, covered with a thin coat of solder. In soldering, hold the iron *under* the joint or tap and heat the conductors until the solder applied to the *opposite* side of the conductors flows freely. Do not use an excessive amount of solder, and *do not move* the joint while the solder is hardening. When hard, the solder should have a shiny appearance. If the hardened solder has a dull-gray appearance, resolder the joint or tap.

Improperly soldered connections will exhibit high electrical resistance and will be weak mechanically. Soldering is an art that you can master only with practice. Also, make sure that you use an iron of the correct wattage. An iron of low wattage may not provide sufficient heat; one of excessive wattage may damage insulation. If you must use a torch-heated iron, make sure that you have proper instruction in its use before attempting any finish work.

New Insulation

The final step in making a joint or tap is to reapply equivalent insulation to the exposed conductors. Plastic electrical tape is recommended for this purpose. Start on top of the original insulation at one end and wrap the tape diagonally so that successive wraps overlap slightly. Keep the tape stretched so that successive wraps will fuse to each other. From the opposite end, work back to the start with the layers at approximately right angles to each other. Repeat until the tape is as thick as the original insulation.

Soldering Lugs

First tin the exposed conductor carefully. Then melt a quantity of solder into the lug and, while it is still in liquid form, insert the previously tinned conductor. Make absolutely certain that the lug attachment is mechanically sound. Limitations on the use of soldering lugs are noted in NEC references 230-81 and 250-115.

Solderless Connectors

Solderless (pressure) connectors are, in general, intended for applications where there is no strain on a connection. In one type, the stripped conductor ends are simply twisted together. Called a *fixture joint*, it has an insulated shell that is screwed over the twisted wires. Another type has a brass insert and set screw. The wires are clamped in the insert by the set screw, and an insulating cap is screwed on over the insert. Another type, handy for making a tap to a continuous wire, has a body with a U-shaped cross section and a nut that slips over the threaded leads of the U.

Special connectors, marked AL for aluminum or CU/AL for copper/aluminum, are required when working with aluminum conductors.

EXERCISES

4–1. Why is an understanding of the NEC important to anyone undertaking electrical wiring?

4–2. Does the NEC take preference over local ordinances? Explain.

4–3. Does the NEC set maximum or minimum standards?

4–4. What is the function of a nationally recognized testing laboratory?

4–5. Does a testing laboratory approve electrical items? Explain.

4–6. What is the area of a round conductor having a diameter of 100 mils?

4–7. What sizes of conductors are covered for ordinary wiring by the American wire gauge? How are these sizes designated?

4–8. What is a wire gauge, and how is it used?

4–9. As applied to metallic conductors, what is meant by the expression *skin effect*? How does it affect the resistance of a conductor?

4–10. Describe the NEC color scheme for (*a*) wires and (*b*) terminals.

4–11. In terms of size, what is meant by *(a)* wire and *(b)* cable?

4–12. What is the lightest-sized wire permitted by the NEC for ordinary residential wiring?

4–13. What does the NEC mean by *(a)* dry, *(b)* damp, and *(c)* wet locations?

4–14. Describe a wire insulation designated *(a)* THHN and *(b)* RHW.

4–15. What is the maximum permissible operating voltage for the insulations described in Section 4–4?

4–16. What does the cable designation 10–6 indicate?

4–17. What types of metal-clad cable does the code permit? Which type is quite common in residential wiring and by what name is it usually called?

4–18. What is meant by the cable designations AC, ACT, and ACL, and in what types of locations can they be used?

4–19. What color insulations will be found in a three-conductor BX cable?

4–20. What types of nonmetallic-sheathed cable does the code permit?

4–21. What location restrictions, if any, are placed on *Romex*?

4–22. What is meant by the term *ampacity*?

4–23. What factors affect the ampacity of a conductor?

4–24. What is the maximum temperature rating of insulation type THHN?

4–25. What factors must always be considered when selecting conductors for a given application?

4–26. What are the maximum voltage drops that the NEC recommends for *(a)* branch circuits and *(b)* feeders?

4–27. In general, how does the size of aluminum conductors compare to that of copper conductors for a given application?

4–28. Define the terms (a) *conduit*, (b) *electrical metallic tubing*, (c) *surface raceway*, and (d) *wireway*.

4–29. What is *greenfield*?

4–30. In what lengths are EMT and rigid electrical conduit manufactured?

4–31. Does the term electrical trade size denote actual physical dimensions? If not, give three examples to illustrate the difference.

4–32. How is the insulation removed from a conductor to permit its attachment to a screw-type terminal?

4–33. How is a joint made using large-stranded wires?

4–34. What is a tap and how is one made using solid wires?

4–35. What precautions must be taken in soldering either a joint or a tap?

4–36. Using the approximation technique described in Section 4–2, determine the resistance of 500 ft of size AWG no. 6 copper wire at room temperature.

4–37. Using either the tables contained in Section 4–7, or code Tables 310-16 through 310-19, as required, determine the ampacity of size AWG no. 12 conductors with THHN insulation, three in a cable. The maximum ambient temperature is 104°F.

4–38. A load takes 48 A. What size copper conductors (type RHW insulation) must be used in a three-wire cable when the maximum ambient temperature is 110°F?

4–39. Convert 115°F to the equivalent degrees Celsius.

4–40. Convert 37°F to the equivalent degrees Celsius.

4–41. Copper conductors with type T insulation are being used to supply a 60-A load, the conductors are in a metallic raceway, the load is located 250 ft from a single-phase, 120-V, 60-Hz ac source, and the maximum ambient temperature is 104°F. What size conductors are required to limit the voltage drop to 2 percent of the applied EMF?

4–42. What size conductor is required for Exercise 4–41 if the source voltage is 240 V, all other factors being unchanged?

4–43. What size conductors would be required for Exercise 4–41 with a 100-A load if all other factors remain unchanged?

4–44. Determine the size of conductor that you would use in Exercise 4–41 if consideration were given to ampacity alone. What does this determination emphasize?

5

Circuit Components

5–1 SWITCHES AND THEIR USE AS CONTROL DEVICES

In Section 1–4 we noted that all practical electric circuits contain four parts: *source, load, conductors*, and one or more *control devices* to turn the circuit on and off under normal conditions. In this section we are concerned with the control devices or, as they are more commonly known, *switches*, and the ways in which they are connected.

SPST Switches

The simplest form of switch is designated *single-pole single-throw* (SPST) to indicate that it opens or closes *one* conducting path and is placed in a designated position by a single *throw* or switching movement. Common forms of the SPST switch include *knife, toggle, push-button*, and *pull-chain* switches. All have *two* terminals. The knife switch has an open mechanism; but in the toggle, push-button, and pull-chain switches, the switching mechanism is enclosed for improved safety and mechanical protection.

Since an SPST switch can either open or close only one conducting path, it must be connected in *series* with the circuit or device to be controlled. Thus, when the switch is *closed* (*on*), the circuit is completed, current passes from the source, through the switch and the load device or devices, and back to the source. When the switch is *open (off)*, the circuit is no longer continuous, circuit current is zero, and no electrical energy is supplied to the load device or devices.

Should the switching mechanism become defective—and this is true of any type of switch—it may leave the circuit either permanently open or permanently closed. In either case, the switch no longer performs its intended function and must either be repaired or replaced. Except for knife switches, replacement is generally the better course of action.

Two methods of connecting an SPST switch are illustrated in the schematics in Figures 5–1(a) and (b). As noted earlier, the switch, in both cases, is connected in series with the load to be controlled. In Figure 5–1(a), however, each component of the load is also series connected; in Figure 5–1(b) the load consists of three parallel-connected lamps. The source may, of course, be either alternating or direct current. Because an ac source is most common in residential wiring, however, this type of source is indicated throughout the remainder of this text, except in Chapter 13, which relates to motors.

Notice the symbol for an SPST switch. Also note that in *any schematics* in this text the physical connection of two or more conductors is always indicated by a heavy dot at the point of intersection between

FIGURE 5–1. Wiring arrangements for SPST switches. In each case, the switch is series connected. The load, however, may consist of either (*a*) series-connected devices or (*b*) parallel-connected devices.

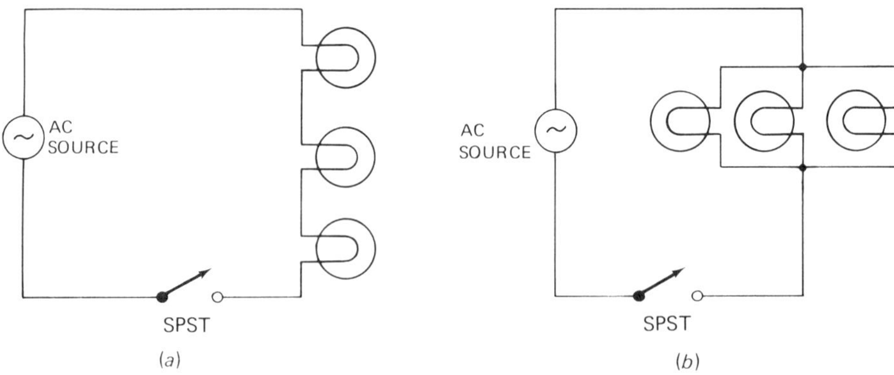

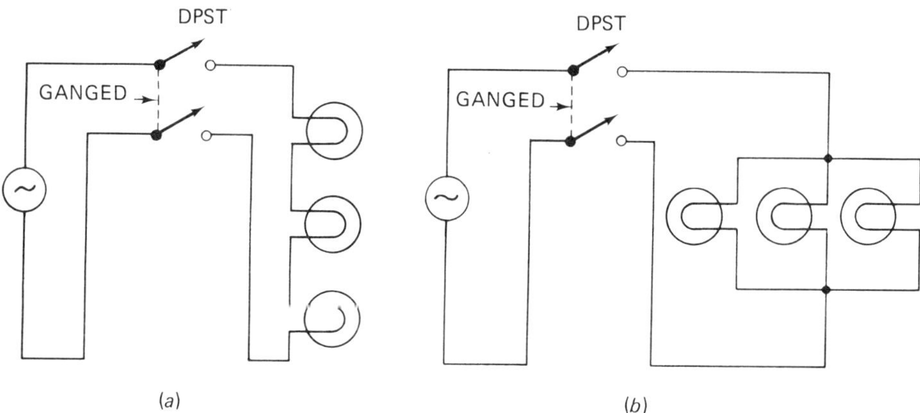

FIGURE 5–2. Wiring arrangement for DPST switches. The two switch mechanisms are mechanically ganged so that both wires are either opened or closed simultaneously. The load may consist of either series-connected devices, as in (*a*), or parallel-connected devices, as in (*b*).

lines representing the conductors. Where lines cross and the heavy dot is absent from the schematic at the point of intersection, there is no physical connection of conductors. In some literature, however, a physical connection is understood to exist wherever two lines intersect (no dot), whereas the absence of connection is indicated by a bridge-like symbol (—⌒—).

DPST Switches

A *double-pole single-throw* (DPST) switch has two *ganged* (mechanically joined) switching mechanisms that simultaneously open or close *two* conducting paths with a *single* throw of the switch. The fact that the switching mechanisms are ganged is indicated schematically by a dashed line. Again, the load may consist of either series- or parallel-connected devices. See Figures 5–2(a) and (b).

Use is made of DPST switches, instead of the SPST variety, when it is necessary to open two *ungrounded* conductors. (The subject of grounding is taken up in Chapter 6.)

Switches of the DPST type have four terminals and, on toggle switches, the words *on* and *off* appear on the handle (throw lever).

SPDT Switches

When it is either necessary or desirable to control a load from *two* locations (for example, upstairs and downstairs in a home), two *single-*

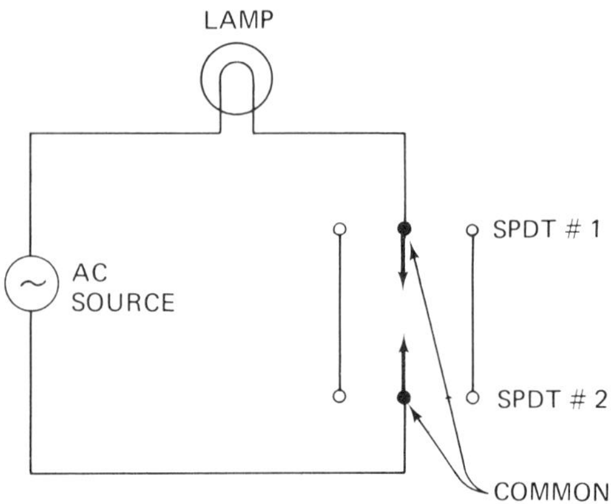

FIGURE 5–3. Two SPDT switches used to control a lamp from two locations, such as the upstairs and downstairs of a home. Switches of this type are also called three-way switches.

pole double-throw (SPDT) switches may be employed. A typical arrangement is shown in Figure 5–3. With both switches in the same position, the circuit is closed and current is supplied to the lamp. If either switch is then thrown to its opposite position, the circuit is open.

As indicated in the Figure 5–3 schematic, an SPDT switch, also referred to in the electrical trade as a *three-way* switch, has *three* terminals. The *common* terminal usually has a dark or oxidized finish. Notice that the common terminal of one switch connects to one end of the load, while the common terminal of the other switch connects to the other end of the source. The remaining four terminals are connected as *two-terminal pairs*, and the wire between each pair is called a *runner*, or sometimes a *traveler*. Mechanical differences in the switches produced by different manufacturers may cause some slight difficulty in connecting the switches properly, but no danger exists if the wrong terminals are chosen initially. The circuit will not work properly, of course, until the correct connections are made.

The words *on* and *off* do not appear on the handle of a three-way switch.

DPDT Switches

To control a load from *three* locations, use is made of *one double-pole double-throw* (DPDT) switch (also called *four-way* switch) in conjunction with *two* SPDT switches. As in any switch having more than one switching mechanism, the switching mechanisms are ganged and, therefore, work in unison.

Typical wiring arrangements for controlling a light from *three* locations are shown in Figures 5–4(a) and (b). Notice that the internal switching action and, therefore, the wiring of the DPDT switch used in (a) differs from that used in (b). Moreover, the common terminal of one SPDT

FIGURE 5–4. Controlling a light from three locations. The different internal switching actions of the DPDT switches necessitate the different wiring arrangements shown in (*a*) and (*b*).

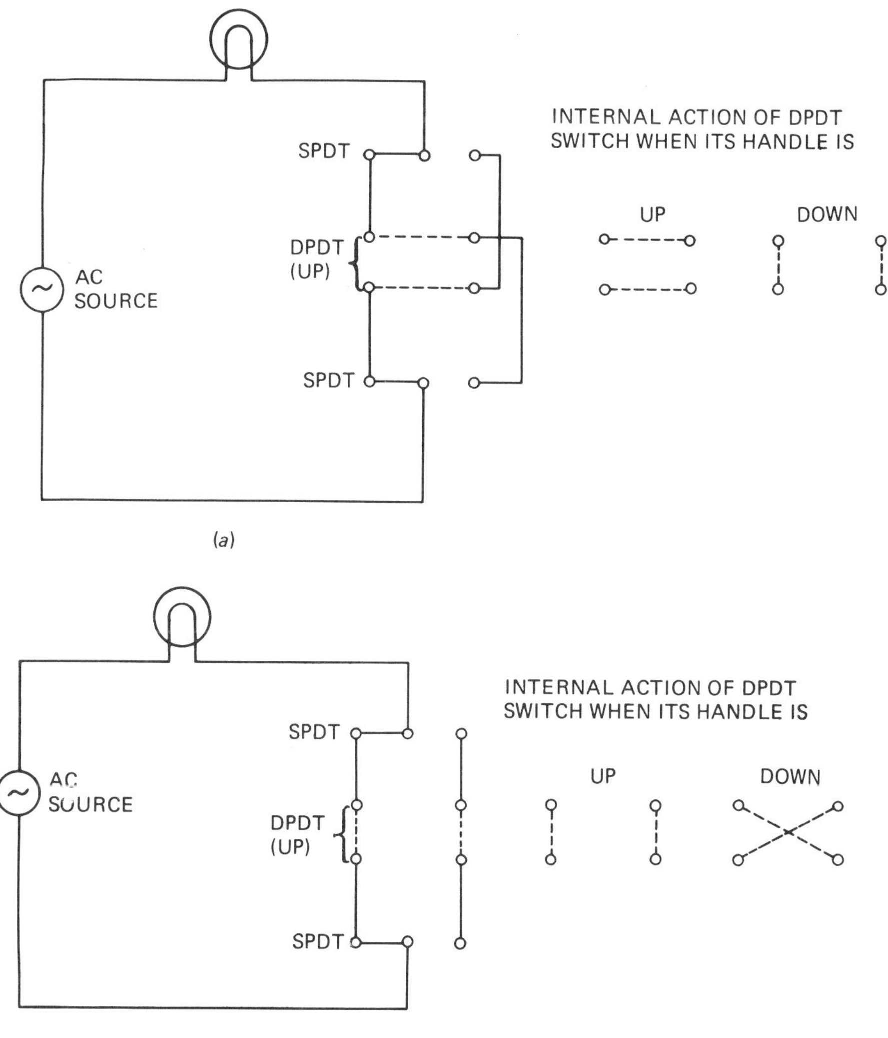

switch connects to one end of the load, the common terminal of the second SPDT switch connects to one end of the source, and the DPDT switch is inserted between the two SPDT switches.

In Figure 5–4(a) the light is *off*. When any one of the three switches is thrown to its opposite position, however, the circuit is completed and the light turns *on*. In Figure 5–4(b) the light is *on* and moving the lever of any one of the three switches turns the light *off*. Examination of either circuit with the switches in any combination of positions will show that the desired control is, indeed, achieved.

When load control from more than three locations is required, two SPDT switches are again used with their common terminals connected to one side of the load and source, respectively; the required number of DPDT switches is connected between them—one additional DPDT switch for each additional location. Two wiring arrangements for controlling a load from four locations are shown in Figures 5–5(a) and (b).

Checking the Internal Action of DPDT Switches

The internal action of DPDT switches can be checked quickly and easily by means of a *continuity checker*. Although many different forms of such testers are available, my own preference is the ohmmeter section of a combination volt-ohmmeter. A wide variety of inexpensive instruments of this type are currently available. All are small in size, lightweight, and completely safe in operation.

As indicated in Figures 5–4 and 5–5, a DPDT switch has four terminals. On the one hand, if continuity (essentially the zero reading on an ohmmeter) is indicated when the test prods of the tester are first placed across the upper terminals and then across two terminals located on the same side of the switch, the switch has the internal action shown in Figure 5–5(a). If, on the other hand, continuity is indicated when the test probes of the tester are first placed on the two terminals located on one side of the switch and then on terminals located at diagonal corners, the switch has the internal action shown in Figure 5–5(b).

With either type of switch, of course, physical movement of the switch handle is required to obtain the correct readings.

5–2 BASIC CODE PROVISIONS ON SWITCHES

Article 380 of the code relates to switches and their application. In Article 100 a *general-use switch* is defined as one intended for use in general distribution and branch circuits. It is rated in amperes and is capable of interrupting the rated current at its rated voltage.

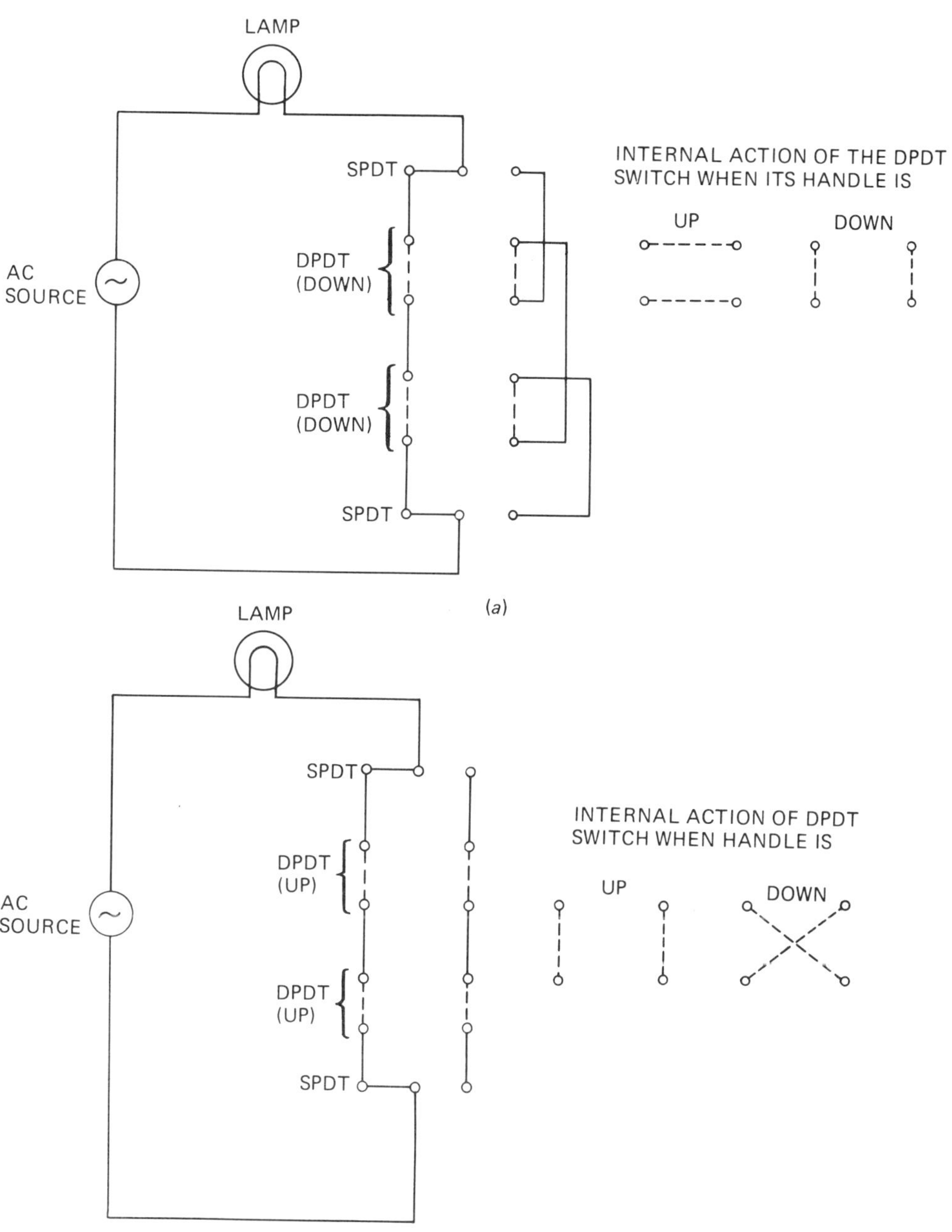

FIGURE 5–5. Controlling a load from four locations. The different internal switching actions of the DPDT switches necessitate the different wiring arrangements. (*a*) Schematic symbols for thermal circuit breakers; (*b*) magnetic circuit breakers.

A *general-use snap switch* is defined as a form of general-use switch so constructed that it can be installed in flush device boxes or on outlet box covers, or otherwise used in conjunction with wiring systems recognized by the code. (Wiring methods are covered in Chapter 7.)

Two types of general-use snap switches are considered in Sec. 380-14, ac and ac–dc. The ac type is readily identified by the letters AC stamped on the switch strap. The absence of any such designation indicates an ac–dc switch.

All switches are marked with the rated current and voltage and, if horsepower rated, the maximum rating for which they are designed.

AC General-Use Snap Switches

Switches of this type are suitable only for use on ac circuits for controlling the following: (1) resistive and inductive loads, including electric-discharge lamps such as the fluorescent type, not exceeding the ampere rating of the switch at the voltage applied; (2) tungsten filament lamp loads not exceeding the ampere rating of the switch at 120 V; (3) motor loads not exceeding 80 percent of the ampere rating of the switch at its rated voltage; (4) snap switches rated 20 A or less directly connected to aluminum conductors, which must be approved for the purpose and marked CO/ALR.

AC–DC Switches

Switches of this type are suitable for use on either ac or dc circuits for controlling the following: (1) resistive loads not exceeding the ampere rating of the switch at the voltage applied; (2) inductive loads not exceeding 50 percent of the ampere rating of the switch at the applied voltage (switches rated in horsepower are suitable for controlling motor loads within their rating at the voltage applied); (3) tungsten filament lamp loads not exceeding the ampere rating of the switch at the applied voltage, when T rated.

T-Rated Switches

As noted in Section 3–2, the tungsten filament is the principal light source in incandescent lamps. The *cold* resistance of a tungsten filament is very small compared to its *hot* resistance. As a result, a very large current flows when the light is first turned on.

The high *inrush* current, although short in duration (less than 0.2 s), necessitates heavy contacts on the switches used in incandescent lighting circuits. Switches of this type are called *T-rated* switches, and their listed current rating is for the lamp's transient inrush current. A 100-W tungsten filament, for example, draws a normal (hot) current of about 0.9 A at 120 V, but its inrush or starting current is about 9 A. Such a lamp requires a switch with a 10-A T rating. When several incandescent lamps are connected in parallel and are controlled by a single switch, the T rating of the switch should exceed the total inrush-current requirement.

The accessibility, grouping, and mounting of general-use snap switches is considered in Chapter 10.

Today, the ac-type switch is used in practically all new wiring and as replacement for the older ac–dc type. The widespread distribution of alternating current by electric utility companies and the lower cost and quieter operation of ac snap switches account for their popularity.

Switch Miscellany

An *isolating* switch is one intended to isolate an electric circuit from the source of power. It has no interrupting rating, and is intended to be operated only after the circuit has been opened by some other means.

A *motor-circuit* switch is rated in horsepower; it is capable of interrupting the maximum operating overload current of a motor of the same horsepower rating as the switch at the rated voltage.

Surface-type switches used with open wiring on insulators must be mounted on subbases of insulating material that separate the conductors at least ½ in. from the surface wired over.

A switch or circuit breaker in a wet location or outside a building must be enclosed in a weatherproof enclosure or cabinet placed in such a way as to prevent the entrance and accumulation of moisture or water. At least ¼ in. of air space is required between the enclosure and the supporting surface.

5–3 RECEPTACLES AND FACEPLATES

In Article 100 a *receptacle* is defined as a contact device installed at an outlet for the connection of an attachment plug. The outlet itself is a point on the wiring system at which current is taken to supply a load.

Today most receptacles are of the duplex type and accommodate two plugs. When purchased, the two receptacles are connected in parallel. In better quality receptacles, however, a small metal strip can be broken out to permit isolation of the two sections. With this arrangement, the top receptacle can be wired so that it is permanently *live* and convenient for supplying a temporary load, such as a vacuum cleaner. The bottom half can then be used for lamps and is switch controlled. Receptacles installed for the attachment of portable cords must be rated at not less than 15 A, 125 V, or 10 A, 250 V.

For safety reasons, switches and receptacles are covered with *face-plates* after installation. Faceplates may be either metallic or nonmetallic. Cutouts are provided for various switch and/or receptacle combinations, and the color can be matched to the room decor.

5–4 LAMP HOLDERS (SOCKETS)

Probably the most common form of *lamp holder* is found in residential wiring. Usually referred to as a *socket*, it is the brass-shell type and ordinarily contains either a key, a push-through, or a pull-chain switching mechanism to control the lamp. The metal shell and cap are lined with insulating material to keep them from becoming part of the electrical circuit (Sec. 410-50).

Lamp holders with a porcelain shell and pull-chain switching mechanism are often used on top of outlet boxes. A weatherproof type is available for outdoor use.

5–5 DEVICES FOR OVERCURRENT PROTECTION

Any electrical system must include safeguards to assure that the ampacity of each conductor is not exceeded. An excess current, called *overcurrent* (or, sometimes, *fault current*), may range from a small overload to a complete short circuit, depending on the location of the circuit fault.

Since the electric power producing heat is equal to I^2R, a given current through a conductor that is double its ampacity creates heat in that conductor four times its normal value. When short-circuit conditions exist, even for a few seconds, the conductor melts and the insulation ignites and sets fire to any surrounding materials that are not fireproof. Overcurrent protection assures that the current will be interrupted before excessive current causes damage either to the conductors themselves or the load that they supply.

There are basically two classes of devices used in residential wiring for overcurrent protection: *fuses* and *circuit breakers*. Neither device can be arranged or installed in parallel, except those circuit breakers that are assembled in parallel and tested and approved as a single unit. Moreover, no overcurrent-protection device can be placed in a permanently grounded conductor, except where it simultaneously opens all conductors of the circuit. In residences, overcurrent devices must be located where they are readily accessible, not exposed to physical damage, and not in the vicinity of easily ignitible materials.

Fuses: General

Fuses are overcurrent devices that destroy themselves when they interrupt the circuit. They are made of a low-melting-temperature metal and calibrated so that they melt when a specific current rating is reached. Because such fuses are connected in series with the load, they open the circuit when they melt. All fuses have an *inverse time characteristic*. Thus a fuse rated at 30 A should carry 30 A continuously; with about a 10 percent overload it should melt in a few minutes, and with a 20 percent overload in less than a minute. A 100 percent overload should cause a correctly rated fuse to blow or melt in a fraction of a second.

Plug Fuses (Sec. 240–E)

In plug fuses of the Edison-base type, a fusible link is enclosed in a housing that prevents the splatter of molten metal when the fuse blows. The condition of the fuse is determined by looking through a window located on top of the plug assembly. If the window (or other prominent feature) is *hexagonal*, the fuse is rated at 15 A or less. If the window is *round*, the fuse is rated at more than 15 A, up to a *maximum* size for fuse plugs of 30 A.

Code Sec. 240-50(a) states that plug fuses and fuse holders shall not be used in circuits exceeding 125 V between conductors, except in circuits supplied by a system having a grounded neutral, and having no conductor at over 150 V to ground. Thus the use of plug fuses *is permissible* between the two hot wires of a 120/240 V installation with a grounded neutral since the voltage from either conductor to ground is 120 V.

Type S Fuses (Sec. 240–54)

Type S fuses and fuse holders were designed specifically to prevent either accidental or intentional replacement of a given fuse with one of a higher rating. When a type S fuse is used with an Edison-type fuse holder, a nonremovable adapter is screwed into it and locked in place. When a type S fuse is inserted in the adapter, it should be screwed in tighter than usual to ensure that the spring under the shoulder of the fuse flattens.

Type S plug fuses and fuse holders are classified at not over 125 V; 0 to 15 A, 16 to 20 A, and 21 to 30 A. Fuses of the 16- to 20-A and 21- to

30-A classifications will not fit fuse holders or adapters of a lower ampere classification. The fuses, fuse holders, and adapters from all U.S. manufacturers are interchangeable.

Cartridge Fuses (Sec. 240–F)

Whenever the circuit current exceeds 30 A, it is necessary to use a cartridge-type fuse and fuse holder. The voltage and current ratings for 0- to 600-A cartridge fuses and fuse holders are shown in Table 5–1.

Fuse holders must be designed so that it is difficult to put a fuse of a given class into a fuse holder designed for a lower current or higher voltage than that of the class to which it belongs. Fuses must also be plainly marked to show ampere rating, voltage rating, interrupting rating where other than 10,000 A, *current limiting* where applicable, and the name or trademark of the manufacturer. (*Exception*: the interrupting rating may be omitted on fuses used for supplementary protection, that is, fuses used with appliances or other equipment to provide individual protection for specific components or circuits within the equipment itself.)

The current-limiting fuse referred to above is a device that, when interrupting a specific circuit, consistently limits the short-circuit current to a value substantially less than that which would exist if a solid conductor of comparable resistance were used in place of the fuse. Current-limiting fuses cannot be installed in non-current-limiting fuse holders.

Circuit Breakers (Sec. 240–G)

Article 100 defines a circuit breaker as a device designed to open and close a circuit by nonautomatic means, and to open the circuit automatically on a predetermined overload of current, without injury to itself when properly applied within its rating.

To energize a circuit breaker, the *on–off* switch or button is thrown to the on position. This closes the contacts of the circuit breaker, which

TABLE 5–1. Current and Voltage Classifications for 0- to 600-Ampere Cartridge Fuses and Fuseholders

Not over 250 V (A)	*Not over 300 V (A)*	*Not over 600 V (A)*
0–30	0–15	0–30
31–60	16–20	31–60
61–100	21–30	61–100
101–200	31–60	101–200
201–400		201–400
401–600		401–600

are held closed by a latch. When the on–off switch is turned off, the latch is released and the contacts are reopened by spring action.

Some circuit breakers have a thermal element connected in series with the contacts and are called *thermal* circuit breakers. Their function is to protect against *gradual* overload conditions. The current passes through the series-connected thermal element and causes it to heat. When excessive heat is produced, as a result of increasing overload condition, a bimetallic strip bends and releases the contact-holding latch automatically. The bimetallic strip is made by having thin strips of two different metals in contact with one another.

Because it takes time for a bimetallic strip to heat, the tripping or opening of a thermal circuit breaker does *not* occur the instant a current exceeds the rated value of the circuit breaker. The manufacturer supplies an operating characteristic in graphic form with each circuit breaker. This curve shows how long it takes for the breaker to trip for a given current. Whenever you select a thermal circuit breaker for a given application, do so by referring to the characteristic curve. Needless to say, a thermal circuit breaker should never be used in any application where a very fast response to overload conditions is required.

Another type of circuit breaker contains an *electromagnet* connected in series with the circuit to be protected. Normal circuit current has no effect on the electromagnet; however, when a short circuit occurs, the electromagnet trips the breaker immediately. Thus a magnetic circuit breaker provides *instantaneous* protection against a short circuit.

Combination thermal–magnetic circuit breakers are also available to protect against both gradual overload and short circuits. Such circuit breakers can be adjusted to open instantaneously for a short-circuit condition and to open at a lower abnormal current after a time delay.

Location of Overcurrent-Protection Devices (Sec. 240–21)

In general, an overcurrent-protection device is required at the point where each conductor receives its supply. Eight exceptions to the general rule are listed in the code. To avoid confusion, however, these exceptions will be treated where applicable.

5–6 OUTLET, JUNCTION, AND SWITCH BOXES

For safety reasons, appropriate boxes or receptacles (Article 370) must be used at every point where conductors are joined or connected to the terminal of outlets or switches.

These boxes may be octagonal, square, or round. Their materials may be metallic or nonmetallic and, according to their depth, either *deep* or *shallow*. Any box less than 1½ in. deep is considered a shallow box. Easily removed knockouts are provided to make holes for the insertion of conductors, but any holes not used must be closed.

Round boxes cannot be used where conduits or conductors requiring the use of locknuts or bushings are connected to the side of the box.

Nonmetallic boxes, not over 100 in.^{3}, are generally used with nonmetallic sheathed cable and with approved nonmetallic conduit. Boxes over 100 in.^{3}, manufactured with bonding means between all raceway and cable entries, may be used with metal raceways and metal-sheathed cable.

Metallic boxes used with nonmetallic-sheathed cable and in contact with any metal surface must either be insulated from their supports and from the metal surface or they must be grounded.

Weatherproof boxes and fittings are required for use in wet locations.

Specially designed boxes are required for floor-mounted receptacles and for lighting-fixture outlets. Most often the boxes are supported from a structural member of the building either directly or by using an approved metallic hanger or a wooden brace not less than 1 in. thick.

In new walls where no structural members are provided, or in existing walls in previously wired buildings, boxes not over 100 in.^{3} in size, specifically approved for the purpose, can be affixed with approved anchors or clamps.

Threaded boxes or fittings not over 100 in.^{3} in size that do not contain devices or support fixtures can be supported by two or more conduits threaded into the box wrench tight and supported within 3 ft of the box on two or more sides.

Outlet boxes for concealed work must have an internal depth of at least 1½ in., except that where the installation of such a box will result in injury to the building structure or is impracticable, a box of not less than ½-in. internal depth may be installed.

In completed installations each outlet box must have either a cover or a fixture canopy. All pull boxes, junction boxes, and fittings must have an approved cover.

Covers of outlet boxes having holes through which flexible cord pendants pass must have either bushings designed for the purpose or smooth, well-rounded surfaces on which the cords may bear. Hard rubber or composition bushings cannot be used.

Number of Conductors in a Box (Sec. 370–6)

Boxes must be large enough to provide free space for all conductors enclosed in the box. Before noting code limitations, let us define three terms.

- *Fitting*—An accessory such as a locknut, bushing, or other part of a wiring system that is intended primarily to perform a mechanical rather than an electrical function.
- *Device*—A unit of an electrical system that is intended to carry but not utilize electric energy. In this section of the code, the word device is taken to mean a switch box.
- *Hickey*—A threaded coupling for attaching a fixture to an outlet box.

Fixture wires excepted, the maximum numbers of conductors permitted in outlet and junction boxes must be as in Table 5–2.

TABLE 5–2. Boxes

Box Dimension (in.) Trade Size or Type	*Min. Capacity (in.³)*	*Maximum Number of Conductors*				
		No. 14	*No. 12*	*No. 10*	*No. 8*	*No. 6*
4 × 1¼ round or octagonal	12.5	6	5	5	4	0
4 × 1½ round or octagonal	15.5	7	6	6	5	0
4 × 2⅛ round or octagonal	21.5	10	9	8	7	0
4 × 1¼ square	18.0	9	8	7	6	0
4 × 1½ square	21.0	10	9	8	7	0
4 × 2⅛ square	30.3	15	13	12	10	6*
4 11/16 × 1¼ square	25.5	12	11	10	8	0
4 11/16 × 1½ square	29.5	14	13	11	9	0
4 11/16 × 2⅛ square	42.0	21	18	16	14	6
3 × 2 × 1½ device	7.5	3	3	3	2	0
3 × 2 × 2 device	10.0	5	4	4	3	0
3 × 2 × 2¼ device	10.5	5	4	4	3	0
3 × 2 × 2½ device	12.5	6	5	5	4	0
3 × 2 × 2¾ device	14.0	7	6	5	4	0
3 × 2 × 3½ device	18.0	9	8	7	6	0
4 × 2⅛ × 1½ device	10.3	5	4	4	3	0
4 × 2⅛ × 1⅞ device	13.0	6	5	5	4	0
4 × 2⅛ × 2⅛ device	14.5	7	6	5	4	0
3¾ × 2 × 2½ masonry box/gang	14.0	7	6	5	4	0
3¾ × 2 × 3½ masonry box/gang	21.0	10	9	8	7	0
FS—Minimum internal depth 1¾ Single cover/gang	13.5	6	6	5	4	0
FD—Minimum internal depth 2⅜ Single cover/gang	18.0	9	8		6	3
FS—Minimum internal depth 1¾ Multiple cover/gang	18.0	9	8	7	6	0
FD—Minimum internal depth 2⅜ Multiple cover/gang	24.0	12	10	9	8	4

* Not to be used as a pull box. For termination only.

TABLE 5–3. Volume Required per Conductor

Size of Conductor	*Free Space within Box for Each Conductor (in.^{3})*
No. 14	2.00
No. 12	2.25
No. 10	2.50
No. 8	3.00
No. 6	5.00

Table 5–2 applies where no fittings or devices such as fixture studs, cable clamps, hickeys, switches, or receptacles are contained in the box, and where no grounding conductors are part of the wiring within the box. Where one or more fixture studs, cable clamps, or hickeys are contained in the box, the number of conductors must be one less than shown in the tables, an additional reduction of one conductor must be made for each strap containing one or more devices, and a further reduction of one conductor must be made for one or more ground conductors entering the box. A conductor running through the box is counted as one conductor, and each conductor originating outside the box and terminating inside the box is counted as one conductor. If no part of a conductor leaves the box, it is not counted. For example, a wire from the green terminal of a receptacle to a grounding point on the box is not so "counted." The volume of a box is the total volume of all assembled sections.

For combinations or conductor sizes not shown in Table 5–2, those in Table 5–3 apply. Exceptions to the above noted code specifications will be dealt with as the need arises.

EXERCISES

5–1. With respect to the load, how must a switch be connected in any circuit? Why?

5–2. When is a DPST switch used in preference to the SPST type?

5–3. Which types of switches would you select to control a load from five locations? How would you connect them? Draw the circuit.

5–4. What types of general-use snap switches are available? In what types of circuits may each type be used?

5–5. Which type of general-use snap switch is most often used in practice? Why?

5–6. What is a *T-rated* switch?

5–7. With what types of load must a T-rated switch be used?

5–8. What is an *isolating* switch?

5–9. What is a *motor-circuit* switch?

5–10. What is a *surface-type* switch?

5–11. What is meant by the statement that all fuses have an *inverse time characteristic*?

5–12. What is the maximum permissible ampere rating of a plug fuse?

5–13. How are plug fuses rated at 15 A or less identified?

5–14. How are plug fuses rated at more than 15 A identified?

5–15. Does the code permit the use of plug fuses between the two hot wires of a 115/230 V installation? Explain fully.

5–16. How do type S plug fuses differ from the Edison-base type?

5–17. When must a *cartridge-type* fuse be used?

5–18. What is a *current-limiting* fuse?

5–19. How does a *thermal-type* circuit breaker operate?

5–20. How does a *magnetic-type* circuit breaker operate?

5–21. For protection against gradual overload, what type of circuit breaker would you select? Why?

5–22. What code restriction exists on the use of round boxes?

5–23. What is a *shallow* box?

5–24. When does the code permit the use of a shallow box?

5–25. How is the permissible number of conductors in a box determined?

III

SYSTEM GROUNDING AND WIRING

6

System and Equipment Grounding

6–1 GROUNDING METHODS

The words *ground* and *earth* mean the same thing. Thus, when any part of an electrical system is grounded, it is connected, either directly or indirectly, to earth.

For reasons of safety, the *neutral* wire of all residential wiring systems is intentionally grounded. Why this is necessary is examined in detail in Section 6–2. Such grounding is illustrated in Figure 6–1, which shows a typical overhead-type residential supply installation.

A *service drop*, supplied and installed by the utility company, is made up of three *service conductors* that extend from the street main or transformer to the building being served. Notice that the service conductors terminate at insulators mounted on the side of the building. The neutral service conductor is grounded at the distribution transformer.

Three *service-entrance conductors* attach to the service conductors near their point of termination. The service-entrance conductors are often

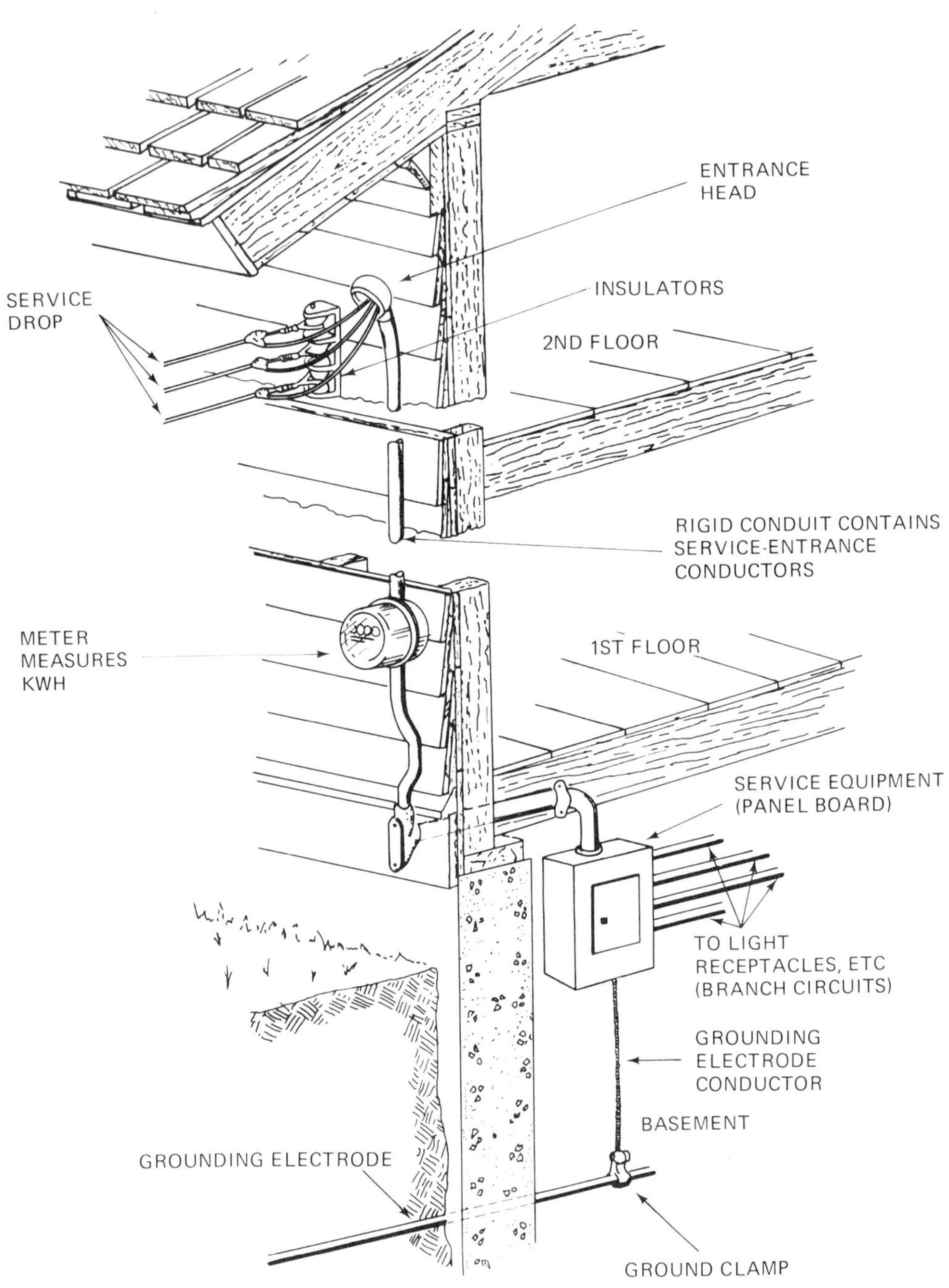

FIGURE 6–1. Typical electric-service entrance.

made up in the form of a cable or are, as in Figure 6–1, enclosed in conduit. Either form of conductor enclosure is called the *service raceway.*

The service-entrance conductors first connect to a meter, which records the number of watthours of electric power consumed. From the meter these conductors extend through the wall of the building and connect to a service switch and fuses (or to a circuit breaker) and accessories, which constitute both the main control and a means for cutoff of the electric supply. Collectively, these items are called the *service equipment*; for safety reasons, they are mounted in a metal enclosure.

At the point where the neutral wire of the service-entrance conductors attaches to the service switch, a *grounding-electrode conductor* is connected. The opposite end of this conductor attaches to a *grounding electrode*, which may be a metal underground water pipe in direct contact with the earth for 10 ft or more (including any metal well casing bonded to the pipe) and electrically continuous to the points of connection of the grounding electrode conductor and the bonding conductors. The grounding electrode may also be the metal frame of a building (where effectively grounded), an electrode encased by at least 2 in. of concrete located within and near the bottom of a concrete foundation or footing that is in direct contact with the earth, or a ground ring encircling the structure and in direct contact with the earth at a depth of at least 2½ ft. (See code Sec. 250-81 for complete information on acceptable grounding electrodes.)

Generally, the service drop, service-entrance conductors, watthour meter, service switch and fuses (or circuit breakers), and the ground are referred to collectively as the *service entrance*.

Physically, overcurrent protection devices for *branch* circuits, that is, circuits which supply lighting and appliance outlets, are almost always located in the same metal enclosure as the service equipment. For that reason, they are usually considered to be part of the service entrance.

The neutral wires of all branch circuits are connected to the grounded neutral wire of the service-entrance conductors. In this way the neutral wire of all residential branch circuits is grounded. The grounded neutral wire of each branch circuit must be insulated and, most often, the insulation is *white* in color. Although the NEC also permits the insulation of the grounded neutral conductor to be natural gray in color, we shall consider it and refer to it as *white* throughout the remainder of this text.

One notable exception to the NEC identification scheme is as follows: because no. 4 and heavier wires are rarely any color other than black, they may be used for the neutral grounded conductor, *provided* the ends of the black wire are *painted black*.

A most important fact that you must remember is that *in residential wiring, the grounded neutral wire is never interrupted by a fuse, circuit breaker, switch, or other device.* Without exception, the white neutral wire always runs directly from the point of attachment to the grounding electrode to the outlet from which electric power is finally channeled for consumption.

The grounded neutral service-entrance conductor and the grounded neutral (white) conductor of each branch circuit form the *system ground* of an electrical installation.

In addition to the system ground, the NEC requires that all metallic equipment—raceways, boxes, fittings, enclosures for fuses or circuit breakers, and so forth—associated with an electrical system be grounded. This type of grounding is called *equipment grounding*.

6–2 WHY GROUNDING IS NECESSARY

The principal reason for system grounding is to limit the *voltage to ground* from rising above a safe value because of a fault *outside* the building. If, for example, outside wires are struck by lightning, the ground acts like a lightning rod to get rid of the very high electrical charges built up on the wires. Furthermore, if neutral service-entrance conductors are not grounded, a faulty power-supply transformer could also cause voltages inside that building to rise far beyond their normal values.

Equipment grounding is necessary to prevent electric shock to persons coming into contact with metallic objects that, either intentionally or accidentally, form part of the electrical system. If, for example, an underground (hot) conductor in the system should accidentally touch any metallic equipment, junction box, motor frame, or fitting, there is voltage between that equipment and ground. If that equipment is not grounded, a very serious shock hazard exists.

To illustrate this problem, study Figure 6–2, which shows an ungrounded raceway connected to a box containing a switch with a wall plate attached to that switch. If the hot (black) conductor of the branch circuit accidentally comes into contact with the switch box, both box and wall plate are at a potential of 120 V above ground. Should anyone in contact with a grounded object now touch the wall plate, his body would

FIGURE 6–2. As a result of accidental contact between a hot conductor and the switch box, a potential of 120 V exists between the entire raceway system and the ground.

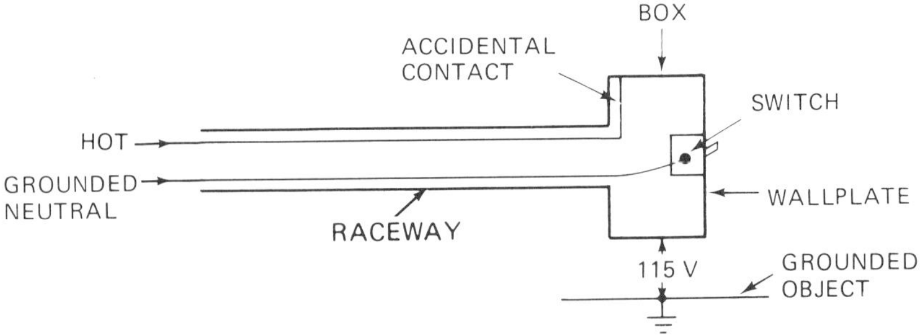

immediately complete the circuit to ground. The resulting shock would be most unpleasant—at best. Depending on the resistance of the person's body and the condition of the ground (dry, damp, or wet), the shock could even prove fatal.

The type of dangerous shock described above is avoided by grounding the raceway and, therefore, the box and wall plate to which it is connected. In this case, if the hot conductor accidentally comes into contact with the box, it is the same as having the black and white conductors come into contact with each other. The resulting short circuit causes the overcurrent protection device of the branch circuit to *blow*. The circuit is then *dead*, since the blown protective device creates an open circuit. The current that causes the protective device to become an open circuit is usually called a *fault current* since it flows when, and only when, there is a fault in the circuit.

The NEC requires that all equipment be grounded to the same electrode (usually a water pipe) that grounds the system (see Fig. 6–3).

6–3 ARE ALL SYSTEMS GROUNDED?

Insofar as residential wiring is concerned, the general answer is "yes," all systems are grounded. Because the code tends to be rather involved on this point, however, further clarification should prove beneficial.

Sec. 250-5(b)(1) states that ac systems of 50 to 1000 V supplying premises wiring shall be so grounded that the maximum voltage to ground on ungrounded conductors *does not exceed* 150 V.

As you now know, a 120/240 V arrangement is generally used when a single-phase source is required. The choice of which conductor to ground should be obvious. As shown in Figure 6–4, if we were to ground one of the hot conductors instead of the neutral, the voltage to ground from one side of the line would be 240 V, and this is a *code violation*.

In the case of a 120/208 V, four-wire, three-phase system (Sec. 2.6), the neutral conductor is again grounded. See Figure 6–5. Since each load then has a grounded conductor, the voltage to ground from any hot conductor is 120 V. This meets code requirements.

Another NEC requirement is that, when single-phase loads are connected from any line wire to ground, each single-phase load have a grounded conductor. All single-phase systems supplying lighting circuits must be grounded.

To sum up, then, if the voltage to any other grounded object can be kept under 150 V, the system must be grounded. This includes all residential electrical systems and/or any system that uses incandescent lamps.

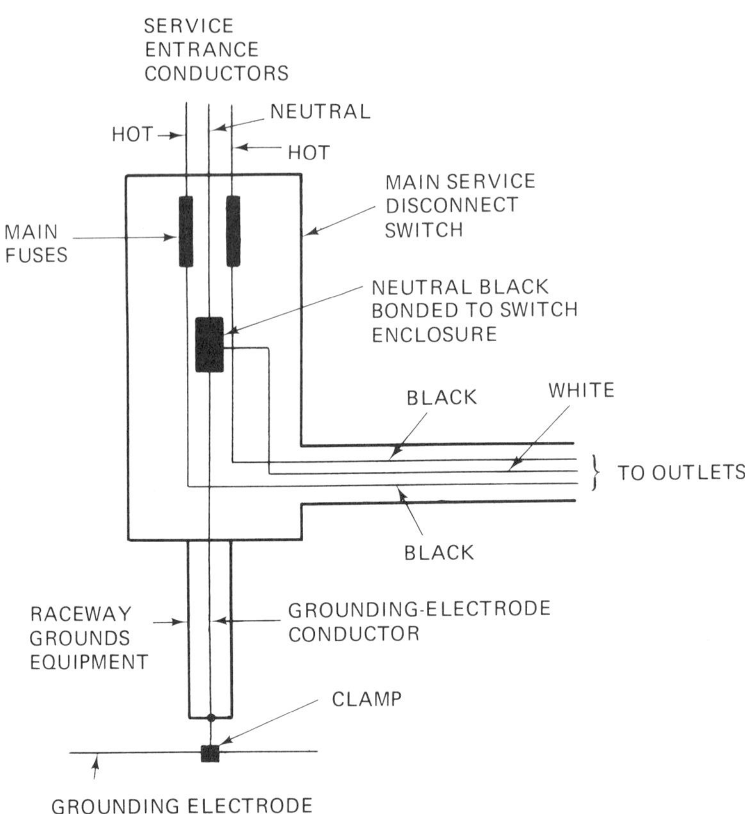

FIGURE 6–3. System and equipment grounding.

6–4 VOLTAGE TO GROUND

We have used the expression *voltage to ground* several times without defining it. Voltage to ground means simply the potential from any point in the system to any other object that itself is grounded.

After a system is grounded, the voltage to ground from any ungrounded conductor must be the same as the voltage to the grounded conductor. Figures 6–6(a) and (b) illustrate this for 120/240 V, single-phase and 120/208 V three-phase systems, respectively.

6–5 MORE ON SHOCK HAZARDS AND THEIR ELIMINATION

As noted earlier in Section 6–3, with a 120/240 V, three-wire supply, all lighting systems must have one circuit of the conductor grounded. Why?

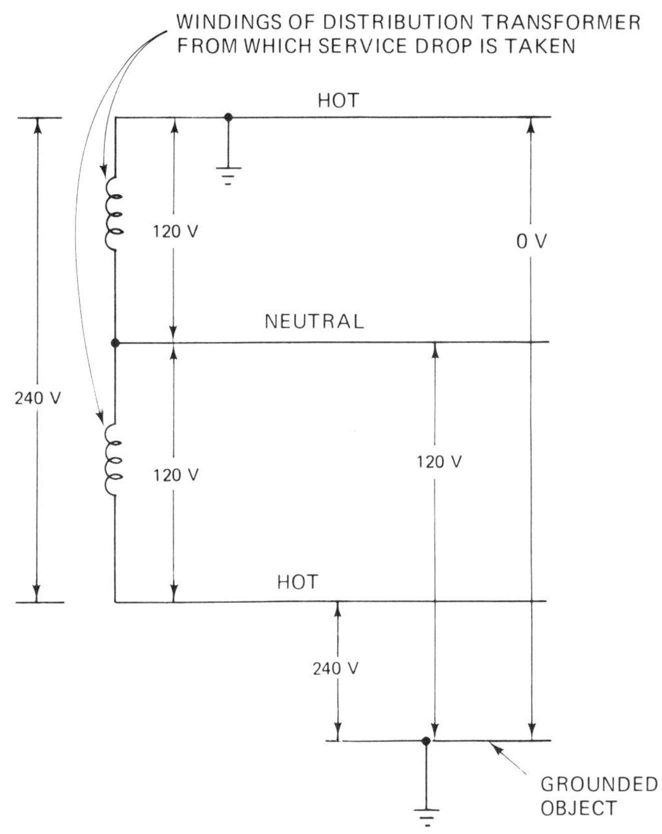

FIGURE 6–4. When a hot line is grounded, rather than neutral, a potential of 240 V exists between one side of the line and the ground.

FIGURE 6–5. Voltage to ground in a 120–208 V three-phase supply with the neutral conductor grounded.

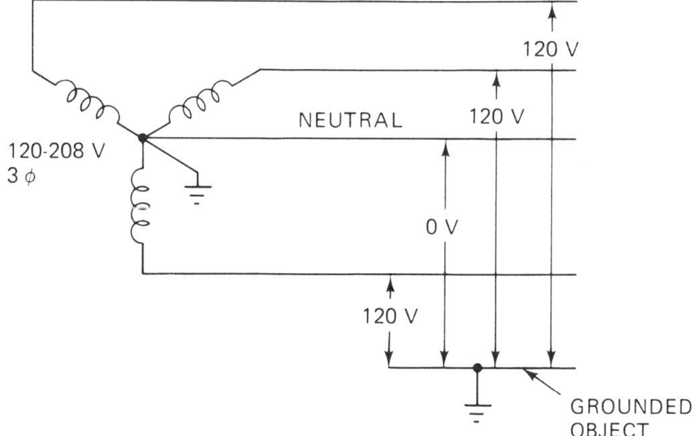

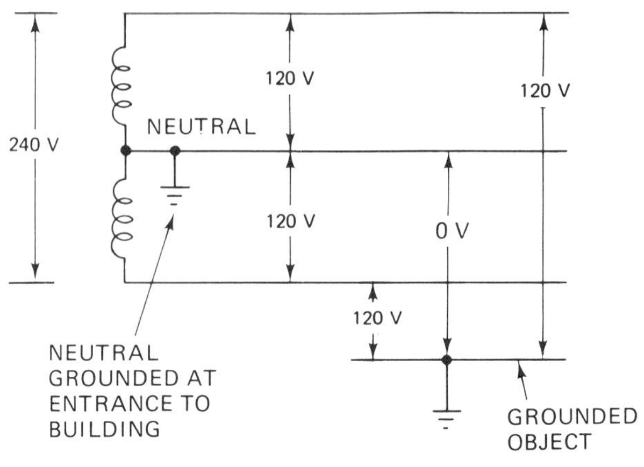

(*a*)

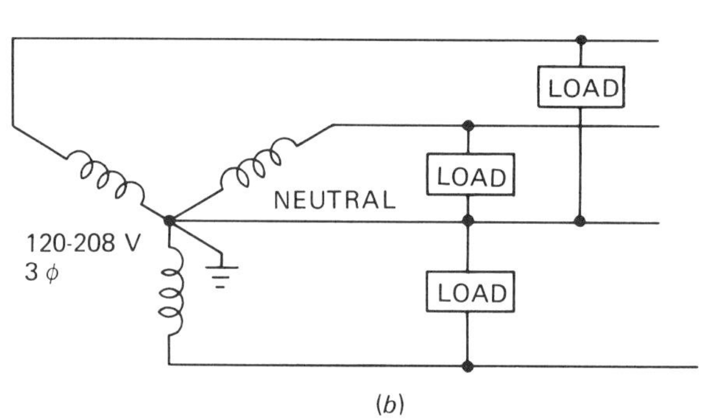

(*b*)

FIGURE 6–6. In properly grounded single-phase, three-wire systems (*a*) and three-phase, four-wire systems (*b*), the voltage to ground from any ungrounded conductor is the same as the voltage to the grounded neutral.

As indicated in Figure 6–7, if the screw-shell terminal of an incandescent lamp is incorrectly connected to an ungrounded conductor, a potential of 120 V exists between the shell and any grounded object. Suppose that a person changing the lamp simultaneously contacts a grounded object and the shell. His body is then the path by which current returns to ground. The resultant shock he receives may even prove fatal.

It should be obvious from the above discussion that the screw shell of any incandescent lamp should be connected to the grounded conductor. Moreover, the grounded conductor can only be grounded at the point where the service-entrance conductors enter the building. Any additional ground violates the NEC, creating hazards that will be discussed shortly.

The grounded terminal of a lampholder is identified by a nickel- or zinc-colored screw. If a lampholder has wire leads instead of screw terminals, the shell is identified by a white or a natural-gray conductor.

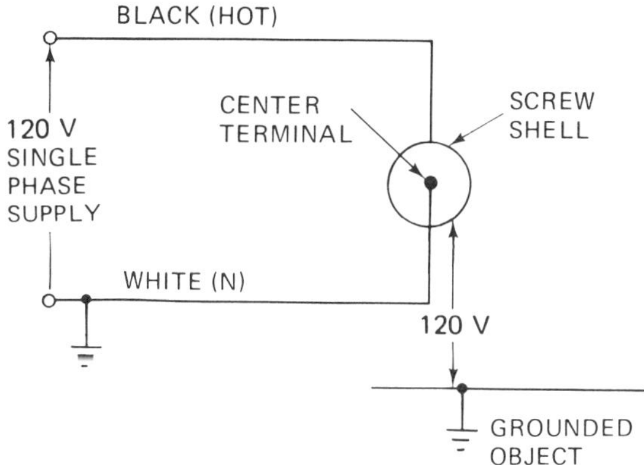

FIGURE 6–7. When the screw shell is connected to the ungrounded conductor, a potential of 120 V exists between the screw shell and any grounded object, and a definite shock hazard exists.

In Section 6–1 we learned that the grounded neutral in a residential wiring system should never be opened or interrupted by a fuse, circuit breaker, switch or other device. Let us see what happens if this rule is violated.

Refer to Figure 6–8(a), which shows a single-pole switch connected in series with the grounded conductor. Notice that even with the switch open, a 120-V potential still exists between the screw shell of a lamp holder and ground. The same situation would hold true, of course, for an open circuit breaker or a blown fuse. Anyone in simultaneous contact with the screw shell and any grounded object would receive a stiff "jolt."

The correct method of installing a single-pole switch in series with the hot conductor is shown in Figure 6–8(b). Installed this way, no continuous path to ground can possibly exist between the screw and ground when the switch is open. Therefore, there is no possibility of an electric shock.

In some installations other than residential you will find circuit breakers that have a sufficient number of poles to disconnect all conductors (both hot and neutral) simultaneously. This may lead to the erroneous conclusion that you can (or should) use fuses in both the hot and grounded neutral conductors of a residential wiring system. Do *not*. The hazard of any such double fusing is shown in Figure 6–9. If the fuse in the hot lead blows, all well and good. If the fuse in the ground lead opens, however, while that in the hot lead does not, a potential difference of 120 V again exists between the screw shell and any grounded object; it is dangerous!

In summary, keep this rule in mind. Always use a *solid*, unbroken neutral with any three-wire, single-phase supply. The same condition also

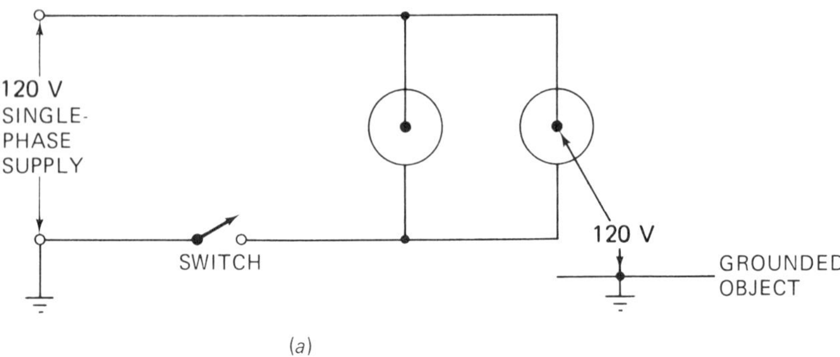

(a)

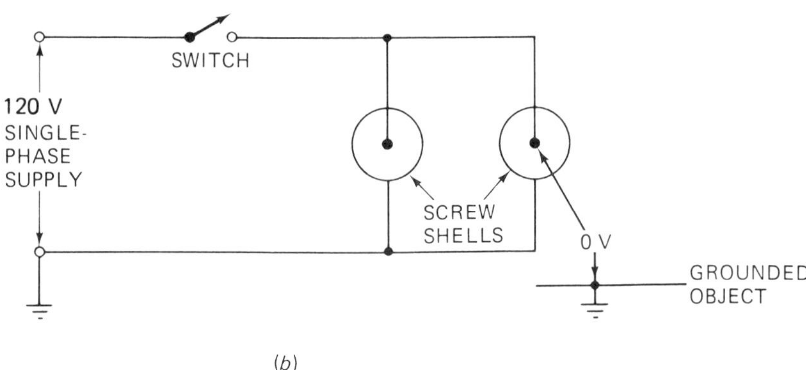

(b)

FIGURE 6–8. When an open single-pole switch is connected in series with the grounded neutral conductor (*a*), a potential of 120 V exists between the center terminals of the screw shells and any grounded object. In (*b*) this hazard is removed by placing the switch in the hot side of the circuit.

FIGURE 6–9. It is dangerous to have fuses connected in both sides of a single-phase circuit with the neutral grounded.

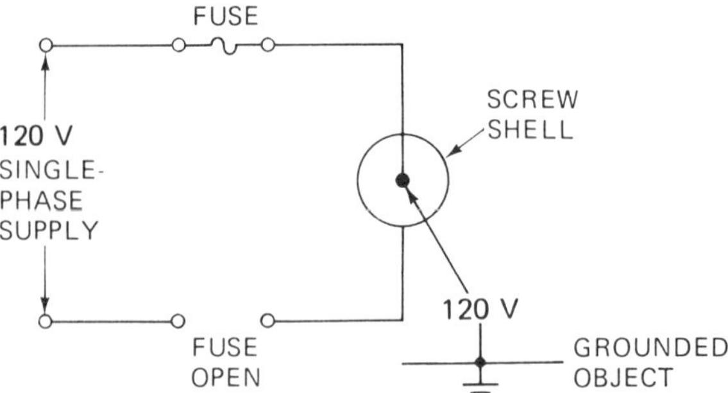

applies to a three-phase, four-wire system. Based on the facts that we have analyzed up to this point, review the proofs that support the validity of this last statement.

6–6 MORE ON EQUIPMENT GROUNDING

As noted earlier, the grounding of all metal that may contact an ungrounded conductor, either intentionally or unintentionally, is known as equipment grounding; it is essential to prevent metal equipment from maintaining a voltage above ground (zero potential). In Section 6–2 we discussed what could happen with an ungrounded raceway connected to a box containing a switch. Figure 6–10 depicts a similar arrangement, except that the raceway ground equipment and the system ground are connected. Should an accidental ground occur on the ungrounded conductor, the resulting fault current follows the path indicated by the arrows.

To permit overcurrent devices to open the circuit, the resistance of the equipment-ground path must be low. Since the resistance of a conductor decreases as its cross-sectional area increases, the use of *thick* conductors ensures the required low resistance. This method of providing a low-resistance ground can be expensive, however, so that construction economics often mean that the use of such conductors must be somewhat limited.

As noted earlier, the system and equipment grounds are joined at the main disconnect (service) switch. Figure 6–11 shows how this is done. Notice that the grounding-electrode conductor is encased in a metallic

FIGURE 6–10. When the raceway system is connected to the system ground, fault current will blow the fuse and remove any shock hazard.

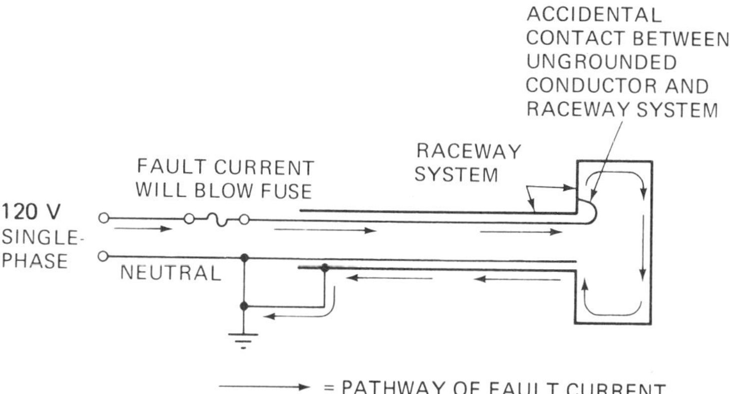

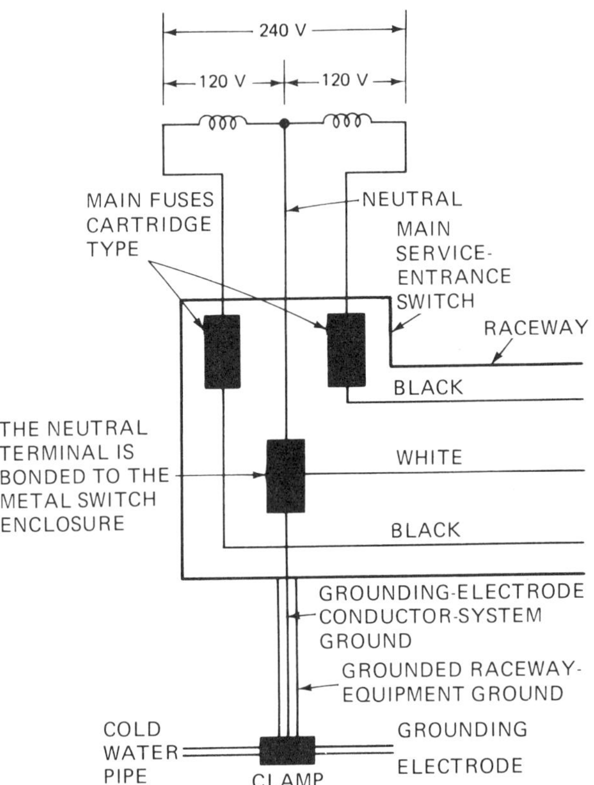

FIGURE 6–11. System and equipment grounding terminals in service-entrance switch.

raceway and that raceway and conductor are attached to the cold-water pipe by means of a heavy clamp. The equipment ground, however, connects only to the raceway of the grounding-electrode conductor, so that there the raceway has a dual grounding function.

To ground a wiring system, calculate the size of the grounding-electrode conductor that you need from the data in Table 6–1. In those cases where you use made electrodes, however, the grounding electrode conductor need not be larger than no. 6 copper wire or its equivalent in ampacity.

When the equipment ground is either conduit or EMT, the low-resistance path that they form permits fault currents to operate overcurrent protective devices easily, so the raceways are maintained at ground potential.

In addition to raceways, every metallic part that is not necessary for the operation of an electrical load must be securely grounded. Thus all

TABLE 6–1. Service and Grounding Conductors for Grounded Systems

Size of Service Conductor		*Size of Grounding Conductor*	
Copper	*Aluminum*	*Copper*	*Aluminum*
2 or smaller	0 or smaller	8	6
1 or 0	00 or 000	6	4
00 or 000	0000 or 250 MCM	4	2
Over 000 to 350 MCM	Over 250 to 500 MCM	2	0
Over 350 to 600 MCM	Over 500 MCM to 900 MCM	0	000
Over 600 to 1100 MCM	Over 900 MCM to 1750 MCM	00	0000
Over 1100 MCM	Over 1750 MCM	000	250 MCM

fittings and locknuts for boxes, switches, faceplate, motors, and the like must be tight.

Armored cable itself provides a path for fault currents. To reduce the resistance of the ground-return path, a continuous uninsulated aluminum ribbon lies between the paper-covered conductors and the armor. The parallel resistance of the armor and the ribbon is, of course, lower than the resistance of the armor itself. Armored cable must be securely fastened to outlets with connectors or clamps and must not be subject to any strain that would loosen the fastening over a period of time.

In nonmetallic cable an extra conductor is provided for equipment grounding. This conductor may be bare, but if it is insulated, the color of the insulation must be *green*.

Portable appliances, such as floor lamps, that do not use grounding-type receptacles and plugs are always a shock hazard. If the screw-shell insulation breaks down and contacts a metal frame, anyone making simultaneous contact with the frame and any grounded object will be shocked.

Grounding by use of a three-wire cord is illustrated in Figure 6–12. The receptacle includes a separate grounding prong. The grounding of the

FIGURE 6–12. Resistance reduces the fault current and the main fuse does not blow. Both shock and fire hazards exist.

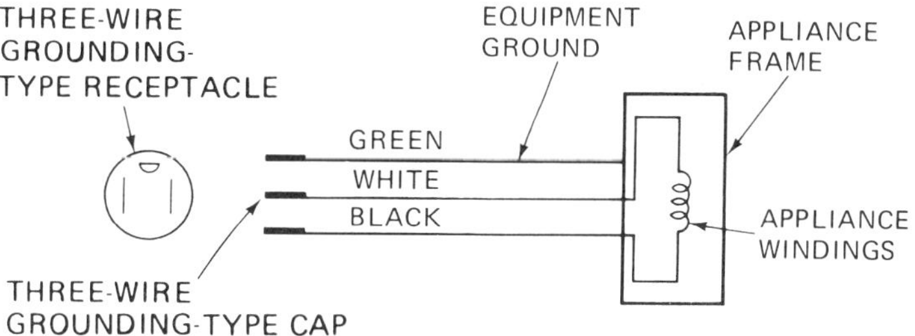

appliance is completed through the receptacle to its box and then through the raceway back to the source.

In the special case of major household appliances—dishwashers, electric ranges, garbage disposals, and the like—the appliance frame is connected to the neutral conductor. This grounding arrangement is not permitted with smaller appliances.

6–7 MORE ON FAULT CURRENTS

A loose equipment-ground connection prevents a fault current from doing its intended job of activating overcurrent devices. To learn why, study Figure 6–13, where a poor ground connection between a raceway and box establishes an equivalent resistance of 2.5 Ω. The three-wire, 120/240 V supply is fused at 50 A. A fault current is produced when a hot connector

FIGURE 6–13. A loose equipment–ground connection prevents a fault current from activating overcurrent devices.

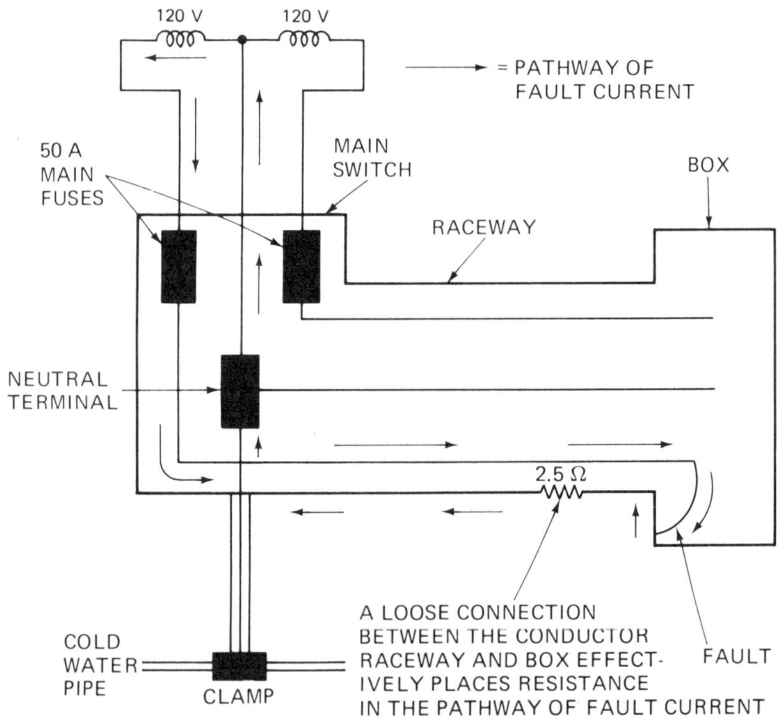

FIGURE 6–14. Lightning-protection points shown along dotted lines are: (*1*) terminals at least 20 ft apart along ridges and within 2 ft of ridge ends; (*2*) down-lead conductors; (*3*) two or more grounds, at least 10 ft deep for the house, less for clothesline; (*4*) roof projections, such as a weather vane, tied into conductor system; (*5*) trees within 10 ft of house connected to ground; (*6*) at least two chimney terminals; (*7*) dormer terminals; (*8*) arrester on antenna, connected to main conductor; (*9*) tie in to system on gutter within six feet of conductor; (*10*) arrester on power lines. (Courtesy of Underwriters' Laboratories).

comes into contact with the box. The maximum current that can pass through the poor (high resistance) ground connection is, by Ohm's law,

$$I = \frac{E}{R} = \frac{120}{2.5} = 48 \text{ A}$$

Under these conditions the 50-A fuse will not blow. As a result, a shock hazard exists between the poor connection and ground. A fire hazard also exists if combustible materials are nearby, since any electrical arcing at the poor connection causes flashover, which, in turn, produces unsafe heat.

6–8 RESIDENTIAL LIGHTNING PROTECTION

Property damage and casualties from lightning generally occur when a bolt makes a direct strike to the roof, chimney, dormer, or other elevated part of a building. Lightning is powerful. A bolt may have a potential ranging from 10 to 100,000,000 V and discharge currents ranging from 1000 to 300,000 A.

A lightning protection system listed by Underwriters' Laboratory is shown in Figure 6–14. The possibility of a home being struck by lightning is greater in rural areas than in urban areas, since high-rise buildings, electric-distribution and telephone poles, and similar objects usually divert lightning bolts to ground.

EXERCISES

6–1. Explain what is meant by the following terms: *(a)* service drop, *(b)* service conductors, *(c)* service-entrance conductors, *(d)* service raceway, *(e)* service equipment, *(f)* grounding-electrode conductor, *(g)* grounding electrode, and *(h)* made electrode.

6–2. What colors are permitted for the grounded neutral wire in a residential wiring system?

6–3. What exceptions, if any, does the NEC permit for the color scheme noted in Exercise 6–2?

6–4. What is the *system ground* in a residential wiring system? What is the *equipment ground* in a residential wiring system?

6–5. In residential wiring, why must the grounded neutral wire be uninterrupted?

6–6. Why is a system ground important?

6–7. Why is an equipment ground important?

6–8. In a residential wiring system, what danger is inherent in double fusing, that is, fusing in both the hot and neutral lines?

6–9. Does the NEC permit separate system and equipment grounds?

6–10. Since there are three service-entrance conductors, why does the NEC require grounding of the neutral line?

6–11. What would happen if, in a residential wiring system, we were to ground a hot conductor rather than the neutral conductor?

6–12. In a 120/208 V, four-wire, three-phase system, is any conductor grounded? If so, which conductor and why?

6–13. Under what conditions must a system always be grounded?

6–14. Precisely what is meant by the expression *voltage to ground*?

6–15. How is the grounded terminal of a lamp holder identified?

6–16. If a lamp holder has wire leads instead of screw terminals, how is the shell identified?

6–17. In a grounded system using no. 2 copper service conductors, what size of copper grounding conductor is needed?

6–18. What step is taken during manufacture to reduce the resistance of the ground-return path of armored cable?

6–19. If an insulated grounding conductor is included in nonmetallic cable, what is the required color of the insulation on the grounding wire?

6–20. Explain how portable-appliance grounding is accomplished with a three-wire cord.

6–21. How is an electric range grounded?

6–22. Explain how a loose equipment-ground connection may prevent a fault current from activating an overcurrent device.

6–23. If an ungrounded cable contacts a cold-water pipe intermittently, does this constitute a serious hazard? Explain your answer.

7

Wiring Methods

7–1 INTRODUCTION

Rigid metallic and nonmetallic conduit, thin-wall conduit, flexible conduit, nonmetallic-sheathed cable, and armored cable are the most often employed residential and farm wiring methods recognized by the NEC. Before any given method is used, however, check your local code; it may limit the NEC specifications.

7–2 RIGID METALLIC CONDUIT

Rigid metallic conduit looks like the plumber's pipe used to carry steam, gas, or water, but it is made from specially annealed steel for easy bending. Its inside surface is also very smooth to facilitate the easy pull-through of wires without damage to their insulation.

Available in 10-ft lengths, rigid conduit may have either a black enamel or galvanized finish. Black-enamel conduits are limited to indoor usage, but the galvanized conduits may be used indoors, outdoors, aboveground or underground. When they are used in cinder fill exposed to permanent moisture, however, they must be encased in 2-in.-thick noncinder concrete. Alternatively, they may be buried 18 in. under the fill (NEC, Sec. 346-3).

Electrical conduits range from ½ to 6 in. in trade sizes; however, the actual physical dimensions differ appreciably from those trade sizes. See Table 7–1.

Cutting

Use an ordinary pipe cutter or a hack saw with 18 teeth per inch. To prevent damage to insulation, use a suitable reamer to remove all burrs on the cut edge afterwards.

Bending and Threading

Conduit benders are standard equipment for an electrician. Generally, the manufacturer provides complete instructions for the use of a particular bender. Read and follow them carefully.

Table 7–2 shows the minimum permissible radius of bends in conduit conductors either with or without lead sheathing. (See Sec. 346-10.) A run of conduit between outlet and outlet, between fitting and fitting, or between outlet and fitting shall not contain more than the equivalent of four quarter-bends (360°), including those bends located immediately at the outlet or fitting.

TABLE 7–1 Dimensions of Rigid Metallic Conduit

Trade Size (in.)	*Internal Diameter (in.)*	*Internal Area (in.^{2})*	*External Diameter (in.)*
0.50	0.622	0.30	0.840
0.75	0.824	0.53	1.050
1.00	1.049	0.86	1.315
1.25	1.380	1.50	1.660
1.50	1.610	2.04	1.900
2.00	2.067	3.36	2.375
2.50	2.469	4.79	2.875
3.00	3.068	7.38	3.500
3.50	3.548	9.90	4.000
4.00	4.026	12.72	4.500
4.50	4.506	15.94	5.000
5.00	5.047	20.00	5.630
6.00	6.065	28.89	6.620

TABLE 7–2. Minimum Permissible Radius of Bends in Conduit

Size of Conduit (in.)	Minimum Radius of Bend (in.)	
	Conductors without Lead Sheath	*Conductors with Lead Sheath*
0.50	4	6
0.75	5	8
1.00	6	11
1.25	8	14
1.50	10	16
2.00	12	21
2.50	15	25
3.00	18	31
3.50	21	36
4.00	24	40
4.50	27	45
5.00	30	50
6.00	36	61

To thread rigid conduit, use a die having a taper of ¾ in./ft. Connections to boxes are made with locknuts and bushings after the conduit is threaded.

Permissible Number of Wires

Table 1, Chapter 9, of the NEC, reproduced in this text as Table 7–3, gives the percent of cross section of conduit and tubing that can be occupied by conductors. Notice that a separate listing is given for lead-covered conductors and that the table does not apply to fixture wires.

Tables 3A through 3C, Chapter 9, of the NEC are derived from the same Table 1 and specify the maximum number of specific types and sizes of conductors that can be enclosed in the ½- to 6-in. electrical trade sizes of conduit or tubing.

TABLE 7–3. Percent of Cross Section of Conduit and Tubing for Conductors*

Number of Conductors	*1*	*2*	*3*	*4*	*Over 4*
All conductor types except lead-covered (new or rewiring)	53	31	40	40	40
Lead-covered conductors	55	30	40	38	35

*See Table 2, Ch. 9, NEC for fixture wires.

When more than three wires are placed in conduit, their ampacity must be derated, as explained earlier in Section 4–7.

Support

Conduit must be firmly supported within 3 ft of each outlet box, junction box, cabinet, or fitting. Either single-hole or two-hole straps can be used to support the conduit. On straight runs, the distance between supports is determined by the electrical trade size of the conduit. These requirements are computed in Table 7–4 (NEC Table 346-12).

7–3 RIGID NONMETALLIC CONDUIT

Rigid nonmetalic conduit also comes in 10-ft lengths, and one coupling is furnished with each length. Materials that have the physical characteristics suitable for conduit of this type are properly formed and treated fiber, asbestos cement, soapstone, rigid polyvinyl chloride (aboveground use), and high-density polyethylene (underground use). Electrical trade sizes again run from ½ to 6 in. nominally.

Use Permitted (Sec. 347–2)

Rigid nonmetallic conduit and fittings approved for the purpose may be used in the following conditions and where the potential is 600 V or less.

- In walls, floors, and ceilings.
- In locations subject to severe corrosive influences (Sec 300-6) and where subject to chemicals for which the materials are specifically approved.

TABLE 7–4. Supports for Rigid Metallic Conduit

Conduit Size (in.)	*Maximum Distance between Rigid-Metal Conduit Supports (ft)*
½–¾	10
1	12
1¼–1½	14
2 –2½	16
3 and larger	20

Reprinted with permission from the 1978 National Electrical Code, Copyright 1977, National Fire Protection Association, Boston, Mass.

- In cinder fill.
- Where walls are washed frequently, the entire conduit system must be watertight. All bolts, straps, screws, and the like, must be corrosion resistant or protected with approved corrosion-resistant materials.
- In dry and damp locations not prohibited by Sec. 347-3.
- For exposed work not subject to physical damage (if approved for the purpose).

Use Prohibited (Sec. 347–3)

Rigid nonmetallic conduit shall not be used in the following conditions:

- In hazardous locations except as covered in Secs. 514-8 and 515-5.
- For the support of fixtures or other equipment.
- Where subject to physical damage unless approved for the purpose.
- Where subject to ambient temperatures exceeding those for which the conduit has been approved.
- For conductors whose insulation temperature limitation would exceed those for which the conduit is approved.

Miscellany

In general, the bending requirements for rigid nonmetallic conduit are the same as those given for the metallic variety.

To compensate for thermal expansion and contraction, expansion joints are sometimes required.

The number of conductors permitted in a single conduit follows the same guidelines given earlier for rigid metallic conduit.

In addition to a support within 3 ft of each box, cabinet, or other conduit termination, all rigid nonmetallic conduit must be supported as required in Table 7–5 (NEC Table 347-8).

TABLE 7–5. Support for Rigid Nonmetallic Conduit

Conduit Size (in.)	*Maximum Spacing between Supports (ft)*
½–1	3
1¼–2	5
2½–3	6
3½–5	7
6	8

Reprinted with permission from the 1978 National Electrical Code, Copyright 1977, National Fire Protection Association, Boston, Mass.

7–4 THIN-WALL CONDUIT

Thin-wall conduit is the name commonly used in the trade to designate electrical metallic tubing (EMT). It is similar to rigid metallic conduit except that its walls are thinner, which makes it lighter and easier to handle. The largest size is 4-in. tubing. It is cut and reamed by using the same methods as for rigid conduit, but the walls are not usually threaded except by special fittings approved for the purpose. Threadless connectors, which consist of a body and a split ring, provide tremendous pressure and electrical continuity when the nut is securely tightened. Adapters permit the use of thin-wall conduit in fittings made for rigid conduit.

All information given for rigid metallic conduit concerning the number of conductors, internal sizes, lengths, bends, and supports is applicable to EMT.

Article 348 indicates that EMT may be used for both exposed and concealed work. The use of EMT is prohibited, however, in the following cases:

- Where, during installation or afterward, it may be subject to severe physical damage.
- Where protected from corrosion by enamel only.
- In cinder concrete or fill where the tubing would be subject to permanent moisture unless protected on all sides by a layer of noncinder concrete at least 2 in. thick or unless the tubing is at least 18 in. under the fill. Where practicable, the use of dissimilar metals throughout the system shall be avoided to eliminate the possibility of galvanic action.
- Ferrous or nonferrous EMT, elbows, couplings and fittings must not be installed in concrete or in direct contact with the earth, or in areas subject to severe corrosive influences unless they are made of a material judged suitable for the condition, or unless corrosion protection approved for the purpose is provided.

7–5 FLEXIBLE METAL CONDUIT

Often referred to as *greenfield*, flexible metal conduit (Article 350) is essentially the same as the outer covering of armored cable.

Installations of flexible conduit must comply with the appropriate (or applicable) provisions of NEC Articles 300 (Wiring Methods), 334 (Metal-Clad Cable), and 346 (Nonmetallic-Sheathed Cable).

Flexible metal conduit must never be used in wet locations, unless the conductors are of the lead-covered type or of some other type specially made for the conditions prevailing in battery rooms or where rubber-covered conductors are exposed to oil, gasoline, or other materials that have a deteriorating effect on rubber. Use of flexible metal conduit in hoistways and in any hazardous location is generally, but not always, prohibited. See Sec. 350-2 for references to determine limitations.

A ⅜-in. nominal trade size of flexible metal conduit may be used in lengths not in excess of 6 ft as a part of an approved assembly or for lighting fixtures. Code Table 350-3 lists the maximum number of insulated conductors acceptable in ⅜-in. flexible metal conduit.

Support is required at intervals not exceeding 4 ft 6 in. and within 12 in. on each side of every outlet box or fitting. For exceptions to this standard, see Sec. 350-4.

Where both the conduit and the fittings are approved for the purpose, flexible metal conduit may be used as a grounding means. *Exception:* Flexible metal conduit may be used for grounding if the length is 6 ft or less, if it is terminated in fittings approved for the purpose, and if the circuit conductors contained therein are protected by overcurrent devices rated at 20 A or less.

In most sections of the country, flexible metal conduit is seldom used. It is very useful, however, in places where it is either difficult or awkward to bend rigid conduit or where a reasonable degree of permanent flexibility is needed.

7–6 PULL BOXES

As you know, the NEC limits the number of bends between outlets in a run of conduit to four quarter-bends—that is, four 90° bends. Thus the total degree of bend—including that of bends located immediately at the outlet—must not exceed 360°. If possible, of course, fewer bends should be employed, since bends only complicate the installation of wires in the raceway system.

Pull boxes (Sec. 370-18) may be installed in raceway systems to reduce the number of bends in conduit. A pull box is simply a sheet-metal enclosure and may be either straight or right-angled, as shown in Figures 7–1(a) and (b), respectively. Straight pull boxes make it easier to pull wires through conduit by providing places where the electrician can reach in and help the wires along.

For raceways ¾-in. trade size or larger and containing conductors of AWG size no. 4 or larger, the minimum dimensions of a pull box (or junction box) installed in a raceway or a cable must comply with the following:

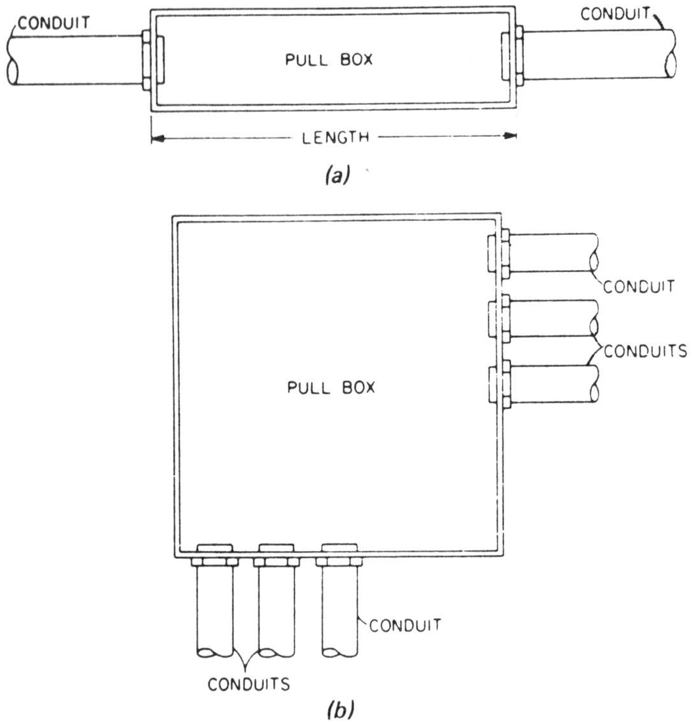

FIGURE 7–1. Typical straight (*a*) and right-angle (*b*) pull boxes.

- In straight pulls, the length of the box must be at least eight times the trade diameter of the largest raceway or cable.

- Where angle or U pulls are made, the distance from the point of raceway or cable entry to the opposite wall must be at least six times the trade diameter of the largest entering raceway or cable. When more than one raceway or cable enters the same side of the box, the distance between the point of entry and the opposite wall must be increased by the sum of the trade diameters of all additional entries. The center-to-center distance between raceway entries enclosing the same conductor must not be less than six times the diameter of the larger raceway.

The rules for angle or U pulls are illustrated by the example shown in Figure 7–2. Here the pull box has two 4-in. and four 2-in. (trade diameter) entering raceways. According to the rules given above, the length of side *D*, and its oposite side, *B*, must be no less than

$$6 \times 4 + 2 + 2 = 28 \text{ in.}$$

Inspection of Figure 7–2 shows that the same dimensions must also hold true for sides C and A.

Assuming that the conductors make a 90° bend inside the box, and the same raceways (positionwise) enclose the same conductors, the center-to-center distance between raceways 3 and 4, which we shall call distance X, is given by

$$\begin{aligned}\text{Distance } X &= \sqrt{17^2 + 17^2} \\ &= \sqrt{289 + 289} \\ &= \sqrt{578} \\ &= 24.04 \text{ in.}\end{aligned}$$

Thus the box shown in Figure 7–2 is entirely adequate.

When the wiring is completed, pull boxes are closed by suitable covers. The covers and internal wiring must be accessible after the building is finished.

7–7 PULLING WIRE

When raceways are completed, a continuous path exists from panelboards to pull boxes, or to junction boxes, or to outlet boxes, or from outlet to

FIGURE 7–2. Right-angle pull box dimensions. A and B represent the lengths of sides; C, the diagonal distance between conduits; and D, the distance of conduit from corner.

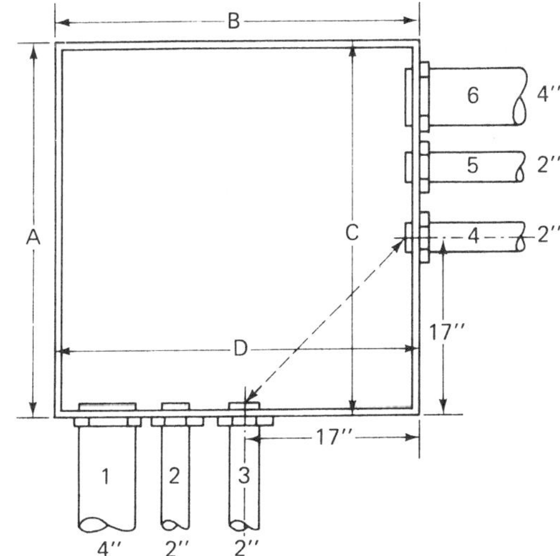

outlet. You can pull the wires into the raceway system with either a snake, a fish wire, or a fish tape. These resilient pulling devices are usually made from tempered steel or plastic rope. Steel fish tapes are often wound on reels, and are used as shown in Figure 7–3. Any wires to be pulled through the conduit can be attached to the hook that is bent onto one end of the steel fish tape.

The best method to use depends, of course, on the overall length of the raceway system, the conduit and conductor sizes, the number of bends, and so forth. Although we are concerned primarily with residential wiring, the methods described for pulling wire are, in some cases, reserved for very large jobs.

The hooked end of the fish tape is pushed into the raceway until it

FIGURE 7–3. Use of steel fish tape wound on a reel. (Courtesy of Ideal Industries, Inc.)

protrudes at an outlet. Wires cut to the correct length are then attached to the fish tape so that they are pulled into the raceway as the fish tape is withdrawn. The wires should be somewhat longer than the raceway, because splices are not allowed inside the raceways. Any excess wire can always be cut off when the installation is completed. The attachment between the wires and fish tape should not be bulky, because a large attachment will make the pulling more difficult. One method of attaching the wires to the fish tape uses a socket-like device called a *basket*. The wire ends are inserted into this basket, which is so constructed that it tightens the ends when tension is applied.

Another wire pulling method employs a compressed air gun that blows a nylon fish line through either rigid or thin-wall conduit. A plastic piston is attached to the line to receive the wire ends.

Wire pulling usually requires two men: one to pull and the other to feed the wires into the raceway and make sure that there are no snarls. Powdered soapstone is often used as a lubricant to make the pulling easier.

Small wires may be attached to the fish tape and pulled by hand; but when the wires are larger, a rope or cable is first attached to the fish tape and pulled through the raceway. Then the wires are fastened to the rope or cable, which in turn is withdrawn to pull the wires into place. A hand pulling machine, which is simply a small hand-operated winch, is often helpful. Turning the handle on the winch winds the cable or rope on a drum. Gears usually lessen the pulling effort by a factor of about 50. Power pulling machines are also available for very heavy loads.

7–8 NONMETALLIC-SHEATHED CABLE

As you know, there are two kinds of nonmetallic-sheathed cable; type NM has the conductors enclosed in an outer fabric braid or plastic sheath; type NMC has the conductors embedded in solid plastic. Code Sec. 336-3 lists the permissible and nonpermissible uses of both types of nonmetallic sheathed cable.

Type NM cable may be used for both exposed and concealed work in normally dry locations. It must never be installed where it will be exposed to corrosive fumes or vapors. Nor should it be embedded in masonry, concrete, adobe fill or plaster, or in a shallow chase in masonry, concrete or adobe, and covered with plaster, adobe, or similar material.

Type NMC cable may be installed for both exposed and nonexposed work in dry, moist, damp, or corrosive locations, and in outside and inside walls of masonry block or tile.

Nonpermissible uses for either type NM or NMC include service-entrance cable, commercial garages, theaters and assembly halls, motion-picture studios, storage-battery rooms, hoistways, any hazardous location, or when embedded in poured cement, concrete, or aggregate.

FIGURE 7–4. Methods of protecting cable: (*a*) cable passes through holes; (*b*) cable on running board; (*c*) cable following building structures; (*d*) cable between guard strips.

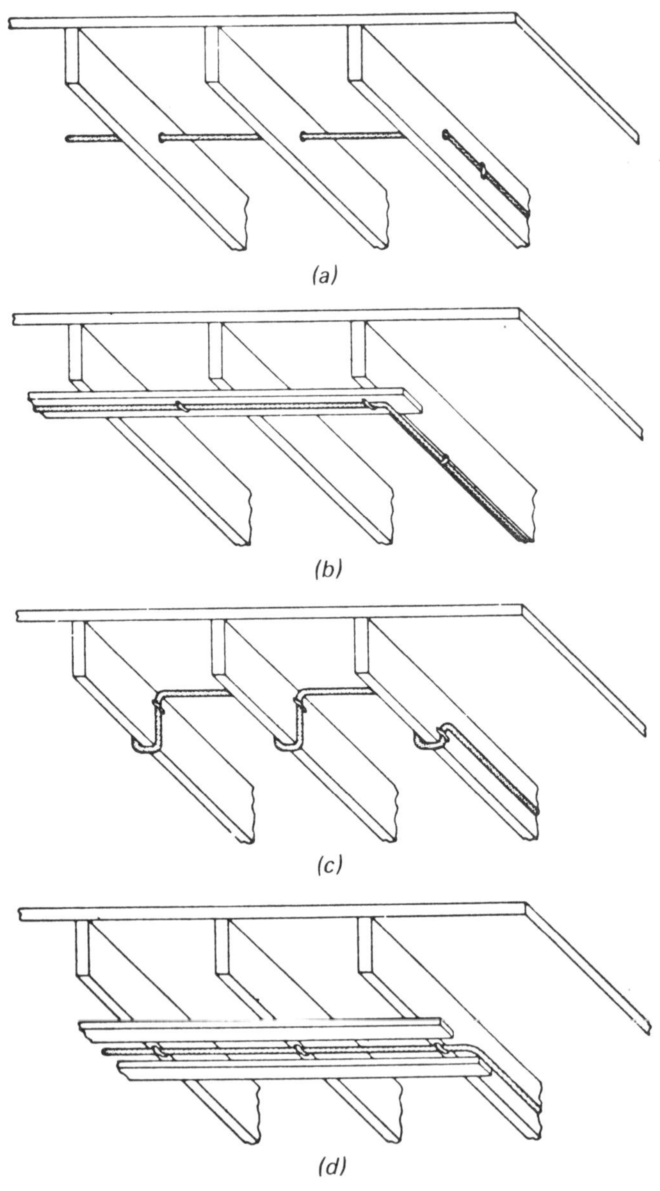

Before nonmetallic-sheathed cable is installed in an oulet box, about 8 in. of the outer covering must be removed by using either a jackknife or a special cable stripper. The stripper saves time, reduces the possibility of damage to insulation, and serves as a wire gauge.

Nonmetallic-sheathed cable is anchored to switch and outlet boxes by the use of special connectors. Many boxes have built-in clamps, which serve the same purpose as these connectors.

Common methods of installing nonmetallic-sheathed cable are shown in Figure 7–4. In Figure 7–4(a) the cable is stapled along the side of the floor joist and then passes through holes near the center of the joists. Wherever possible, holes through the joists (or any lumber) should be located at least 1¼ in. from the edge to avoid mechanical weakening of the joist.

In Figure 7–4(b) a running board is nailed across the joists. The cable is stapled along the bottom edge of the joist and to the running board.

In Figure 7–4(c) the cable follows the building structure. This method uses more materials and cable than necessary, and the voltage drop increases. The methods shown in Figures 7–4(a) and (b) are preferred.

When attics are wired, guard strips like those shown in Figure 7–4(d) are used to protect cables that cross rafters and joists. The guard strips must be at least as high as the cable is thick. Where the cable is more than 7 ft above the attic floor, guard strips are not needed. Protection is needed only for 6 ft around the attic access opening if there is no permanent stairway or ladder.

When installing cables, always follow the approximate contour of the building. Do not take shortcuts across open spaces. When nonmetallic-sheathed cable is run at an angle with the joists in unfinished basements, cables no smaller than two no. 6 or three no. 8 may be fastened directly to the lower edges of the joists. For smaller cables, use the method shown in Figures 7–4(a) and (b). Cables of all sizes run parallel to joists and must be fastened to the sides or edges of the joists.

The cable must be supported by approved staples, straps, or similar fittings at intervals not exceeding 4 ft 6 in., and within 12 in. from every cabinet, box, and fitting. (For exceptions see Article 336-5.) The radius of the curve at the inner edge of any bend must not be less than five times the diameter of the cable.

7–9 ARMORED CABLE

Type AC cable may be used in dry locations. Where walls are exposed to excessive moisture or dampness or are below grade line, type ACL (lead-covered cable) must be used. (See Article 333 for complete information.)

A two-conductor armored cable is shown in Figure 7–5. Note that a bonding strip is in intimate contact with the armor. As explained earlier, the purpose of this bonding strip is to reduce the resistance of the wire. An insulating bushing is slipped inside the armor to protect the insulation from sharp edges on the armor. The cutout shows the inside paper wrap. When the insulating bushing is being installed, keep the bonding strip outside the bushing.

Armored cable may be cut either with a hacksaw or an armor cutter. Any hacksaw used for this purpose should have 32 teeth per inch. Hold the hacksaw at right angles to the armor, not at right angles to the cable. Take care not to damage the insulation of conductors inside the cable. After you cut the armor, twist it to separate the parts. Once the armor is separated, it can be pulled off to uncover the bonding strip and paper wrapping.

EXERCISES

7–1. Describe the appearance and finish of rigid metallic conduit.

7–2. In what way is the use of black-enamel rigid metallic conduit limited?

7–3. In what lengths and in what trade sizes are rigid metallic conduit available?

7–4. How is rigid metallic conduit cut?

7–5. What restriction is placed on the bending of rigid metallic conduit?

7–6. When installed in rigid metallic conduit, must the ampacity of conductors be given consideration? Explain in detail.

7–7. What are the support requirements for rigid metallic conduit?

FIGURE 7–5. Two-conductor armored cable.

INSULATING BUSHING

ARMOR FOR MECHANICAL PROTECTION

BONDING STRIP

CUTOUT SHOWING INSIDE WRAP

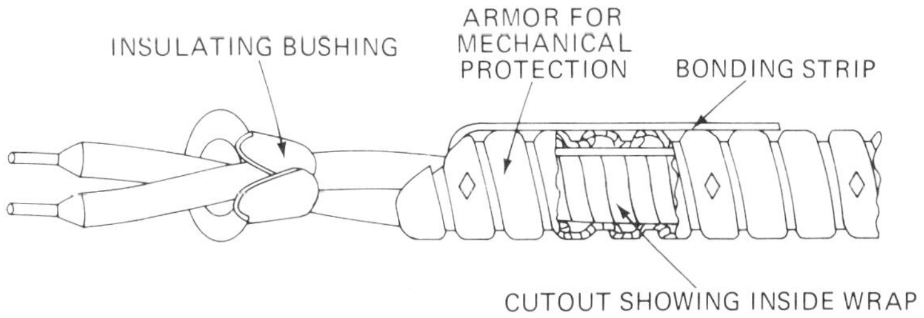

7–8. Where may rigid nonmetallic conduit be used?

7–9. Where may rigid nonmetallic conduit not be used?

7–10. Where thermal expansion and/or contraction are experienced, what precaution must be taken with rigid nonmetallic conduit?

7–11. In what electrical trade sizes is thin-wall conduit available?

7–12. Is EMT threaded in the same manner as rigid metallic conduit? Explain fully.

7–13. Where is the use of EMT prohibited?

7–14. In what types of installations is *greenfield* handy?

7–15. What restrictions are placed on the use of flexible metal conduit?

7–16. How must flexible metal conduit be supported?

7–17. What is the purpose of a pull box?

7–18. If a straight pull box is used with, say, 3-in. conduit, what is its required length?

7–19. A right-angle pull box has one 2-in. conduit and three 1-in. conduits attached on each of two sides. What must be the minimum size of the box? What is the distance between the centers of the two nearest 1-in. conduits? Assuming that the pull box is square and the two conduits are the same distance from the corner, determine distance *D* (Figure 7–2).

7–20. In a small residential installation employing rigid metallic tubing, describe how you would normally pull the conductors into the raceway system.

7–21. What are the two types of nonmetallic-sheathed cable, and how may they be used?

7–22. Describe three techniques for installing nonmetallic-sheathed cable.

7–23. What precautions must be taken when nonmetallic-sheathed cable is used to wire an attic that is less than 7 ft high?

7–24. What is the purpose of the bonding strip that is in intimate contact with the outside sheath of armored cable?

7–25. Describe how you would cut armored cable with a hacksaw.

8

Sizing the Service Entrance and Branch Circuits

8–1 INTRODUCTION

Since the service, service equipment, and systems ground provide protection for the rest of the wiring system, most inspections start at this point. Once the inspection of the service, service equipment and systems ground is completed, the inspector will turn his attention to branch circuits and to feeders—where they exist.

By definition (NEC, Article 100), feeders are the circuit conductors between the service equipment and the branch-circuit overcurrent devices. Since, in a residential wiring system, both the main and branch-circuit overcurrent devices are located in the same cabinet, there are no feeders.

Again, by NEC definition, a branch circuit is that portion of the wiring system between the final overcurrent device protecting the circuit and the outlet or outlets. In determining where a branch circuit actually begins, notice that this definition specifically states "between the final overcurrent device protecting the circuit." Thus a device not approved for branch-

FIGURE 8–1. Common Electrical symbols.

GENERAL OUTLETS

Ceiling Wall

- Outlet
- (D) Drop Cord
- (E) Electrical Outlet
  For use only when circle used alone might be confused with columns, plumbing symbols, etc.
- (F) Fan Outlet
- (J) Junction Box
- (L) Lamp Holder
- $(L)_{PS}$ Lamp Holder with Pull Switch
- (S) Pull Switch
- (V) Outlet for Vapor Discharge Lamp
- (X) Exit Light Outlet
- (C) Clock Outlet (specify voltage)

CONVENIENCE OUTLETS

- Duplex Convenience Outlet
- Convenience Outlet other than Duplex:
  1 = single; 3 = triplex; etc.
- Duplex Convenience Outlet—Split Wire
- Duplex Convenience Outlet—Grounding Type (GR)
- Weatherproof Convenience Outlet (WP)
- Range Outlet (R)
- Switch and Convenience Outlet (S)
- Special Purpose Outlet
- Floor Outlet

SWITCH OUTLETS

S	Single Pole Switch
S_2	Double Pole Switch
S_3	Three-Way Switch
S_4	Four-Way Switch
S_D	Automatic Door Switch
S_K	Key-Operated Switch
S_P	Switch and Pilot Lamp
S_{CB}	Circuit Breaker
S_{WCB}	Weatherproof Circuit Breaker
S_{MC}	Momentary Contact Switch
S_{RC}	Remote Control Switch
S_{WP}	Weatherproof Switch
S_F	Fused Switch
S_{WF}	Weatherproof Fused Switch

SPECIAL OUTLETS

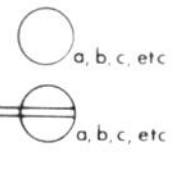

$S_{a,b,c,etc}$

Any Standard Symbol as given above with the addition of a lower-case subscript letter may be used to designate some special variation of Standard Equipment of particular interest.

PANELS, CIRCUITS, AND MISCELLANEOUS

- Lighting Panel
- Power Panel
- Branch Circuit: concealed in ceiling or wall
- Branch Circuit: concealed in floor
- Branch Circuit: exposed
- Home Run to Panel Board
  Indicate number of circuits by number of arrows. *Note:* Any circuit without further designation indicates a two-wire circuit. For a greater number of wires, indicate as follows: —/—/—/— (3 wires), —/—/—/—/— (4 wires), etc.
- Feeders
  Note: Use heavy lines and designate by number corresponding to listing in Feeder Schedule.
- Underfloor Duct and Junction Box: Triple System
  Note: For double or single systems eliminate one or two lines. This symbol is equally adaptable to auxiliary system layout.
- (G) Generator
- (M) Motor
- (I) Instrument
- Controller

AUXILIARY SYSTEMS

- Push Button
- Buzzer
- Bell
- Annunciator
- Outside Telephone
- (T) Bell-Ringing Transformer
- Standard Battery
- Auxiliary System Circuits
  Note: Any line without further designation indicates a two-wire system. For a greater number of wires, designate with numerals in a manner similar to, "— — — 12 No. 18W 3/4" Conduit (c)" or designate by number corresponding to listing in Schedule.
- $\square_{a,b,c}$ Special Auxiliary Outlets
  Subscript letters refer to notes on plans.

circuit protection, such as a thermal cutout or motor-overload protective device, is not considered to be the starting point of a branch circuit.

To plan a wiring system, it is necessary to make certain calculations, such as those demonstrated in this chapter. First, however, let us take a look at the symbols used in wiring diagrams and the types of building plans that you will encounter.

8–2 SYMBOLS

Electrical symbols that commonly appear on the architectural drawings of residences are shown in Figure 8–1. Study these symbols carefully and refer to them as needed. Notice in particular that round and square symbols designate outlets served by full voltage and low voltage, respectively.

8–3 GETTING FAMILIAR WITH BUILDING PLANS

The use of symbols is illustrated in the very simple wiring plan for a two-car garage shown in Figure 8–2. Duplex convenience outlets, lamp holders, and switches are indicated by their respective symbols. The dashed lines

FIGURE 8–2. Garage wiring plan.

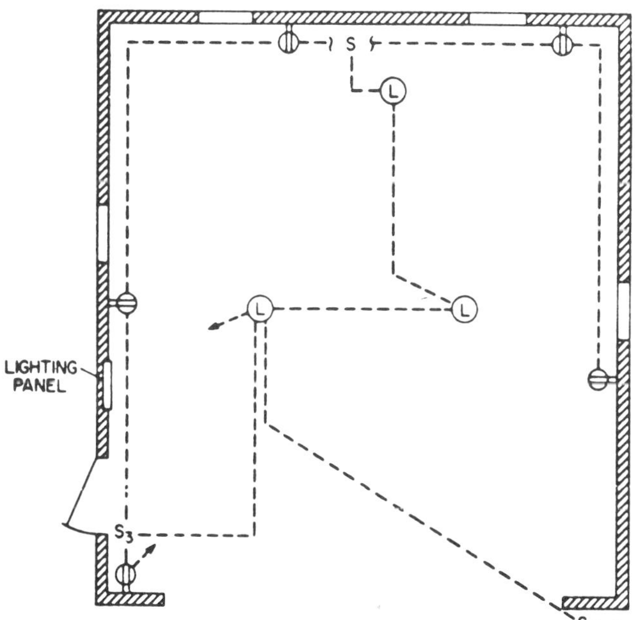

connecting the devices indicate that exposed wiring is used for the connections between the devices. The lighting panel is shown without any indication of whether it contains switches and fuses or circuit breakers; therefore, wiring plans should include notes giving additional information such as wire sizes, number of circuits, and type of equipment.

Study the wiring layout for a small house with a basement in Figure 8–3. The laundry facilities and lighting panel are assumed to be in the basement, and, for simplicity, the basement plan with its wiring layout is not shown.

In the kitchen, receptacles are provided for a refrigerator, clock, iron, toaster, and other small appliances. A special outlet, marked S_p, is provided for ironing; it has a switch and a pilot lamp. Special outlets are provided for the range (R), range hood (RH), garbage disposal (GD), and dishwasher (DW). Note that the two small-appliance circuits are shared by the kitchen and dining room and that they do not enter other rooms.

Adequate receptacle outlets must be installed in the other rooms. They are spaced at approximately equal distances (always less than 6 ft apart). Each bedroom has a combination switch–receptacle outlet and a ceiling light for general illumination. Each closet has a light controlled by a pull-chain switch. The receptacle at the entrance door is convenient for connecting a vacuum cleaner or other small appliance.

The bath has a ceiling light for general illumination and special lights at the mirror. It also has a separate exhaust fan.

Near the front door there is a split-wired duplex receptacle. The top half of the receptacle is wired to two three-way switches, and the bottom half is always alive. The terrace has two weatherproof receptacles for portable lamps, decorative lighting, and the like.

Three-way switches are used in the hall, kitchen, and living room to reduce the amount of walking required for control. Notice the push-button switches at both doors. These switches control the hall chimes.

The lines connecting the outlets in Figure 8–3 represent runs of cable. The arrowheads on the circuits indicate the *home runs*, which are the cables leading to the service entrance. The number of circuits can be determined by counting the arrowheads. The solid lines indicate circuits concealed in the ceilings or walls, and the broken lines indicate exposed wiring in the basement.

At first, reading and understanding wiring diagrams may be somewhat difficult, but with practice, you will find that this part of the electrician's job is not difficult. A good practice exercise is to make two wiring diagrams of your own home, one as it actually exists, and one as you would like to have it.

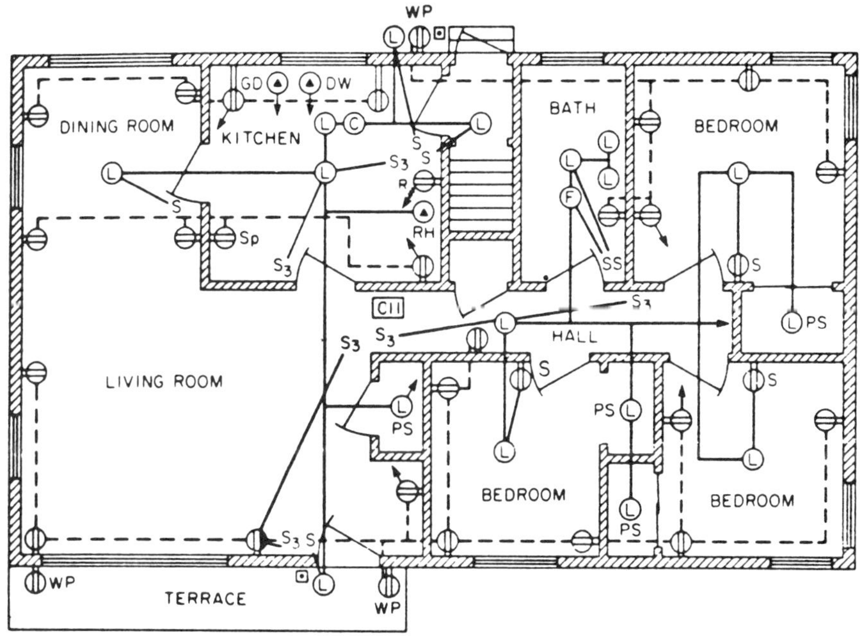

FIGURE 8–3. Wiring layout for a small house.

8–4 DETERMINING THE LOAD: GENERAL

When the wiring for a house is planned, the requirements of the NEC and any local code must be used as a guide. Keep in mind that these requirements represent a *minimum*. Good wiring practice may require additional or higher-capacity circuits. The computed load should represent all the power required for lighting, small appliances, and other loads, such as water pumps, water heaters, space heaters, clothes driers, and air conditioners. It is also very important to provide for future additional appliances that will increase the load on the electrical system.

8–5 LIGHTING LOAD

Table 220-2(b) of the NEC specifies that the lighting load be calculated on the basis of 3 W/ft^2 of occupied space. The floor area is computed from the outside dimensions of the building, apartment, or area involved, and by the number of floors. Open porches, attached garages, and unfinished or unused spaces (unless adaptable for future use) are not included here.

The 3 W/ft^2 unit value noted above is based on *minimum* load conditions and 100 percent power factor, and may not provide sufficient capacity for the installation contemplated.

In view of the trend toward higher-intensity lighting systems and increased loads due to more general use of fixed and portable appliances, each installation should be calculated for the load *likely* to be imposed, along with a capacity-increase factor to ensure safe operation.

Where electric-discharge lighting systems are to be installed, a high-power-factor-type light should be used to avoid the need for conductors having greater ampacity.

The *continuous* load—that is, a load where the maximum current is expected to continue for three or more hours—supplied by a branch circuit must not exceed 80 percent of the branch-circuit rating. *Exceptions:* (1) Where the assembly, including the overcurrent device protecting the branch circuit, is approved for operation at 100 percent of its rating, the continuous load supplied by the branch circuit may equal the ampacity of the branch-circuit conductors. (2) Where branch circuits are derated, branch-circuit loads must not exceed the derated ampacity of the conductors. (Review Section 4–7.)

To determine the installation requirements of a typical residential load, assume a one-story house with a full basement. The outside dimensions of the ranch-type house are 30 by 60 ft, and it is considered likely that the basement will someday be finished. Then the total area on which to base your calculations is $30 \times 60 \times 2 = 3600$ ft^2. Allowing 3 W/ft^2, the lighting load would be $3 \times 3600 = 10{,}800$ W. To determine amperage, transpose the power equation to $I = P/E$. Using the load requirement determined above, and assuming a supply voltage of 115 V,

$$I = \frac{P}{E} = \frac{10{,}800}{115} = 93.9 \text{ A}$$

For practical purposes, the amperage requirement is taken as 94 A.

Each circuit is usually designed to carry either 15 or 20 A. If the local code prohibits the use of 15-A circuits, you would use 20-A circuits. The number of circuits needed is determined by dividing the total current by the required amperage and rounding the answer off to the next higher whole number. Assuming 15-A circuits and a total current requirement of 94 A, the "sample" house would need

$$\frac{94}{15} = 6.3 \text{ circuits}$$

Rounded off to the next higher whole number, 6.3 becomes 7. Thus seven

15-A circuits would be needed. Alternatively, the number of 20-A circuits required is

$$\frac{94}{20} = 4.7 \text{ circuits}$$

Rounded off to the next higher whole number, five circuits are needed.

With 15-A circuits you should use AWG no. 14 copper wire; with 20-A circuits you must use AWG no. 12 copper wire.

The load should always be distributed over several circuits to prevent the whole house from being darkened by the blowing of one fuse or the tripping of one circuit breaker. Also, to the greatest extent practical, the load should be divided evenly between branch circuits.

8–6 SMALL-APPLIANCE AND LAUNDRY LOADS

The small appliances considered here do not include fixed appliances, such as garbage disposals, built-in dishwashers, and garbage compactors.

For the small-appliance load in kitchen, pantry, family room, dining room, and breakfast room of dwelling occupancies, two or more 20-A branch circuits must be provided for all receptacle outlets, in addition to those required for lighting. Such small-appliance branch circuits must have no other outlets (except for an electric clock).

Receptacle outlets supplied by at least two appliance-receptacle branch circuits must be installed in the kitchen.

Each small-appliance branch circuit must be able to supply a minimum of 1500 W.

A laundry load of 1500 W is assumed for a typical residence. The load is supplied by a separate 20-A branch circuit, which terminates with one duplex receptacle located within 6 ft of the intended location of the washing machine. If an electric clothes dryer is to be used, another circuit is installed in the laundry area.

8–7 ELECTRIC RANGES

Electric ranges are supplied by a separate 240-V circuit. Before we say more about ranges, however, let us define one very important term: *demand factor*.

Demand factor is the ratio of the maximum demand of the system, or any part of the system, to the total connected load of the system or of that part of the system under consideration.

Because all loads on a system are rarely turned on simultaneously, the maximum demand is considered to lie somewhere between the maximum connected load and the actual usage. The demand factor is expressed as a percentage of the maximum load, and is used in practice to determine the size of conductors and/or overcurrent devices needed for a particular application.

The demand factor is determined by a series of tests; the resulting information appears in Table 8–1 (NEC Table 220-19), to which the following notes apply.

Note 1. Over 12-kW through 27-kW ranges all of the same rating: for ranges individually rated more than 12 kW but not more than 27 kW, the maximum demand in column A shall be increased 5 percent for each additional kilowatt of rating or major fraction thereof by which the rating of individual ranges exceeds 12 kW.

Note 2. Over 12-kW through 27-kW ranges of *unequal ratings*: for ranges individually rated more than 12 kW and of different ratings but none exceeding 27 kW, an average value of rating shall be computed by adding together the ratings of all ranges to obtain the total connected load (using 12 kW for any range rated less than 12 kW) and dividing by the total number of ranges; and then the maximum demand in column A shall be increased 5 percent for each kilowatt or major fraction thereof by which this average value exceeds 12 kW.

Note 3. Over 1¾ kW through 8¾ kW: in lieu of the method provided in column A, it shall be permissible to add the nameplate ratings of all ranges rated more than 1¾ kW but not more than 8¾ kW, and multiply the sum by the demand factors specified in column B or C for the given number of appliances.

Note 4. Branch-circuit load: it shall be permissible to compute the branch-circuit load for one range in accordance with Table 220-19. The branch-circuit load for one wall-mounted oven or one counter-mounted cooking unit shall be the nameplate rating of the appliance. The branch-circuit load for a counter-mounted cooking unit and not more than two wall-mounted ovens, all supplied from a single branch circuit and located in the same room, shall be computed by adding the nameplate rating of the individual appliances and treating this total as equivalent to one range.

Note 5. This table also applies to household cooking appliances rated over 1¾ kW and used in instructional programs.

8–8 MOTOR LOADS

Loads such as air conditioners are considered as motor loads. The nameplate current rating of the electric motor is changed to an equivalent

TABLE 8–1. Demand Loads for Household Electric Ranges, Wall-Mounted Ovens, Counter-Mounted Cooking Units, and Other Household Cooking Appliances over 1¾-kW Rating

	Maximum Demand	*Demand Factors*	
Number of Appliances	*Column A: Not over 12-kW Rating (kW)*	*Column B: Less than 3½-kW Rating (%)*	*Column C: 3½- to 8¾-kW Rating (%)*
1	8	80	80
2	11	75	65
3	14	70	55
4	17	66	50
5	20	62	45
6	21	59	43
7	22	56	40
8	23	53	36
9	24	51	35
10	25	49	34
11	26	47	32
12	27	45	32
13	28	43	32
14	29	41	32
15	30	40	32
16	31	39	28
17	32	38	28
18	33	37	28
19	34	36	28
20	35	35	28
21	36	34	26
22	37	33	26
23	38	32	26
24	39	31	26
25	40	30	26
26–30	15 kW plus 1 kW for each range	30	24
31–40		30	22
41–50	25 kW plus ¾ kW for each range	30	22
51–60		30	18
61 & over		30	16

Column A is to be used in all cases except as otherwise permitted in Note 3 (at left).
Reprinted with permission from the 1978 National Electrical Code, Copyright 1977, National Fire Protection Association, Boston, Mass.

number of kilowatts using the formula $P = EI$. An air conditioner with a nameplate rating of 7 A at 240 V, for example, requires

$$7 \times 240 = 1680 \text{ W} = 1.68 \text{ kW}$$

Sometimes it is convenient for your calculations to use the current drawn by the motor. If a branch circuit supplies only one motor, then branch-circuit ampacity should be at least 1.25 greater than the motor current. When the branch circuit supplies several motors, multiply the largest current by 1.25 and add the result to the sum of the other motor currents. Consider the case when a branch circuit supplies three motors drawing respectively 6, 9, and 12 A. The 12-A current is multiplied by 1.25 before adding it to the other currents:

$$12 \times 1.25 = 15 \text{ A} \quad \text{for the 12-A motor}$$
$$6 + 9 + 15 = 30 \text{ A} \quad \text{total current}$$

Therefore, a 30-A branch circuit is needed to supply these three motors. The same method may be used to calculate current and determine the size of service-entrance equipment.

Other common motor loads that should be considered are garbage disposals, water pumps, heating system motors, and permanent attic fans.

8–9 OTHER LOADS

The power used by appliances, such as clothes dryers, dishwashers, and electric space heaters, is calculated according to the nameplate ratings of the appliances. Only the more common appliances have been mentioned in load calculations. Other loads that may have to be considered include those used for hobbies and/or in a small business.

8–10 ALTERNATE LOADS

In calculating the size of service-entrance wires, you will, of course, consider the fact that certain load combinations are not likely to occur simultaneously. Air conditioners, for example, are not likely to be operating at the same time as space heaters. Use the larger of the two or more loads then to calculate load requirements. If the air conditioner uses, say, 2 kW and the electric space heater uses 10 kW, you would only consider the space-heater load. If a space heater supplements a heat pump, however, remember that the space heater *plus* the heat pump make up the total load since they would be used simultaneously.

8–11 SIZING THE SERVICE ENTRANCE

Once the total load in kilowatts is determined, the service-entrance size is calculated. First, from the total number of kilowatts compute the total current requirement. This requirement determines the smallest permissible sizes for the service-entrance wires and for the *panelboard*, the metal enclosure that houses the fuse-switch arrangements or circuit breakers. Because all loads are not likely to be used at the same time, a demand factor like that used for electric ranges can be applied to the total load. One approved method is to add up all the loads, excluding the electric range. For the first 3 kW, use a 100 percent demand factor and for the remainder, a 35 percent demand factor. To this total, add the range load computed in the manner explained in Section 8–7. A second approved method is to add all loads, including the electric range's nameplate rating, and then use a demand factor of 100 percent for the first 10 kW and a 40 percent demand factor for the remainder.

The computation of service-entrance size is best illustrated by the examples presented below. Chapter 9 of the NEC includes similar example calculations. You should study them and become thoroughly familiar with the methods used for load determination.

For uniform application of the provisions of Articles 210, 215, and 220, nominal voltages of 115 and 230 V are used in example problems to compute the ampere load on the conductors.

8–12 EXAMPLE PROBLEMS

Problem 1

A single-family, ranch-style home has a floor area of 2000 ft^2 and is built on a cement slab; an oil burner used in the heating system has two ¼-hp motors, each drawing 6 A. What is the minimum size requirement for the service-entrance conductors?

Solution

The lighting load, based on 3 W/ft^2 is $2000 \times 3 = 6000$ W = 6 kW. A small-appliance load of 3 kW and a laundry load of 1.5 kW are assumed to meet with the minimum requirements of the NEC. Each furnace motor uses a power of $(6 \times 115)/1000 = 0.69 \approx 0.7$ kW. The total load is reckoned thus:

Load	*Power (kW)*
Lighting	6.0
Small appliances	3.0
Laundry	1.5
Furnace	1.4
Total	11.9

For the first 3 kW of the load, the demand factor is 100 percent; for the remaining 8.9 kW, the demand factor is taken as 35 percent, or $0.35 \times 8.9 = 3.1$ kW. Then the total load is 6.1 kW.

Since 6.1 kW = 6100 W and $I = P/E$, the current in the service-entrance conductors for a 115/230 V three-wire system is

$$\frac{6100}{230} = 26.6 \approx 27 \text{ A}$$

Although the current is only 27 A, a 100-A service entrance must be used because the allowable minimum number of circuits is five.

Problem 2

Now consider the residence described in Problem 1 as a total-electric house using 12 kW of electric space heating with individual thermostats in five rooms, a 5-kW clothes dryer, a 1.5-kW dishwasher, a 2-kW water heater, a 12-kW electric range, and an air-conditioning load of 6 kW; what is the minimum size that can be used for the service-entrance conductors?

Solution

Since the air-conditioning load is less than the electric space-heating load, it is set aside, and the total is reckoned as follows:

Load	*Power (kW)*
Lighting, at 3 W/ft^2	6.0
Small appliances	3.0
Laundry	1.5
Electric space heating	12.0
Clothes dryer	5.0
Dishwasher	1.5
Water heater	2.0
Electric range (nameplate rating)	12.0
Total	43.0

The demand factor for the first 10 kW is 100 percent and, for the remaining 33 kW, 40 percent. Since $0.40 \times 33 = 13.2$ kW, the total load is $10 + 13.2 = 23.2$ kW = 23,200 W, and the current in the service-entrance wires is

$$\frac{23{,}200}{230} = 100.9 \approx 101 \text{ A}$$

For such an installation, choose service-entrance equipment and wire of the next larger size. For example, no. 2/0 copper wire with type T insulation rated at 145 A is suitable. Standard panelboard ratings are 140 A for circuit breaker types and 200 A for fusible types.

In Problem 2, the electric space heating was derated because each room had its own individual thermostat. If there are fewer than *four* separate thermostats, no derating is used. If, for example, only three separate thermostats were used, the load would be 10 kW at 100 percent demand factor, plus 12 kW for the heating at 100 percent demand factor, plus 8.4 kW (21 kW at 40 percent demand factor), for a total of 30.4 kW. The service-entrance current would be

$$\frac{30{,}400}{230} = 131.2 \approx 131 \text{ A}$$

A 140-A service entrance could be used in both cases, but note that the nonderated heating unit increases the current by about 31 percent.

8–13 OTHER SELECTION FACTORS

The 100-A service entrance is the smallest size normally used. The service entrance must have a current capacity equal to or greater than the calculated load. A capacity that is larger than the calculated load is preferred, because an increased load may be added later. A 60-A service entrance may be used only when fewer than six two-wire branch circuits are used and the computed load is less than 10 kW.

Because most houses have occupied areas greater than 1000 ft^2, a minimum of three circuits is needed for general lighting and appliances. (Remember that you must allow two circuits for small appliances and one for a laundry.) Even without considering heating or air conditioning, you can see that six circuits are needed. That number exceeds the five circuits permitted with a 60-A service entrance. Consequently, the 100-A service is the smallest size in common use. Local electrical codes must also be considered because some do not permit the use of service entrances with ratings of less than 200 A.

8–14 BRANCH CIRCUITS

The number of branch circuits affects the size of the service entrance because the service-entrance conductors must match the service-disconnect and the service-overcurrent devices. If many branch circuits are used, a large panelboard is required. Understandably, large panelboards have high current ratings.

The usual procedure is to provide a branch circuit for each of the listed loads:

- Electric range
- Automatic clothes washer
- Clothes dryer
- Electric water heater
- Dishwasher
- Garbage disposal
- Water pump
- Motor on oil-burning furnace
- Motor on blower in furnace
- Permanently connected motors with ratings greater than ⅛ A hp
- Permanently connected appliances with ratings greater than 1000 W
- Any automatically started motor such as a fan

The circuits can be either 115 or 230 V. Circuits rated at 230 V are preferred because smaller wires can be used. Also, the voltage drop in the wires is lower.

When more than one receptacle is used on a branch circuit, the circuit capacity is limited. For example, a 15-A branch circuit cannot supply a portable appliance requiring more than 12 A or 1380 W. A 20-A circuit is limited to 16 A or 1840 W. The portable appliance should not use more than 80 percent of the circuit capacity. When a branch circuit supplies both portable and permanently connected appliances, the fixed load must not exceed 50 percent of the circuit rating. Therefore, the highest fixed load for a 15-A circuit is 7.5 A or 863 W. A 20-A circuit can supply fixed appliances using up to 10 A. This limits the circuit to fixed-appliance loads up to 1150 W.

Branch circuits rated at 30, 40, or 50 A cannot be used for lighting or for small-appliance circuits in residences. Their use is limited to fixed appliances such as electric water heaters and fixed space heaters. The NEC specifies that the maximum load on a 30-A circuit shall not exceed 80 percent of circuit ampacity. Thus a 30-A circuit can supply a load that requires up to $0.8 \times 30 = 24$ A. The 40- and 50-A circuits are restricted to supplying electric ranges.

How many branch circuits should you install in a house? As you can see from the detailed considerations above, there is no standard answer because the number of circuits depends upon the size of the house and the expected electrical load. As a general rule, there should be more branch circuits than are minimally required by the NEC. The presence of extra circuits permits the addition of other electrical equipment in the future. Extra circuits also make the electrical system more convenient, because each circuit carries less load and that reduces the chances of blown fuses or tripped circuit breakers. Finally, the load is better balanced between circuits, so that the power waste for heating wires is reduced.

8–15 FEEDER CIRCUITS

The service-entrance conductors of single-family residences can be considered as feeders for certain loads. A 75 percent demand factor may be applied when four or more fixed-appliance loads are connected to these conductors. However, the appliances must be other than electric ranges, clothes dryers, air conditioners, or space heaters. If a residence contains four loads, such as water heaters, fixed dishwashers, garbage disposals, furnace motors, or water pumps, the demand factor can be used. The same demand factor can be also applied to similar loads in multifamily dwellings. The following example problem will clarify the use of the demand factor.

Problem 3

What is the current in the ungrounded line when a 230-V, 2500-W water heater, a ½-hp, 115-V garbage disposal, a ½-hp, 230-V water pump, a ¼-hp, 115-V furnace motor, and a ¼-hp, 115-V attic fan are fed by a 115/230 V feeder? The motor nameplate currents are 4.8 for the ¼ hp motors, 6.6 A for the garbage disposal, and 3.6 A for the water pump.

Solution

The current drawn by the water heater is

$$I = \frac{P}{E} = \frac{2500}{230} = 10.9 \text{ A}$$

Because the garbage disposal draws the largest motor current, its current is increased by 1.25 to determine the line current:

$$1.25 \times 6.6 = 8.25, \quad \text{or } 8.3 \text{ A}$$

To balance the load, the furnace motor and the attic fan should be connected to different ungrounded conductors. It is sufficient to compute the current in the ungrounded line that has the larger load. In this example, the line feeding the garbage disposal has the larger load. The total current in this line is calculated as follows:

Load	*Current (A)*
Water heater	10.9
Garbage disposal	8.3
Furnace motor	4.8
Attic fan	0.0
Water pump	3.6
Total	27.6

When the demand factor of 75 percent is considered, the line current is

$$0.75 \times 27.6 = 20.7 \text{ A} \quad \text{Answer}$$

NOTE. When feeder circuits are used, the line current calculated for the service-entrance conductors is reduced—from 27.6 to 20.7 A in this example. When the size of the service entrance is determined, the other loads, such as lighting, small appliances, and electric range, are calculated as before. The feeder current is added to the total current of other loads to determine the total current in the service entrance.

8–16 SELECTING THE PANELBOARD

Once the total current and the number of circuits have been determined, the panelboard can be selected. Its current rating cannot be less than the computed load current. Also, it must protect the service-entrance conductors from excessive current. A typical panelboard in newer homes uses circuit breakers. The main circuit breaker, which has a rating equal to that of the panelboard, is located at the top of the panelboard. Below the main circuit breaker are the two-pole circuit breakers for protecting 240-V circuits, and then single-pole circuit breakers for protecting 120-V circuits. For safety, all the circuit breakers are mounted in a dead-front cabinet. That means that there are no exposed terminals or uninsulated wires showing, and the cabinet door is labeled with an index on which the use of each circuit breaker is written after installation. Sometimes two circuit breakers are connected in parallel on the watthour-meter side to provide special ratings. For example, two 70-A circuit breakers can be used to make a 140-A panelboard. Service circuit breakers are rated at 30, 40, 50, 60, 70, 100, 125, 150, 175, and 200 A. Equipment larger than 200 A is available for larger installations.

A panelboard may also have pull-out switches and plug fuses. The standard ratings are 30, 60, 100, and 200 A. The main lighting switch is a pull-out switch that holds two cartridge fuses. Depending upon the internal connections in the panelboard, the main switch may protect the other pull-outs and the plug fuses or only the plug fuses. There may also be pull-outs for the range, water heater, and clothes dryer. Then come fuse holders for plug fuses. The plug fuses protect 120-V branch circuits used for lighting, laundry, and small appliances. Remember that, in new installations, adapters for type S fuses must be inserted in the fuse holders designed for plug fuses.

Each pull-out switch holds two cartridge fuses for a particular size of fuse. For example, a fuse holder made for fuses between 35 and 60 A will not accept fuses 30 A and smaller or 70 A and larger. Each pull-out can be inserted in two positions. In the correct position, the current flows through normally, whereas in the reverse position the pull-out acts as an open switch. It is easy to determine the positions because of the markings molded into the pull-out switch.

When you select the panelboard for a house, be wise and select one larger than is actually needed at the time of installation so that there will be room for spare branch circuits. No one can foresee all the equipment that a householder may wish to add in the future, but some can be anticipated in this way. Also, the panelboard should conform to local code custom. Contact your power supplier or electrical inspector to determine what equipment is acceptable.

EXERCISES

8–1. How would you determine where a branch circuit begins?

8–2. In building plans, how does the architect indicate *(a)* exposed and *(b)* covered wiring used for connection between devices?

8–3. What symbols are used to designate outlets served by *(a)* full voltage and *(b)* low voltage?

8–4. What letter symbols are used to designate a *(a)* range, *(b)* dishwasher, *(c)* garbage disposal, and *(d)* range hood?

8–5. In building plans, how are cables leading to the service entrance indicated? How can the number of branch circuits be determined?

8–6. For wiring a house, is it wise to use NEC and/or local code requirements? If so, why?

8–7. If the lighting load for a house is based on 3 W/ft^2, does this ensure that the system will provide sufficient capacity for the installation contemplated? Explain your answer.

8–8. What precaution should be taken in planning a house-wiring layout if the use of electric-discharge lighting is contemplated? Explain your answer.

8–9. What is meant by the expression *continuous load*?

8–10 In general, what is the continuous-load limitation on a branch circuit?

8–11. In a two-story house having a finished basement, the outside dimensions are 32 by 58 ft. What is the lighting load? What is the total current requirements with a supply voltage of 115 V?

8–12. For the house in Exercise 8–11, how many *(a)* 15-A and *(b)* 20-A branch circuits are required for the lighting system?

8–13. Why should the lighting load always be distributed over several branch circuits?

8–14. What are the branch-circuit requirements for the small-appliance load in the kitchen, pantry, family, dining, and breakfast rooms of dwelling occupancies?

8–15 How much wattage must each small-appliance branch circuit be able to supply?

8–16. What laundry load is assumed in a typical residence and how must this load be supplied?

8–17. What provision must be made if the use of an electric clothes dryer is also contemplated?

8–18. What is meant by the expression *demand factor*?

8–19. What is the voltage requirement for an electric range?

8–20. How is demand factor *(a)* expressed and *(b)* determined?

8–21. For a range rated at 10 kW, what maximum demand is assumed?

8–22. For a range rated at 7.5 kW, what maximum demand is assumed?

8–23. For a total range load of 11 kW, what is the total demand?

8–24. What load is imposed by an air conditioner carrying a nameplate rating of 6 A at 230 V?

8–25. Suppose that a branch circuit supplies one motor drawing 8 A. What is the required ampacity of the branch circuit?

8–26. Suppose that a branch circuit supplies two motors drawing respectively 8 and 10 A. What is the required ampacity of the circuit?

8–27. In calculating the total load on a wiring system, would you include both a washer and dryer? Explain your answer.

8–28. Describe two methods for determining total load requirements for a residential wiring system.

8–29. A ranch-style home built on a cement slab has a floor area of 1800 ft^2; an oil burner used in its heating system has two ¼-hp motors. What minimum size is required for the service-entrance conductors?

8–30. Assume that the home in Exercise 8–29 is "all electric" and uses 20 kW for electric heating, with individual thermostats in four rooms, a 6-kW clothes dryer, a 1.5-kW dishwasher, a 3-kW water heater, a 9-kW range, and a 7-kW air conditioner. What is the minimum size required for the service-entrance conductors?

8–31. When may the heating load be derated for electric space heating?

8–32. Normally, what is the smallest size allowed for a service entrance? Why?

8–33. Under what special circumstances may a service entrance smaller than 100 A be used?

8–34. Name 10 loads for which a separate branch circuit is normally used. With such loads, is it better to use a 115- or a 230-V supply source? Give the reason for your choice.

8–35. For portable appliances, what is the maximum wattage that can be supplied by *(a)* 15-A and *(b)* 20-A branch circuits supplied by a 115-V source?

8–36. When a branch circuit supplies both portable and permanently connected appliances, how much of the circuit rating, expressed as a percentage, can apply to the fixed load?

8–37. What is the maximum permissible load on a 30-A branch circuit?

8–38. If the service-entrance conductors of a single-family residence are considered as feeders for certain loads, and four or more fixed-appliance loads are connected to these conductors, what demand factor may be applied? What appliances must be excluded from such calculations?

8–39. Describe typical panelboards using *(a)* circuit breakers and *(b)* pull-out switches and plug fuses.

8–40. For residential use, what are the standard ratings for circuit breakers?

8–41. What are the standard ratings for pull-out switch and plug-fuse-type panelboards?

8–42. What factors should be considered in selecting a panelboard?

IV

INSTALLATION

9

Installing the Service Entrance and Ground

9–1 PREPARATION FOR WIRING

Before any electrical wiring is started, the complete system is designed by either an architect, a contractor, or an electrician. If the design is completed before the electrician starts a job, it is only necessary for him to follow the construction and wiring plans in order to install the specified wires and equipment at the correct locations. For many small installations, however, the design is either nonexistent or sketchy, and the electrician is called upon to design a system that will meet the requirements of whatever code prevails—either local or NEC.

Schematic Diagrams

Because buildings are seldom wired alike, it is not possible to draw a general plan that will apply to all installations. Certain main features are common to all wiring plans, however, regardless of the use to which the building is put.

Wiring plans for the service entrance of a typical residence are shown in Figure 9–1(a) in combined pictorial–schematic form and in Figure 9–1(b) in pure schematic form. In each case a 120/240 V, single-phase, three-wire circuit is assumed. Figure 9–1(a) is self-explanatory. Notice in Figure 9–1(b), however, that a single straight line indicates both the service drop and the service-entrance conductors. The fact that there are three conductors is indicated by the bias lines. Also in Figure 9–1(b), the kilowatthour meter is simply indicated by a circle, and no internal details of the

FIGURE 9–1. Wiring plans for the service entrance of a typical residence: (*a*) detailed plan; (*b*) schematic form.

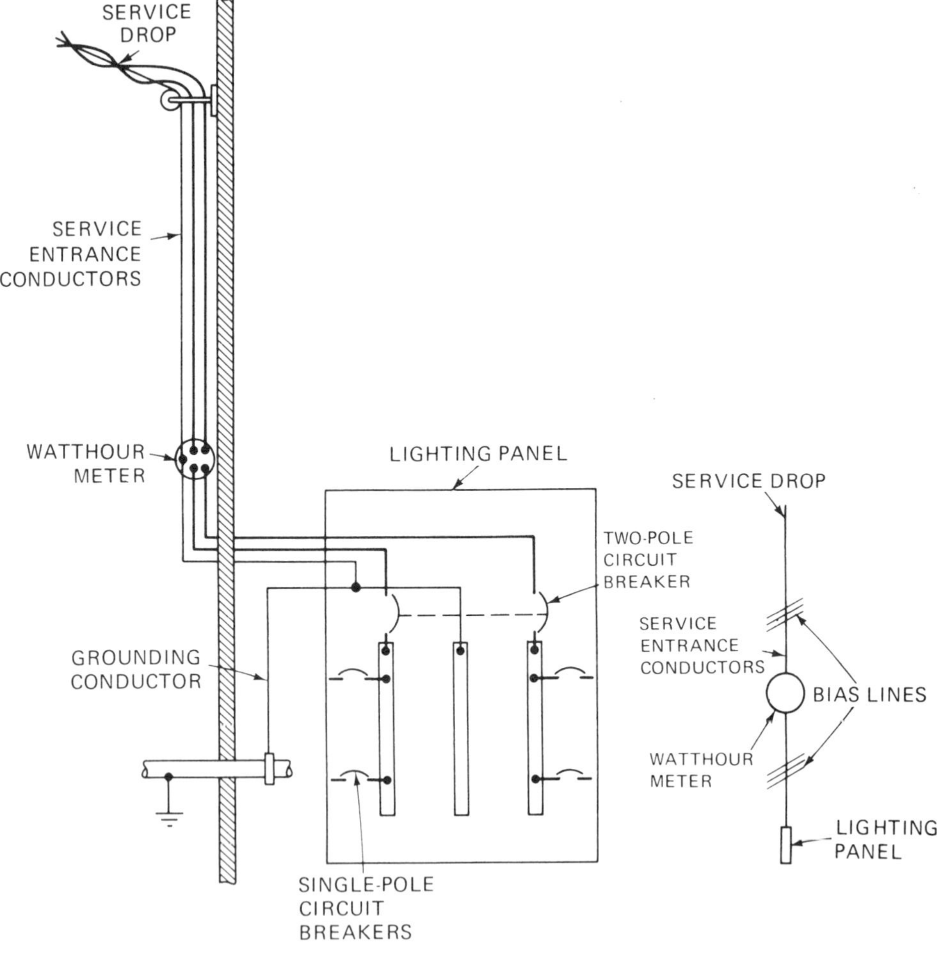

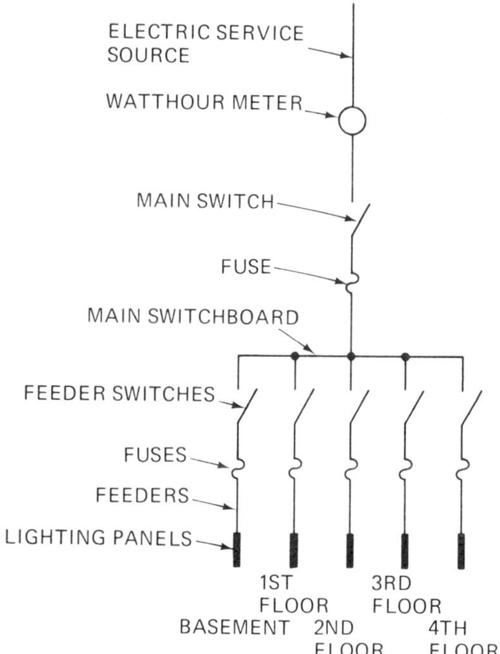

FIGURE 9–2. Modified form of one-line schematic diagram.

panelboard and systems are shown. Diagrams of the type shown in Figure 9–1(b) are often referred to as one-line schematic diagrams. Unfortunately, the recommended uniform method is not always followed in the preparation of such diagrams. In some one-line schematic diagrams, for example, the bias lines are omitted and other methods are used to specify the number of wires.

A different form of one-line schematic diagram is shown in Figure 9–2. In it the building has three floors and the electric service is assumed to be a 120/240 V, single-phase, three-wire circuit. The source is connected through the kilowatthour meter to the main service-disconnect switch and fuse to the individual circuits included in the panelboard. Because the lighting panels are remote from the main panelboard, a feeder runs from each branch-circuit overcurrent device to its associated light panel.

Figure 9–2 illustrates a major advantage of one-line schematic diagrams. If all parts, such as wires, switches, and fuses were shown pictorially, the diagram would have to be quite large and would be very difficult to draw. Also, if three wires were shown for each part of the circuit, the amount of detail would tend to be confusing. Although the specifics conveyed by diagrams of this type are limited, these one-line schematics are ideal for preliminary work.

Final Preliminary Steps

Once the wiring plans are completed and reviewed, make a careful survey of the building. Pay particular attention to the wall construction in order to ensure that adequate space has been provided for the mounting of outlet boxes. Sometimes the ceilings and walls are furred; that is, thin strips of wood are provided to create an air space or to permit the installation of insulating materials. If, for example, 1-in. furring strips are designated for use on a masonry wall, remember that a less-than-inch-thick mounting board will be available for outlet boxes because the nominal 1-in. measurement means that the lumber is really only ¾ in. In such a case, a hole may have to be cut into the masonry wall to make room to sink the outlet boxes. Also, be sure to check on the space available for cables or conduit, and note the items that must be avoided, such as water pipes and ductwork for the heating system.

Although a floor plan should show the location of a switch with respect to the walls, it will not show the distance of a switch from ceiling or floor. Only a section (or detail) drawing will give that height, which should then be marked on the drawing to be used during the wiring installation. If the service-entrance location is not shown on available plans, contact the electric supplier since he determines where electric power will enter the building.

Plans are usually drawn to a scale of either ⅛ or ¼ in./ft. This scale must be marked on the plans.

9–2 THE SERVICE DROP

As noted in Section 6–1, the service drop is connected and installed by the utility company staff. Although this part of the installation does not involve the nonindustrial electrician directly, it is recommended that you read Sec. 230-24 of the code to familiarize yourself with the clearance of service drops not exceeding 600 V. These clearances are maintained as follows: two wire, 120 V, single phase; three wire, 120/240 V, single phase; four wire, 120/208 V, three phase; and four wire, 277/488 V, three-phase systems.

Service-drop conductors must, of course, have sufficient ampacity to carry the load. In addition, they must have adequate mechanical strength, and must not, in any case, be smaller than AWG no.8 copper or AWG no.6 aluminum. The only exception to this rule is when the drop supplies only limited loads of a single-phase circuit, such as controlled water heaters. Here the conductors must be no smaller than AWG no.12 hard-drawn copper or equivalent.

9–3 SERVICE-ENTRANCE CONDUCTORS

As the electrician you are responsible for the installation of the service-entrance conductors. But you must not start the installation until the utility company designates the location of the service entrance. Also, be sure to obtain a copy of the power company's rules, which will be in strict accordance with the local code. Follow them carefully. The service-entrance conductors are usually installed on the outside of the building, but they may be installed on the inside as long as NEC and/or local code requirements are met.

Wiring Methods (Sec. 230–43)

The wiring methods that may be used are limited to the following: open wiring on insulators, rigid metal conduit, intermediate metal conduit, electrical metallic tubing, service-entrance cables, wireways, busways, auxiliary gutters, rigid nonmetallic conduit, cable bus, type MC cable, or mineral-insulated, metal-sheathed cable.

Conductors Considered Outside Building (Sec. 230–44)

Conductors are considered to be outside the building when they are buried in at least 2 in. of concrete beneath a building, or when they are installed in a raceway that is enclosed by concrete or brick that is at least 2 in. thick.

Mechanical Protection (Sec. 230–50)

Individual open conductors or cables other than approved service-entrance cables must never be installed within 8 ft of the ground or where exposed to physical damage. Service-entrance cables must be of the protected type or be protected by conduit, EMT, or other approved means wherever liable to contact with awnings, shutters, swinging signs, or installed in exposed places in driveways, near coal chutes, or otherwise exposed to physical damage.

Individual Open Conductors Exposed to Weather (Sec. 230–51[c])

Individual open conductors exposed to weather must be supported on insulators, racks, brackets, or other sustaining means, placed at intervals of not more than 9 ft and so as to separate the conductors at least 6 in. from each other and 2 in. from the surface wired over. Or the supports must be at intervals not exceeding 15 ft if they maintain the conductors at least 12 in. apart. For 300 V or less, conductors may have a separation of not less than 3 in. where the supports are placed at intervals not exceeding 4½ ft and so as to maintain the conductors at least 2 in. from the surface wired over and with a separation of at least 3 in. between conductors.

Individual Conductors Entering Building or Other Structure (Sec. 230–52)

Where individual open conductors enter a building or other structure, they must enter through roof bushings or enter through the wall in an upward slant through individual, noncombustible, nonabsorptive, insulating tubes. U-shaped drip loops must be formed on the conductors before they enter the tubes.

Service Cables (Sec. 230–E and F)

Approved service-entrance cables must be supported by straps or other approved means within 12 in. of every service head, gooseneck, or connection to a raceway or enclosure and at intervals not exceeding 4½ ft. (The service head, also called *entrance cap* and/or *weather head,* is a fitting at the top of the service conduit that keeps rain from entering the conduit. It consists of three parts—a body that attaches to the service conduit, an insulating block to separate the wires, and a cover to keep the rain out and hold the assembly together.)

Cables that are not approved for mounting in contact with a building or other structure must be mounted on insulating supports installed at intervals not exceeding 15 ft and in a manner that will maintain a clearance of not less than 2 in. from the surface over which they pass.

Service-entrance conductors entering buildings are insulated; those on the exterior of buildings must be either insulated or covered. *Exceptions:* A grounded conductor may be (1) bare copper used in a raceway or part of a service cable assembly, (2) bare copper for direct burial when copper is suitable for soil conditions, (3) bare copper for direct burial without regard to soil conditions where part of an approved cable assembly with a moisture- and fungus-resistant outer covering; (4) aluminum or copper-clad aluminum without individual insulation or covering used in a raceway or for direct burial when part of an approved cable assembly with a moisture- and fungus-resistant outer covering.

To determine which type of service-entrance conductor to use, check with your local electrical inspector or power company. The size of wires to use depends, of course, upon the loan that is calculated in the manner demonstrated in Chapter 8 of this book and Chapter 9 of the NEC.

Many times the neutral wire is two AWG numbers smaller than the hot wires because it carries less current. With 240-V loads and balanced 120-V loads, the neutral carries no current.

Generally, on an overhead service, service-entrance cables are never spliced except in the meter socket. Exceptions to this rule are given in Sec. 230-46 of the NEC.

Two popular types of service-entrance cable are shown in Figure 9–3. Notice that both cables have two conductors, which are first covered with insulation and then braid. On top of the braid is a bare neutral made from many strands of small wire spun in a flat layer. The cable shown in Figure 9–3(a) is designated type SE, style A; the A indicates that the cable is armored. The armor is a galvanized steel tape. The armor is in turn

FIGURE 9–3. Two popular forms of service-entrance conductors: (*a*) type SE, style A; (*b*) type SE, style U.

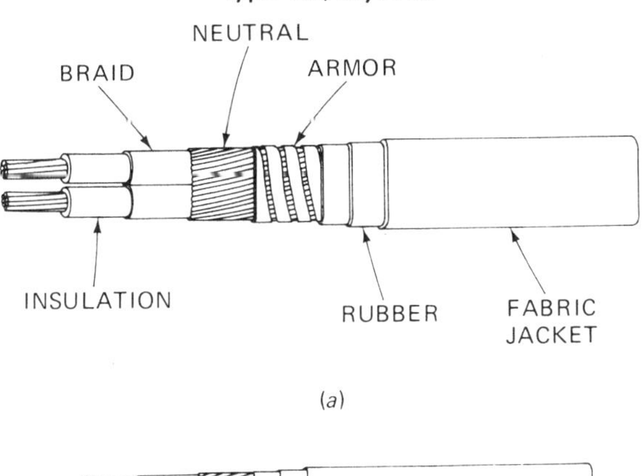

(*a*)

(*b*)

wrapped in rubber tape for protection against moisture. The whole cable is finally jacketed in fabric, usually painted gray. The cable shown in Figure 9–3(b) is designated type SE, style U; the U indicates that the cable is unarmored.

9–4 SERVICE-HEAD INSTALLATION

After suitable locations have been selected for the service bracket and the panelboard, the path of the service-entrance cable can be decided. The service head must be located above the point of attachment of the service-drop conductors to the building. Where it is impracticable to locate the service head above the point of attachment, it may be located not farther than 24 in. from the point of attachment.

A weather head with a removable cover is shown in Figure 9–4. Before installing it, strip the sheath and the outer insulation from the service-entrance cable for a distance of 3 ft. If type SE, style A cable is being used, remove the galvanized steel tape for the same distance. Now unwind the fine wires of the neutral and twist them together to form a single stranded wire. To place the weather head on the service-entrance cable, first remove the cover and loosen the cable clamp. Then the weather head can be slipped over the end of the cable with the three wires extending through the small holes. The cable clamp should grip the service-entrance cable where the outer insulation and jacket have not been removed.

Sometimes a special fitting with the shape of a gooseneck is used instead of a weather head. When the fitting is installed it looks like a vertical U. The service-entrance cable is bent to conform to the shape of the fitting, and the ends of the cable are taped and painted to seal out moisture.

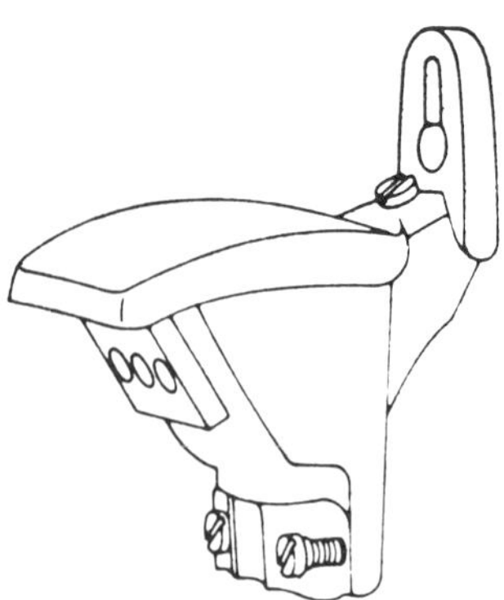

FIGURE 9–4. Typical weatherhead.

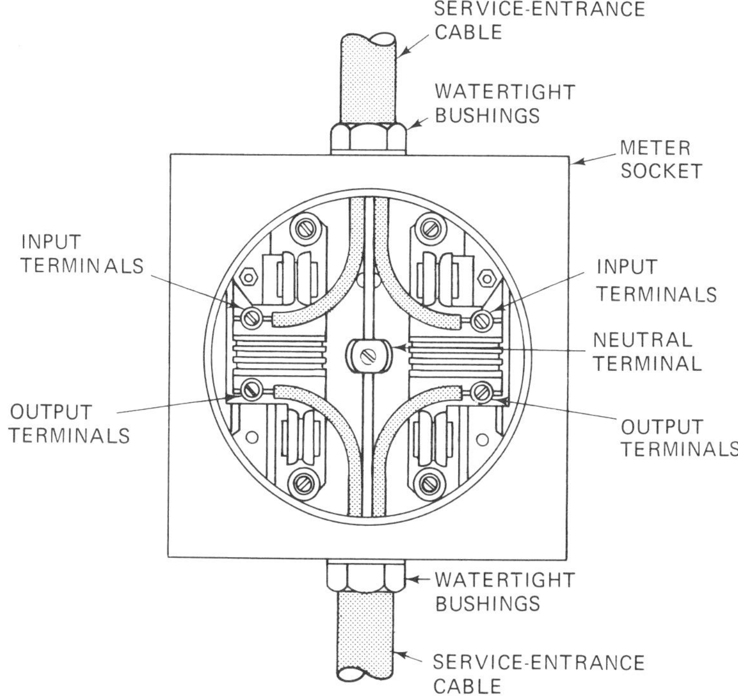

FIGURE 9–5. Meter-socket wiring.

9–5 THE METER SOCKET

The meter socket or base may be supplied by the power company or by the electrical contractor. In either case, the electrician mounts it 5 to 6 ft above ground level. The meter socket is threaded to accept watertight bushings.

The service-entrance cable is prepared in the manner already described for the weather head, except that less of the overbraid, rubber, tape, and steel tape needs to be removed.

The wiring of a meter socket is shown in Figure 9–5. The service-entrance cable enters the meter socket from the top through a watertight bushing. The watertight bushing clamps the service-entrance cable where its sheath and outer insulation have not been disturbed. The neutral terminal is located in the center of the socket. The two hot wires are connected to the input terminals. The service-entrance cable connecting to the panelboard enters the meter socket through a watertight bushing in the bottom of the meter socket. Its neutral wire is also connected to the

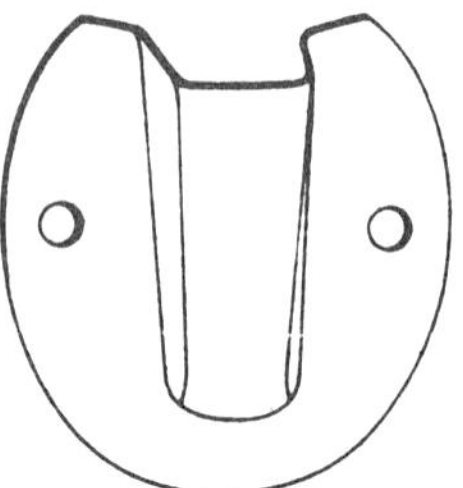

FIGURE 9–6. Typical sill plate.

center terminal and its hot wires to the output terminals. Both red wires of the service-entrance cable should be kept on the same side of the meter socket.

In some parts of the United States, watthour meters are installed inside buildings. They are usually mounted on the same board as the panelboard. When this mount is used, allow space to mount the meter socket above the panelboard. In other areas, a cabinet that is a combination meter socket and panelboard is preferred.

9–6 ENTERING THE BUILDING

At the point where a service-entrance cable enters a building, a *sill plate* is used to keep rain from following the cable into the building interior. A typical sill plate is shown in Figure 9–6(a). Usually, a soft-rubber compound is supplied with the sill plate and is used to seal any opening that might exist.

When the service-entrance conductors are enclosed in conduit, it is customary to use a raintight entrance ell at the point of entry into the building. With the ell's cover removed, it is a simple matter to pull wires around its right-angle corner.

If the building is of wood-frame construction, you can make the entry hole for cable or conduit by using an electric drill and a hole saw. For large-sized holes, use an electric saber saw. Use a star drill if you have to cut masonry, such as brick, concrete block, or concrete. If you cut the hole by hand, use a maul or sledge hammer to strike the star drill. Use a power hammer if you can. It considerably reduces the work of drilling masonry.

9–7 OUTSIDE SERVICE-ENTRANCE MISCELLANY

Typical fittings used in conduit-type, service-entrance installations are illustrated in Figure 9–7: a locknut and insulating bushing are shown respectively in (a) and (b); their use is illustrated in (c). The first step is to screw the locknut toward the conduit when starting the locknut. The conduit with the locknut is slipped through the hole in an outlet box or panelboard where the proper-sized knockout has been removed. If the surface of the panelboard or other device is painted, scrape off the paint so that the locknut will make good contact with the metal. Then screw on the bushing and tighten it with a screwdriver or pliers; finally, tighten the locknut also with a screwdriver or pliers. Because the locknut is inwardly curved next to the outlet box, it will dig into the box. The rounded bushing surface prevents damage to the insulation.

A grounding bushing [Figure 9–7(d)] has a large screw for connecting a grounding wire and a small screw with a sharp point that is tightened after the grounding bushing is firmly in place. The sharp point digs into the metal to ensure a good electrical connection.

In a conduit installation, the weather head, meter socket, entrance ell, and panelboard are joined by pieces of conduit. Nipples (short pieces of threaded rigid conduit) of various lengths up to about 12 in. are sometimes required. Where conduit cannot be run in a straight line—

FIGURE 9–7. Conduit fittings: (*a*) locknut; (*b*) bushing; (*c*) use of locknut and bushing; (*d*) grounding bushing.

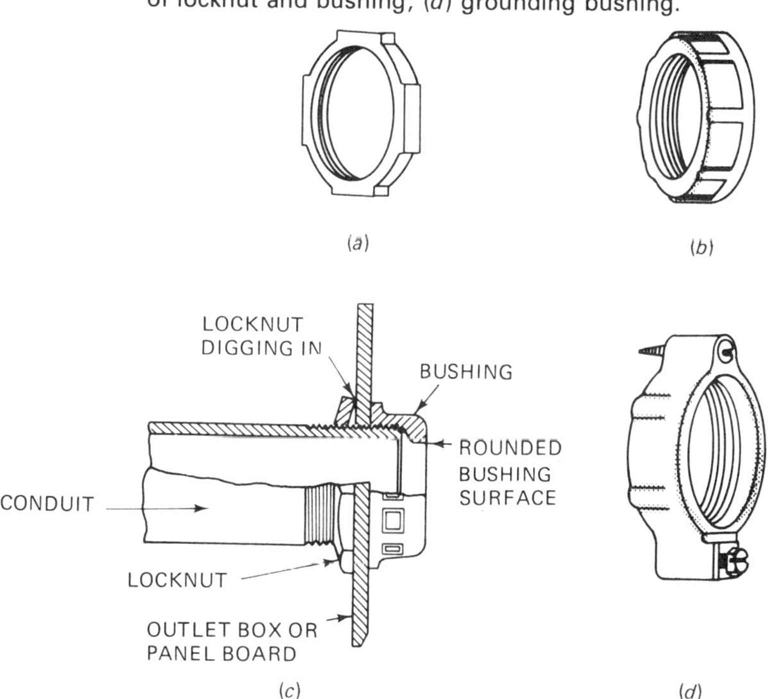

nearly always desirable for appearance's sake—prebent elbows are available.

Start the installation by mounting the service bracket and then cutting a hole for the conduit to enter the building. The size of the conduit depends, of course, on the size of the service-entrance conductors. The meter socket must be made for the same size conduit.

Remember that the weather head should be 12 in. or more above the service bracket and the meter socket 5 to 6 ft aboveground. Assemble the conduit on the ground with the weather head at the top, the meter socket at its correct location, and the entrance ell at the bottom. The entire assembly is often referred to as a *stack*. A short piece of conduit that joins the entrance ell and panelboard is used as a hinge to swing the stack into position and fasten it to the building by means of approved straps or clamps. The conduit is fastened to the panelboard by using a grounding bushing.

9–8 MAST-TYPE RISERS

With ranch-style homes the mast-type riser shown in Figure 9–8 may be needed to obtain adequate clearance between the service drop and ground. The mast is simply a piece of conduit with a diameter of 2 or 2½ in. to provide sufficient rigidity. A weather head is mounted at the top of the mast. Below the weather head is an insulator for connecting the service-drop neutral. The insulator should be at least 1½ ft above the roof, and the mast should be no more than 4 ft from the edge of the roof. Flashing with an adjustable shield is used at the roof surface to prevent water leaks. Two supports fasten the mast to the building by means of bolts. The fitting at the bottom of the mast is used to connect the conduit that runs to the meter socket. Wire the mast-type riser in the same way that you wire a regular conduit-type service entrance.

9–9 UNDERGROUND SERVICE

Underground service is now a common installation method because the availability of relatively inexpensive cable makes the installation economical. In severe winter areas, wind and icing problems are eliminated by the use of underground service. Another reason for the increased use of underground services is that many people consider the overhead type unsightly. In some places, all wires and transformers are underground; in others, only the wires from the power supplier's pole to the residence are underground.

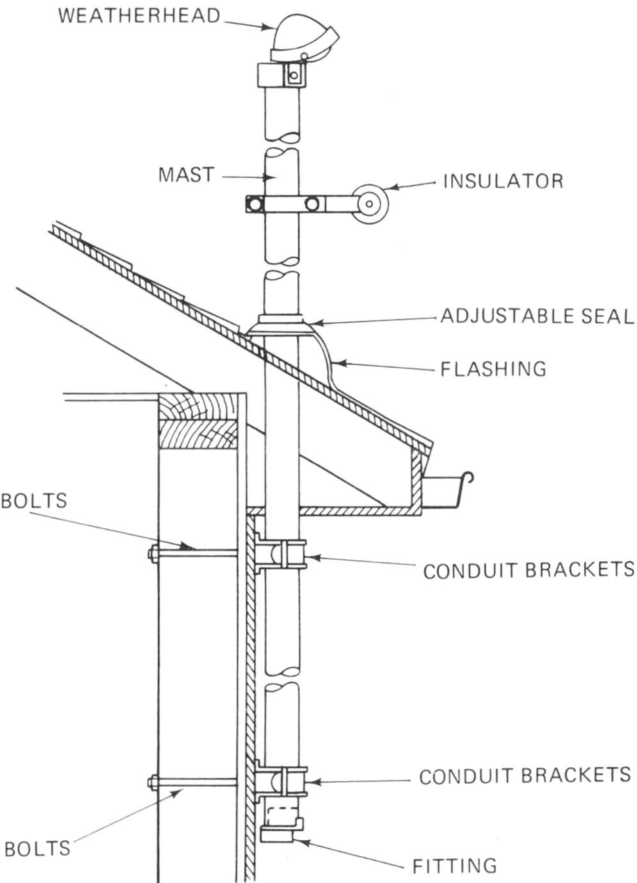

FIGURE 9–8. Mast-type riser.

One form of cable commonly used for underground installation is designated USE (underground service entrance). This cable has an inner layer of type RHW insulation that is especially moisture resistant and an outer layer of insulation that is very sturdy and also water resistant.

Type UF (underground fused) cable has the type TW insulated conductors embedded in a moisture-resistant plastic compound that is also very tough mechanically. It may be buried directly in the ground, *but it must be protected by fuses or circuit breakers at the starting point.* Type UF cable is generally restricted to farm installations where overcurrent protection is provided at the meter pole, which is the starting point. See Article 339 of the NEC for complete details on usage of type UF cable.

Underground cables are buried directly in the ground and no splices are permitted. Both conductors and cables so buried must be at a minimum depth of 18 in. This depth may be reduced to 12 in. provided supplemental protective covering, such as a 2-in. concrete pad, metal raceway, pipe or other suitable protection, is provided.

When a multiconductor cable is installed, the neutral may be bare. When separate conductors are used without conduit, the neutral wire must be insulated like the ungrounded wires. Be sure to properly identify both ends of the neutral so there will be no question when the conductors are connected. Lay the cables in a trench in a loose "snakey" manner. Do not stretch them tightly and do not permit conductors to cross since this may accelerate wear. Individual conductors should be grouped together in a trench.

Metering

A typical outdoor-metering installation for an underground service is illustrated in Figure 9–9. The meter socket holds the watthour meter and a large-sized conduit. The conduit, which projects underground about one

FIGURE 9–9. Outdoor metering for underground service.

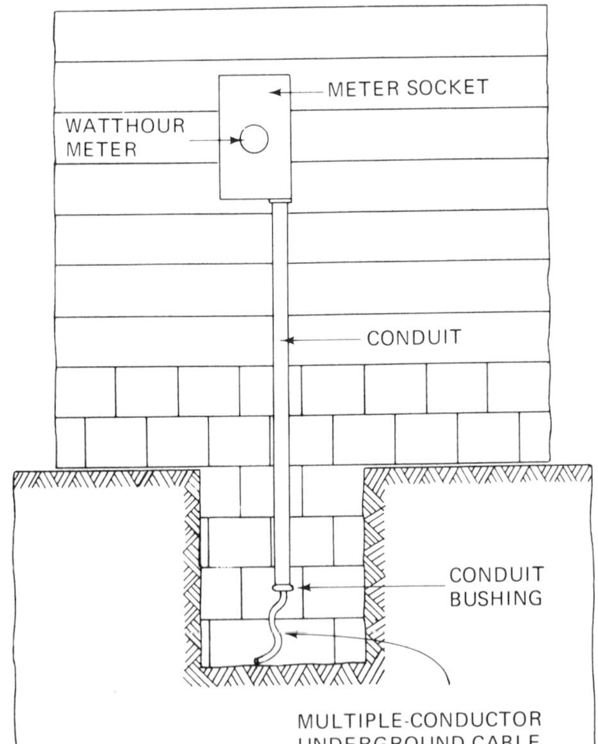

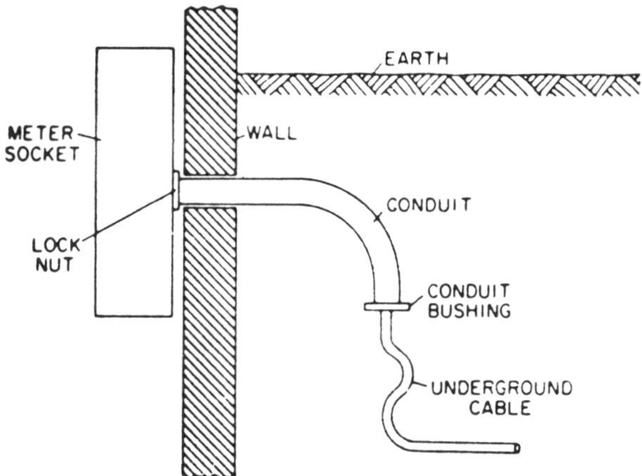

FIGURE 9–10. Underground cable running through a wall.

half the depth of the trench, has a conduit bushing on its bottom end. A large multiconductor cable drops out of the conduit with an S-shaped loop rather than in a direct run to the bottom of the trench. The cable loop reduces the chance of cable damage due to earth movement, especially where the earth freezes and thaws.

When the watthour meter is located inside the building, run a piece of conduit through the wall as shown in Figure 9–10. Be sure to bend the conduit downward. Then seal around both cable and conduit to keep moisture out of the building. The conduit carries the cable directly to the meter socket or to a cabinet that contains both a meter socket and circuit breakers. Although the meter socket may be fastened directly on the wall, the use of a piece of plywood between the wall and socket is preferred. Sometimes the meter socket and panelboard must be mounted a short distance from the place where the underground cable enters the building. In this case, splice the underground cable to the other wires that run to the meter socket. Make splices in a junction box, which is later closed with a blank cover.

Pole Connections

At the power company's pole, the underground cable must be protected for 8 ft or more aboveground. Ordinarily, you can do this with a piece of conduit with a weather head on the top and a conduit bushing on the bottom. Use a piece of conduit long enough to extend underground about one half the depth of the trench. Make an S-shaped loop in the cable where it extends from the conduit to protect it from frost damage.

In some underground electrical services the distribution transformer is placed in a heavy metal vault. You then make connections to the transformer primary inside the vault and connect the service-entrance wires in an aboveground splice box. It is also of heavy metal construction.

9–10 PANELBOARD WIRING

Panelboards are made for surface and flush mounting. They have knockouts on top, bottom, sides, and rear. When the panelboards are surface mounted, the rear knockouts are not used, since they lay against the wall or a piece of plywood. Panelboards have concentric knockouts, so care must be taken in removing these knockouts to keep from making oversized holes. Because the concentric knockouts provide very poor continuity for ground currents, grounding bushings are used with conduit installation.

A typical circuit-breaker-type panelboard contains a heavy copper bar called a *neutral bus,* which has numerous terminals for the connection of the systems ground and the neutral wires of all branch circuits. The ungrounded service-entrance conductors attach to screw-type terminals embedded in the base of the main circuit breaker. On the load side of the main circuit breaker there is another heavy bus with plug-in provisions for branch circuit breakers.

A wiring diagram is included with each panelboard to help the electrician wire the board correctly. Run the service-entrance wires into the panelboard and connect them to the appropriate terminals. Be sure to use only Allen wrenches to tighten the Allen-head set screws. They have a six-sided recess rather than a hexagonal head or a screwdriver slot. Place the wires parallel to the sides of the panelboard so that the wiring will have a neat, workman-like appearance. Avoid sharp bends; use sweeping bends so that the insulation is not pressed against fittings or the panelboard sides. Use raintight fittings and panelboards for exterior installations. When conduit is being installed, use a grounding bushing on the inside of the panelboard. After the set screws are tightened, run wires to the neutral bus or to the separate grounding terminal provided in some panelboards.

9–11 GROUNDING

The usual method of grounding, as noted in an earlier section, is to run the grounding-electrode conductor to the water pipe of the municipal water system. Make sure that the water system has metal pipes with conducting fittings. When there is no municipal water system, the steel casing of a well may be used as a ground, but the drop pipe in a dug well cannot be used as a ground. Sometimes the metal framework of a building or a gas pipe can be used for grounding. Before using anything other than the municipal water system, however, consult the local inspector and get his approval for the grounding system that you propose to use. When the water system is in the ground, the best procedure is probably to run the grounding wire to the nearest cold-water pipe and then place a jumper across the water meter. A jumper is a short piece of wire the same size as the grounding wire. Because a cold-water pipe runs more directly to ground than a hot-water pipe, a cold-water pipe is preferred.

The grounding-electrode conductor may be bare, insulated, or armored. It is always one continuous piece without any splices. A copper grounding wire is never smaller than AWG no. 8. This size serves the purpose if the largest service-entrance conductors are no. 2 or smaller. For either AWG no. 1 or 0, use a no. 6 grounding-electrode wire; use a no. 4 if the service-entrance conductors are no. 2/0 or 3/0. Table 250-94 of the NEC also lists the size of grounding-electrode wire to use with larger service-entrance conductors or with aluminum conductors. If it is necessary to use a grounding rod or ''made'' electrode, the grounding-electrode conductor need never be larger than AWG no. 6 copper or its equivalent in ampacity.

A grounding-electrode conductor or its enclosure shall be securely fastened to the surface on which it is carried. A no. 4 or larger conductor shall be protected if exposed to severe physical damage. Number 6 wire needs no protection if it follows the surface of the building closely and is securely stapled to it; otherwise, it shall be in rigid metal conduit, intermediate metal conduit, EMT, or cable armor. Grounding conductors smaller than no. 6 shall be in rigid metal conduit, intermediate metal conduit, EMT, or cable armor.

Ground Clamps

Select ground clamps carefully, and make sure that the clamp and pipe or rod material are similar. Select a copper or brass clamp, for example, with copper pipe or a copper-coated grounding rod. The ground clamp for armored cable is shown in Figure 9–11(a). It actually has two clamps, a small clamp for the bare grounding wire and a larger clamp for the armor. When a conduit protects the ground wire, use the ground clamp shown in Figure 9–11(b). The conduit is grounded by screwing it into the clamp; be sure the other end is grounded at the panelboard. When either bare or insulated wire is used as the grounding wire, use the ground clamp shown in Figure 9–11(c). Because soldered connections are not permitted in the

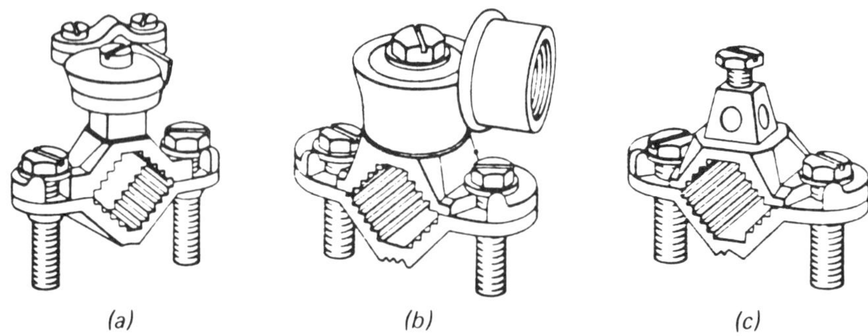

FIGURE 9–11. Ground clamps for (*a*) armored cable, (*b*) conduit, and (*c*) bare or insulated wire.

grounding wire or in the service entrance, all the clamps shown in Figure 9–11 make a pressure-type connection. Note also that ground clamps made with a metal strap are also prohibited for use on the grounding wire or in the service entrance.

EXERCISES

9–1. In a one-line schematic diagram, what are two common methods of designating the use of three conductors?

9–2. What is a major advantage of one-line schematic diagrams?

9–3. To what structural details should you pay particular attention when you survey a building to be wired?

9–4. What type of drawing will show the height between, say, a switch and the floor?

9–5. To what scale are wiring plans usually drawn?

9–6. Normally, what is the smallest size of copper wire that can be used in a service drop? Are there any exceptions to the rule?

9–7. What materials and tools should you have on hand before attempting to install service-entrance conductors?

9–8. What methods may be used for installing service conductors extending along the exterior of a building?

9–9. What conductors are considered to be outside a building?

9–10. What mechanical protection must be provided for service-entrance cables?

9–11. How must individual open conductors exposed to the weather be supported?

9–12. How must approved service-entrance cables be supported?

9–13. In general, service-entrance conductors entering a building are insulated; those on the exterior of buildings must be insulated or covered. Are there any exceptions to this rule? If so, specify the exceptions.

9–14. What is the distinction between type SE, style A and type SE, style U service-entrance cables?

9–15. What is a *weather head?* How is it placed on a service-entrance cable?

9–16. Where must a service head be located?

9–17. What is a *gooseneck* and how is it used?

9–18. How far aboveground should a meter socket be located?

9–19. Describe how a meter socket is wired.

9–20. What is a *sill plate* and how and why is it used?

9–21. What is an *entrance ell* and how and why is it used?

9–22. In a masonry building, describe how you would make the entry hole for service-entrance cable or conduit.

9–23. How are a locknut and insulating bushing used?

9–24. In a run of conduit, what is a *nipple?*

9–25. How would you assemble and install conduit for service-entrance conductors?

9–26. Describe a *mast-type* riser. When and why is it used?

9–27. What are some advantages associated with an underground service?

9–28. Describe USE and UF cable.

9–29. Can type UF cable be used in the normal service entrance?

9–30. What is the minimum depth for underground cable?

9–31. Describe *(a)* outdoor and *(b)* indoor metering installations for an underground service.

9–32. How are pole connections usually made for an underground service?

9–33. What precautions should be taken in working with knockouts on panelboards? Why?

9–34. Describe the usual method of grounding systems and several alternative methods that may or may not be permitted by the local code.

9–35. What is the smallest permissible copper grounding-electrode conductor? What is the limitation for this size of grounding-electrode conductor?

9–36. For a "made electrode," what is the largest size of copper grounding-electrode conductor required?

9–37. If a no. 8 copper grounding-electrode conductor is employed, what precautions, if any, must be taken?

9–38. What precautions, if any, must be taken with a no. 6 copper grounding-electrode conductor?

10

New Work

10–1 GENERALIZATIONS

For you as an electrician, *new work* refers to the installation of wiring systems in buildings under construction. *Old work* obviously refers to wiring installations and/or modifications in buildings where the construction is already completed. In this chapter we are concerned solely with new work; old work will be separately considered in Chapter 11.

Roughing In

The installation of conduit or cable outlet, switch, and junction boxes, and service entrance takes place during the early stages of construction. Collectively, this work is called *roughing in*. Wires are pulled into a conduit system only after the lathing, plastering, papering and similar work are completed. Fixtures, switches, and receptacles are put in place when the building is nearly completed.

Grounding

Although grounding does not have to be done first, it is convenient to complete this part of the job before device connections are made. When armored cable is used, the outlet boxes are automatically grounded to the armor. When nonmetallic-sheathed cable is used, there is a choice between cable with or without a grounding wire. Generally, however, cable with a grounding wire is more convenient.

When the wiring is complete, the grounding wire must be continuous from box to box all the way back to the ground terminal (neutral bus) on the panelboard. The removal of any device from a box must be done in such a way that it will not break the continuity of the grounding wire. This rule is needed to assure good grounding and a safe electrical system.

Metallic Outlet Boxes

It is easy to connect grounding wires properly. One of several different methods is illustrated in Figure 10–1. The bare grounding wire from each cable is joined to a jumper that connects to the receptacle grounding terminal. A grounding clip is used to fasten the grounding wire to the outlet box. Although a screw may be used in place of the grounding clip,

FIGURE 10–1. Proper connecting of grounding wires.

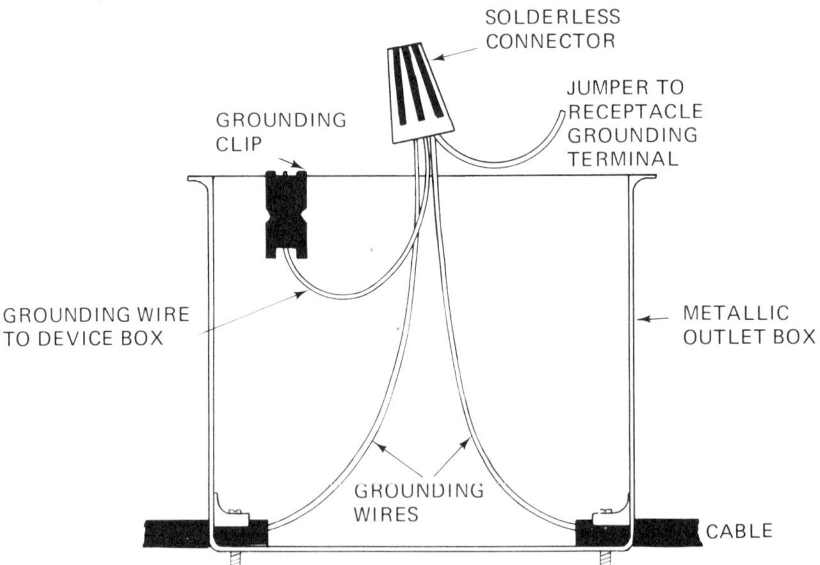

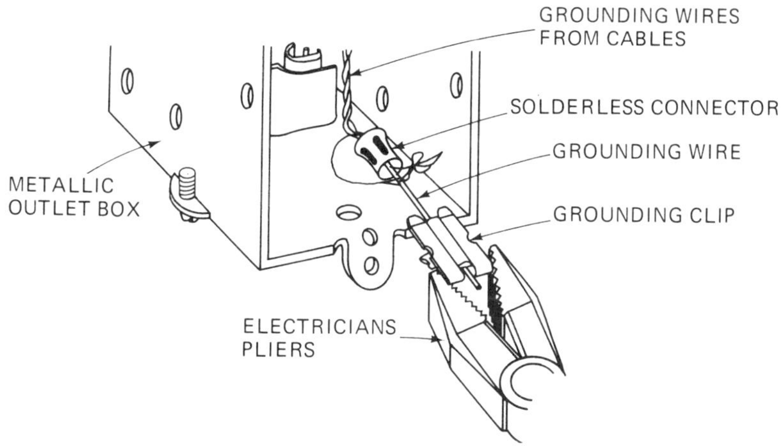

FIGURE 10–2. Installing a grounding clip.

that screw cannot also be used for any other purpose, such as clamping the cable. Some outlet boxes are made with a 6-in. grounding wire installed and fastened by a screw. Then all you have to do is add a grounding wire for each receptacle and splice all the ground wires together. The splice does not have to be insulated; just make sure that you tuck the grounding wires into the bottom of the outlet box carefully. To install a grounding clip, slip it over the end of the grounding wire and then grasp it with a pair of electrician's pliers, as shown in Figure 10–2. Tap it on until it is tight and bend the wire over the edge at the same time. When grounding clips are used, it is best to install them with the grounding wire during roughing in.

When a receptacle is installed directly on a surface-mounted outlet box, the jumper to the receptacle may be omitted because there is direct metal-to-metal contact. When outlet boxes are flush mounted, however, direct contact cannot be depended upon unless the receptacles are designed and approved for installation without a jumper. Always use the jumper unless you are sure that the receptacle is approved and will be grounded satisfactorily without it.

The grounding terminal on a receptacle is colored green and is hexagonal in shape. Wrap the jumper around the grounding wire and tighten securely.

Nonmetallic Outlet Boxes

Receptacles installed in nonmetallic outlet boxes must be grounded even though the outlet box itself is not. Use nonmetallic-sheathed cable with a grounding wire. Remember that the grounding wire must be continuous and return without interruption to the panelboard. If the outlet box has a metal faceplate (cover), run the grounding wire to the box and connect it to the cover.

Terminal and Lead Colors

The terminals and leads on receptacles, lamp holders, lighting fixutres, and other devices are color coded as an aid to correct wiring. Grounding terminals are green, neutral terminals are white, and the terminals for the hot wires are natural brass.

The wires on lighting fixtures usually have black and white insulation; alternatively, both wires may be of the same basic color, but one will include a tracer stripe. The wire with the tracer is neutral. If the fixture uses an incandescent lamp, connect the neutral wire to the outer screw shell. When a grounding wire is included, it is either solid green or green with a yellow tracer.

A single-pole switch has two natural-brass terminals. Since switches are used only to interrupt the hot side of a circuit, the ungrounded wire attaches to both terminals.

10–2 DEVICE LOCATIONS

Receptacles

A convenient recommended height for duplex receptacles is 12 in. above the floor, because the outlet is readily accessible and the cords of small appliances drag less. In the kitchen, bathroom, laundry, and garage, a good height is 48 in. above the floor. Usually, this places the receptacles about 12 in. above kitchen counter tops.

Present practice is to provide receptacles enough so that no point in a room is more than 6 ft from a receptacle. As a result, the distance between receptacles is less than 12 ft. Any wall space greater than 2 ft should be counted when receptacles are located. Sliding panels, such as glass doors leading to a patio, are counted as wall space. When sliding doors are used, special receptacles may be required.

Most often, receptacles are spaced equally; however, the spacing is sometimes varied to allow for the probable placement of furniture. A closer spacing of from 4 to 6 ft is recommended above the kitchen counter top. Also provide a receptacle for the refrigerator and mount it so that it will be hidden when the refrigerator is in place.

To permit the use of power tools and decorative outdoor lighting, install weatherproof receptacles on the exterior house walls at convenient locations. Be sure to keep them at least 18 in. above ground level. Special receptacles with key locks are available to prevent vandalism; for extra convenience, these outdoor receptacles may be controlled by switches mounted inside.

Switches

Switches for outdoor receptacles should be located about 48 in. above the floor, on the lock side of the door, and within 6 in. of the door frame. Select the switch positions carefully so they are convenient for occupants to use in the normal course of passage from room to room. Upon entering the house, for example, a person should be able to turn on a light while standing in the open doorway, and upon leaving a given room to enter either a hallway or another room, he should be able to light the space entered and turn off the light in the space vacated. This can be accomplished, of course, by either three- or four-way switches.

A combination switch and pilot light is often installed when a light, such as one in a basement, cannot be seen from the switch location.

In bedrooms where the branch-circuit wires run from the panelboard to a switch and then to a light, consider the use of a combination switch and receptacle. Then the switch is always alive; that is, it controls the light only. This combination is particularly convenient because it provides a ready connection for vacuum cleaners and the like.

Lighting Outlets

Wall-mounted lighting fixtures are useful at the sides of a bathroom mirror and should be placed about 4½ to 5½ ft above the floor, depending upon their design and intended use. A fluorescent lamp on each side of a mirror is preferable to either incandescent lamps or a single fluorescent lamp because they provide uniform lighting for makeup, shaving, and other activities requiring unshadowed illumination. Quite often, lamps are mounted on a medicine cabinet that also includes a switch and an outlet, but a separately mounted wall switch is usually more convenient. When an enclosed shower stall is planned, the use of a vaporproof light fixture, with a wall-mounted switch located outside the shower stall, should be considered.

The location of special lighting outlets at bookcases, fireplaces, coves, draperies, cornices, and kitchen workspaces calls for special planning. For outdoor use also consider the possibility of installing weatherproof spotlights, floodlights, sidewalk lighting, or post lighting with automatic switching—in addition to any weatherproof receptacles already provided. Remember that for any such exterior lighting control you must use weatherproof switches.

Pendant (suspended) lamps cannot be used in closets. The lamp holder can be mounted either on the ceiling or on the wall above the door. When swinging doors are used, the closet lamp may be controlled by a door switch recessed in a jamb. Then, opening the door automatically turns the light on and closing turns it off.

Signal Systems

A chime controlled by push-button switches is the most common residential signal system that an electrician installs. Systems of this type are supplied from a small transformer that steps down 120 V to between 12 and 20 V. Because the transformer power capacity is very limited, it can be connected directly to a 15-A lighting (general-purpose) circuit without a special fuse. Annunciator wire is used for connections and should be kept at least 2 in. from other wire. Many chimes are designed to sound two notes: one for the front door and one for the back door. Push-button switches control chimes and are mounted on the latch side of the door where they are easy to see and push.

In operation, buzzers and doorbells are similar to chimes, but they operate at lower voltages, between 6 and 10 V.

Always take care to make good, low-resistance splices when working with chimes, buzzers, or bells. A high-resistance splice may make the signal system inoperative.

10–3 MOUNTING OUTLET BOXES

Outlet boxes are mounted while the wall studs and ceiling joists are exposed. Be sure to learn the thickness and type of interior wall before you install the boxes. If the wall is noncombustible—made from plaster or gypsum board, for example—the front edge of the outlet box must be no more than ¼ in. behind the final front surface of the wall. If the wall material is combustible—made of wood paneling or the like—the front edge of the outlet box must be flush with the wall.

Outlet boxes are often sold with internal brackets to make their installation easier. The square outlet box shown in Figure 10–3(a) uses a common bracket. The tabs bent in the bracket help anchor the box. Drive one or two nails through the small holes in the bracket and into a wooden

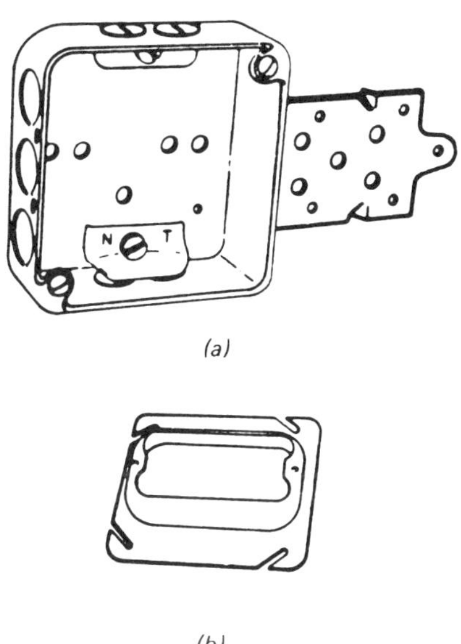

FIGURE 10–3. Square box and raised cover: (*a*) outlet box with bracket; (*b*) cover for switch or duplex receptacle.

stud or joist in order to fasten the box securely. Because of their large volume, square outlet boxes also make good junction boxes. They can be adapted for switches or receptacles by using a raised cover like the one in Figure 10–3(b). Covers are also sold to adapt square boxes to lamp holders. Manufacturers' catalogs show many other cover styles.

A device outlet box with nail holes can be attached to studs. See Figure 10–4(a). Note that the nails are used outside the box proper. The regular device box with removable sides usually has nail holes in the sides. Use 16 penny common nails (16d or 3½ in. long) to nail the box to a stud. Device-outlet boxes are also made with brackets for easier mounting. Where device-outlet boxes are ganged, the device boxes with mounting ears —Figure 10–4(b)— can be fastened on metal box supports, as shown in Figure 10–4(c). Wooden strips at least 1 in. thick (nominal lumber size) and mounted between the studs can replace the metal box supports if desired. Small wood screws are driven through the holes in the ears to fasten the outlet box to the wooden strips.

Octagonal outlet boxes are often used with ceiling lighting fixtures and are suspended from ceiling joists by a bar hanger (Figure 10–5). The center knockout in the bottom of the outlet box is removed so that the stud on the bar hanger can support the outlet box. Adjustable-length bar hangers that fit between studs are also made. Bar hangers are made with various offsets to accommodate different thicknesses of wall material. A

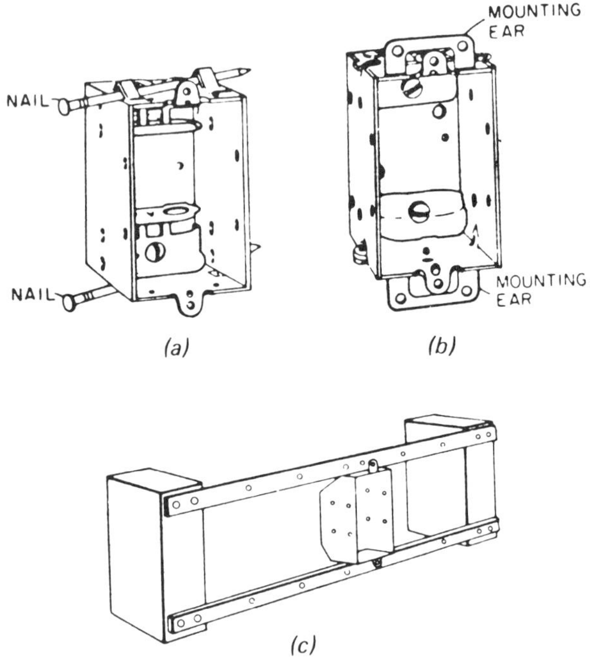

FIGURE 10–4. Device outlet boxes: (*a*) with nails; (*b*) with mounting ears; (*c*) on box-support strips.

1½-in.-deep box is normally used, but boxes as shallow as ½ in. may be used when it is not practical to use deeper boxes, as may be the case in old work particularly.

Because outlet boxes are made with different cable clamps for nonmetallic cable than for armored cable, be sure to obtain boxes that have matching cable clamps. Of course, if bushings are going to be used, boxes without cable clamps are preferred. The approximate route for cables should be plotted beforehand so that an outlet box of the proper size can be installed at each location. Remove the knockouts before fastening the outlet boxes in position, but be careful—unused knockouts must be refilled with metal plugs.

FIGURE 10–5. Bar hanger.

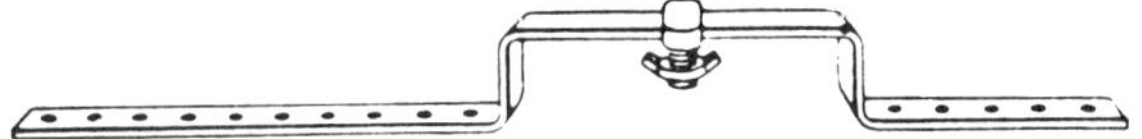

10–4 CABLE INSTALLATION IN OUTLET BOXES

Before inserting armored cable into an outlet box, bend the bonding strip back along the armor. If the outlet box has internal cable clamps, insert the cable into the outlet box where a knockout has been made. Then push the cable into the box until the insulating bushing hits the cable clamp. A correctly installed cable is shown in Figure 10–6. The cable enters the device box from below. The wires protrude through the hole in the cable clamp, which is held in place by a screw. Before you tighten the screw, be sure that the insulating bushing or lead sheath is visible and seated against the cable clamp. Also make sure that the bonding strip is placed where it is firmly gripped by the cable clamp when the screw is tightened. Cut off the excess bonding strip. Do not strip the ends of the wires at this time; just tuck them into the box.

Nonmetallic cable is simply slipped into the outlet box until the sheath can be seen in the box. The cable clamp is then tightened and the wires are tucked into the box. The cable clamp shown in Figure 10–6 is used

FIGURE 10–6. Device-box installation.

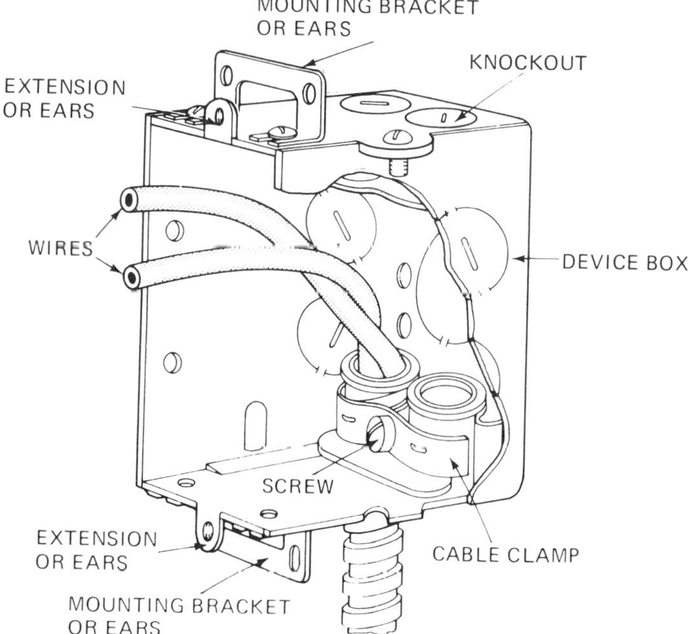

with armored cable. The cable clamp of Figure 10–3(a) is used with nonmetallic cable.

When panelboards and outlet boxes do not have cable clamps, use bushings to fasten the cable. Both nonmetallic and armored cables are prepared in the same way. Slip the bushing on the nonmetallic sheathed cable until the sheath shows through the bushing; then tighten the bushing screws. The smaller-sized cables have flat sides, so you should rotate the bushing until its flat part mates with the flat sides of the cable before tightening the screws. The bushings for armored cables have peepholes, so that the insulating bushing or lead sheath can be observed. Remember to place the bonding strip where it will be firmly gripped when the bushing screw is tightened.

Remove the bushing locknut and slip the connector into the knockout of the outlet box or panelboard. The locknut is dish shaped, not flat; screw it on the bushing so that its teeth dig into the metal. The usual way to tighten a locknut is to use a screwdriver and pliers (see Figure 10–7). If the cable is armored, the locknut must make good contact with the outlet box, because the bushing and armor form parts of the ground path.

10–5 WIRING OUTLETS

Outlets are not wired until the building is nearly completed, so the electrician must be careful not to damage or soil wall surfaces while he works. Before you wire the outlet boxes, be sure that they are clean.

When the walls and ceilings are plastered, the outlet boxes are often splattered or even covered with plaster. Plaster may extend over the edges of the outlet boxes and will have to be carefully cut back. In an extreme case, a careless workman may completely cover over an outlet box. To anticipate this, some electricians stuff paper into the outlet boxes after

FIGURE 10–7. Tightening a locknut.

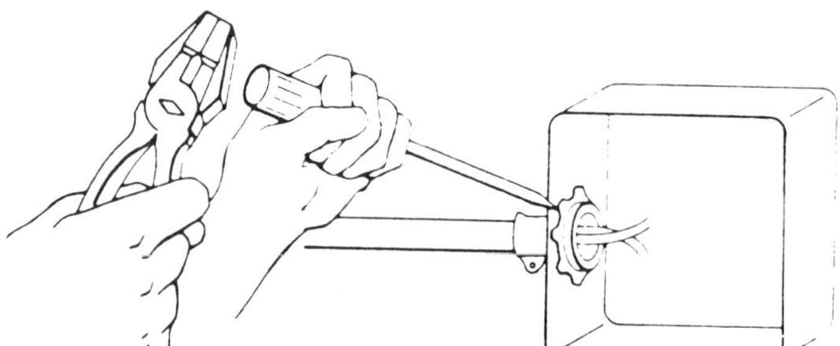

rough wiring in order to keep dirt and plaster out. Clean the boxes thoroughly before you start wiring.

Grounding

There is no special order for connecting the wires in an outlet box, but complete each box before moving on to the next one. Wire one whole circuit at a time rather than sections of several at random. Grounding does not have to be done first, but it is convenient to do it before device connections are made.

In Section 10–4 both nonmetallic-sheathed cable and armored cable were mentioned without specific consideration of grounding. If armored cable is used, the outlet boxes are automatically grounded by the armor. If nonmetallic-sheathed cable is used, choose cable with a grounding wire, since all metallic boxes in kitchens, bathrooms, basements, and in any farm buildings, as well as all outdoor boxes, must be grounded. Metallic boxes must be grounded if they are in contact with metal lath and are located where either the box or the device it contains can be touched by a person standing on the ground or where a person can touch grounded objects such as plumbing. Standing on a concrete floor is the same as standing on the ground. The grounding wire must be run to every box that contains a receptacle. Other information about the wiring of outlets was covered in Section 10–1.

10–6 WIRING RECEPTACLES

A duplex receptacle has two white terminal screws on one side and two brass terminal screws on the other side. See Figure 10 8. A green grounding terminal is on the end. Four terminal screws are provided so that the circuit can be run from one duplex receptacle to the next without splicing.

The duplex receptacle [Figure 10–8(a)] is shown on a wiring plan by the symbol shown in Figure 10–8(b). The black ungrounded wire is connected to a brass terminal screw. The white neutral wire is connected to a white screw terminal. Just remember to connect the white wire to a white terminal screw. The remaining wire is bare and is connected to the outlet box by a screw because nonmetallic-sheathed cable is being used, as illustrated. When armored cable is used instead, install a short bare wire as a jumper to ground the duplex receptacle to the outlet box.

When terminal screws are used, prepare and wrap the wires as described in Section 4–12. Wiring time can be reduced by using copper

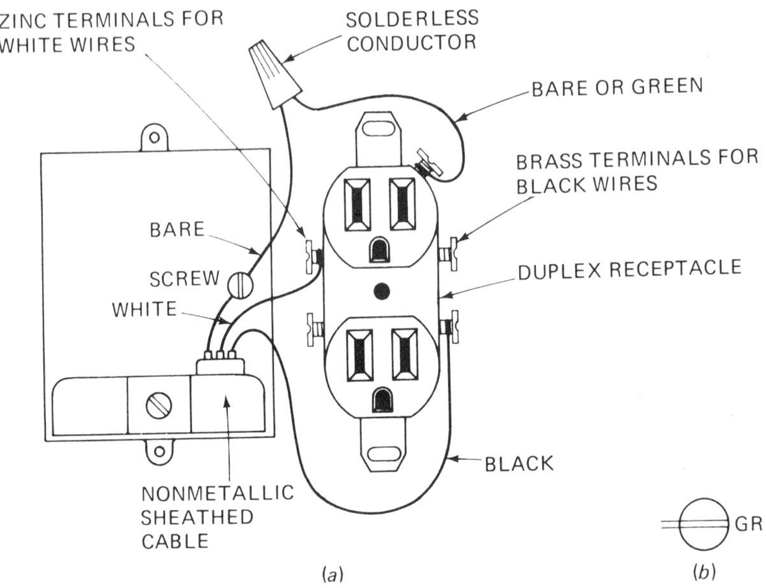

FIGURE 10–8. Wiring a duplex receptacle: (*a*) wiring diagram; (*b*) symbol.

wire in duplex receptacles equipped with pressure terminals. The terminal colors are molded into the plastic, but the markings vary with the manufacturer. One manufacturer, for example, molds the letters WHITE and GR into the device but does not label the ungrounded terminals.

When only one cable enters an outlet box, as in Figure 10–8, some of the pressure terminals or screw terminals are unused of course. Often, however, two cables enter an outlet box. In that case, connect one white wire to each white terminal and one black wire to each brass (ungrounded) terminal. Both terminals of a color are common (connected together) so that the circuit is completed through the duplex receptacle. Terminal screws are meant for and approved for securing one wire only. Never place two wires on one terminal screw. When there are more wires than terminal screws, splice the wires and run a jumper wire from each group up to the appropriate terminal.

Two-Circuit Duplex Receptacle

In a two-circuit duplex receptacle each half is independent and has its own terminals. The ordinary duplex receptacle has two terminals on each side, and both terminals on a side are common. Many better duplex receptacles can be converted easily from the ordinary type to the two-circuit type by using a screw driver to remove the contact link that connects the two terminals.

Two-circuit duplex receptacles are widely used where the top half is to be controlled by a switch and the bottom half is permanently connected. Instead of using a permanent ceiling light, for example, a floor or a table lamp may be plugged into the top half and switched. An electric clock, which you want to run continuously, could be switched into the bottom half. Another efficient use of two-circuit duplex receptacles can be made where two circuits run nearly parallel, as kitchen appliance circuits do. If the two circuits are fed from opposite sides (legs) of the 120/240 V service, only three wires are needed. Then, when two different appliances are plugged into the same duplex receptacle, they will automatically be on opposite legs of the three-wire circuit.

Switched Duplex Receptacles

The wiring of a switched duplex receptacle is shown in Figure 10–9. Note that the terminal screws are shown extended and the grounding wire is not shown—but only in order to make the drawing clearer. Use a convertible duplex receptacle and remove the contact link from the brass (ungrounded) side only. Leave the white (neutral) side connected, and then connect the white wire from the panelboard directly to either white terminal screw and tighten. Splice the black wire from the panelboard to a short piece of black wire and run it to the lower brass terminal on the duplex receptacle. In this way, the lower half is not switched and is, therefore, always ready to supply electric power. Also splice to the black wire from the panelboard a white wire that goes to a terminal on the switch. Then complete the wiring with a black wire from the switch to the top half of the duplex receptacle. The switch now controls the load plugged into the top half of the duplex receptacle.

Previously it was stated that the rule is that the white wire is always reserved for the neutral; yet in Figure 10–9 we can see that the white wire is spliced to a black wire. Here the white wire becomes an ungrounded wire feeding the switch—and violating the rule. This exception connection is permitted by the NEC in switch loops where two-wire cables are used, because the cables are only made with a black wire and a white wire. Care must be taken to switch *only* the ungrounded wire. Note that black wires are connected to the brass terminal screws on the duplex receptacle and the neutral wire is connected to the white terminal. The wiring would be incorrect if the wires connecting the switch were *reversed* so that a white wire connected to a brass terminal.

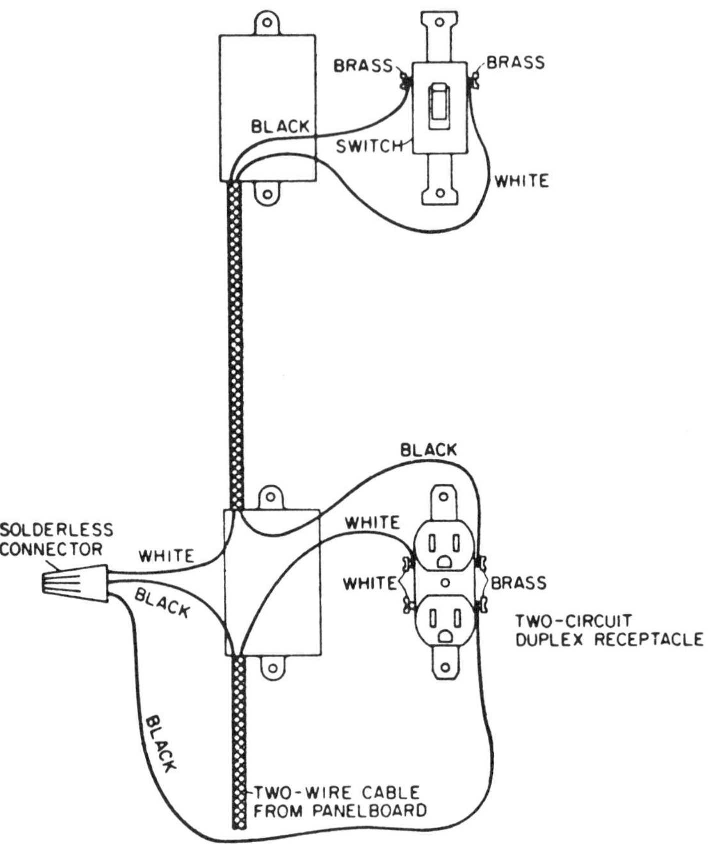

FIGURE 10–9. Wiring of a switched duplex receptacle.

Three Wire Circuits

The circuits discussed so far have contained two wires: one ungrounded wire and a grounded neutral wire. If two such circuits are run in the same general area, there are four wires, as shown in Figure 10–10(a). Note that lines *a* and *b* are ungrounded wires, and the remaining lines *n* are the neutrals, which are connected to the same neutral bus in the panelboard. If lines *a* and *b* are connected to opposite sides of the service entrance, current will flow as indicated by the arrows. "Opposite sides of the service entrance" means that, if line *a* is connected to the black service-entrance wire, line *b* is connected to the red service-entrance wire. The voltage between either lines *a* and *n* or *b* and *n* is 120 V, whereas the voltage between lines *a* and *b* is 240 V.

Since the neutral wires are connected together at the panelboard, they are, in effect, one wire. Why use two wires when one will do? A three-wire 120/240 V circuit will do the work of two two-wire 120-V circuits. Consider the condition when equal loads are used on the three-wire circuit shown in Figure 10–10(b). Assume that each load draws 20 A; then 20 A flows in lines *a* and *b*, but the two arrows on the neutral *n* are in opposite directions. As a result, the neutral currents cancel each other, there is no voltage drop in the neutral, and less voltage drop in each circuit. When the currents are unequal, the neutral current is still reduced. If, for example, line *a* is carrying 20 A and line *b* is carrying 15 A, the neutral *n* is carrying 20 − 15 = 5 A. Although the neutral current is no longer zero, it is much less than either current. The single loads in Figures 10–10(a) and (b) may be replaced by several smaller loads where they are needed if they are about equally distributed between the two sides of the circuit.

A three-wire circuit is wired with a three-wire cable consisting of a white neutral wire, a black wire, and a red wire. Use a convertible duplex socket, and remove the connecting link between the brass terminal screws. Wire the white wire to a white terminal screw as with a two-wire circuit. Wire the red wire to one brass terminal screw and the black wire to the other brass terminal screw. *Never wire the red wire to the white terminal screw* or you will have 240 V across the receptacle. The three-wire cable may be looped from box to box in the same manner as the two-wire cables. Larger outlet boxes may be needed, however, because of the additional wires.

10–7 WIRING PULL-SWITCH LAMP HOLDERS

Pull-switch, or pull-chain, lamp holders are often used in basements, closets, and attics. On a wiring layout a ceiling-mounted pull-chain lamp

FIGURE 10–10. Comparison of (*a*) two-wire and (*b*) three-wire circuits.

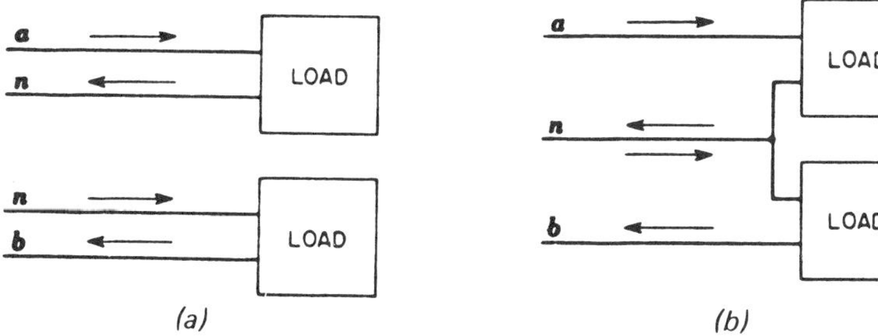

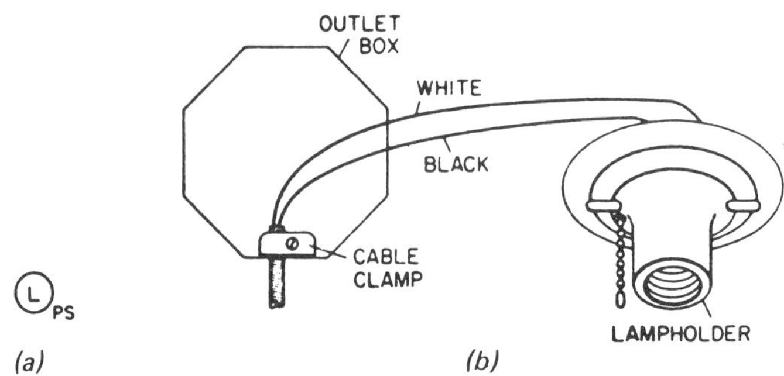

FIGURE 10–11. Pull-chain switch lamp-holder wiring: (*a*) symbol; (*b*) wiring diagram.

holder is indicated by the symbol shown in Figure 10–11(a); the wiring of the last one on a circuit is shown in Figure 10–11(b). The typical lamp holder has one brass terminal screw and one white terminal screw. Connect the black wire to the brass terminal and the white wire to the white terminal.

Not all lamp holders are constructed in the same way, but the wiring does not differ. In some lamp holders the terminals are directly accessible on the top. In a second type, a ring must be unscrewed to give access to the threaded area into which the lamp bulb screws. This switch–socket assembly pulls out so that the wires may be placed on the terminal screws. After the wiring is completed, reassemble the lamp holder; the pull switch must be aligned with the insulating base so that the parts slip together. A third type of lamp holder has pressure terminals for use with copper wire. A fourth type is equipped with 6-in. leads; splice the black lead to the black wire and the white lead to the white wire.

It is usually necessary to splice wires that continue on to another outlet, because most lamp holders have only two terminal screws. With receptacles, splicing is unnecessary, because they have four terminal screws. When the cables continue to another outlet, the outlet is wired by using two short pieces of wire (jumpers), as shown in Figure 10–12. Each jumper should be about 6 in. long and is made usually from a piece of wire of the same size as that used in the cable. Connect the black jumper to the brass terminal of the lamp holder and the white jumper to the two black wires from the cables, and splice the white jumper to the white wires from the cables.

Switched Lamp Holder

A lamp holder or a lighting fixture controlled by a single-pole switch is shown by the symbols in Figures 10–13(a) and (b). A dash below the

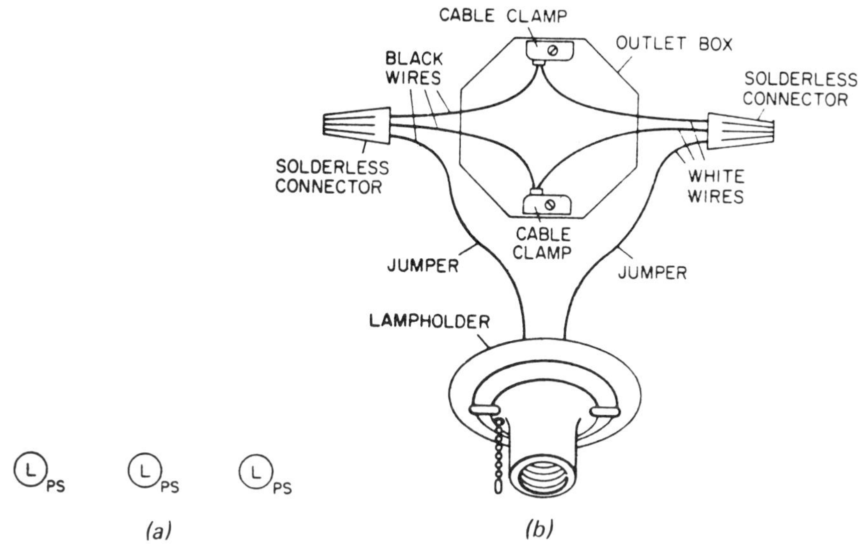

FIGURE 10–12. Wiring of a pull-chain switch lamp holder when cable continues to another outlet: (*a*) symbol; (*b*) wiring diagram.

FIGURE 10–13. Wiring of switched lamp holders: (*a*) symbol for switch feed; (*b*) symbol for lamp-holder feed; (*c*) wiring for switch feed; (*d*) wiring for lamp-holder feed.

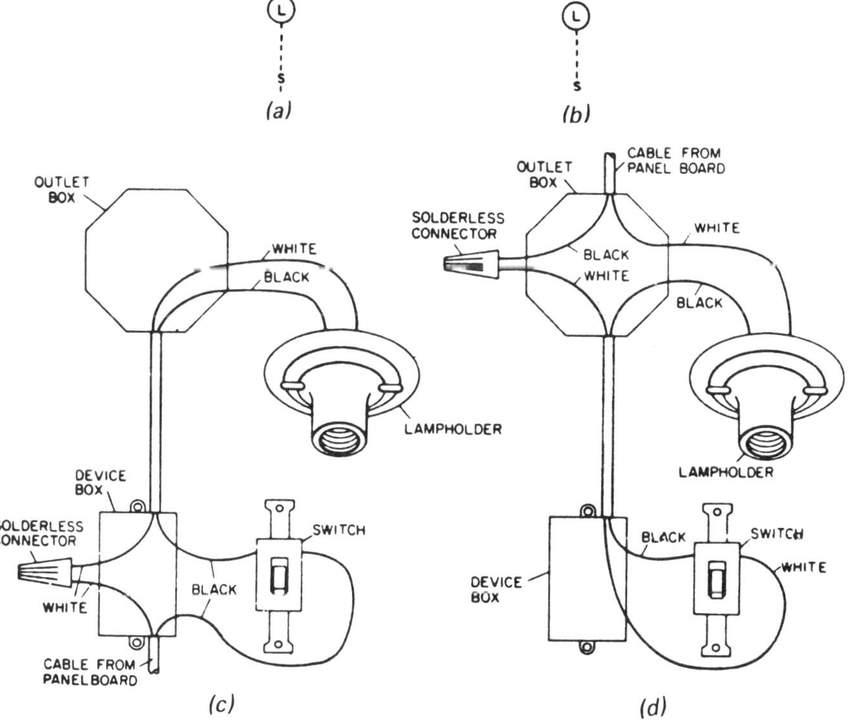

symbol S in Figure 10–13(a) shows that the cable from the panelboard runs to the switch. If there is no dash below the switch symbol, the cable from the panelboard runs to the lamp holder. When the cable runs to the switch first, the wiring is done as shown in Figure 10–13(c). At the device box, the two white wires are spliced together. Each black wire is connected to one side of a toggle switch. At the outlet box the black wire is wired to the brass terminal screw of the lamp holder, and the white wire is wired to its white terminal screw.

When the cable from the panelboard runs directly to the lamp holder outlet box, the connections are made as shown in Figure 10–13(d). The white wire from the panelboard is wired directly to the lamp holder white terminal, but the black wire is spliced to the white wire in the cable running to the switch. The black wire from the switch is connected to the brass terminal on the lamp holder. At the switch the white and black wires are each connected to a switch terminal.

In both Figures 10–13(c) and (d), the lamp holder is connected to one black wire and one white wire. Note, however, the difference in the switch wiring. Both wirings are correct and both are approved by the NEC. In Figure 10–13(c) the switch is connected by two black wires, but in (d) the switch is connected by one black wire and one white wire. This change in wire colors is permitted when two- or three-wire cables are used. When conduit is used, the white wire from the outlet box to the device box would be replaced by a black wire. In both Figures 10–13(b) and (d) only the ungrounded wire is switched; the neutral is not broken by a switch, although it is spliced in (c).

Although a lamp holder is shown in both Figures 10–13(c) and (d), a lighting fixture could be used just as well. An incandescent lighting fixture usually would have one black wire and one white wire. Some fluorescent lighting fixtures have one black wire and two white wires. Just splice all the black wires in one group and all the white wires in another group.

10–8 TWO LAMP HOLDERS CONTROLLED BY ONE SWITCH

Often two or more lighting fixtures are controlled by a single switch. The symbols in Figure 10–14(a) show two lamps controlled by a single-pole switch. Switch feed is used, and the switch is wired the same as in Figure 10–13(c). Again, the white wire from the panelboard is spliced to the white wire feeding the switch. The wiring of the right-hand outlet box is the same as in Figure 10–13(c). At the left-hand outlet box, a cable and two jumpers are used. All the black wires have been spliced together in one

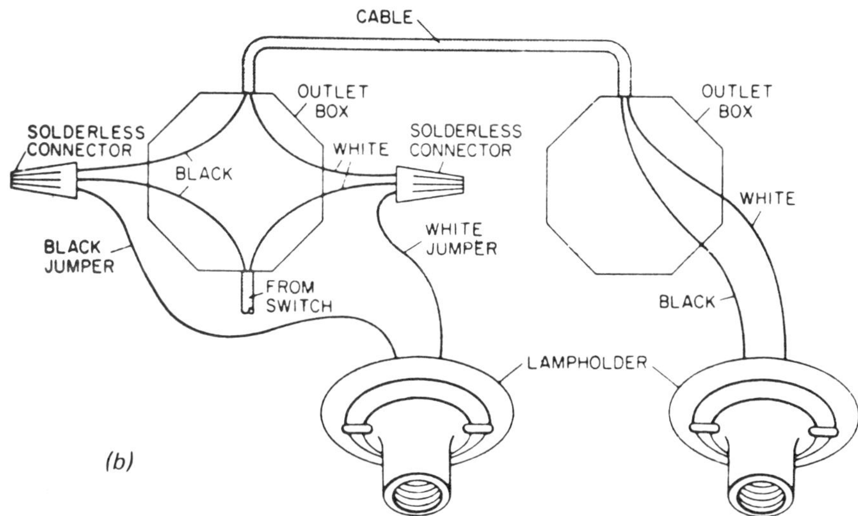

FIGURE 10–14. Wiring for two lamps controlled by one switch and using switch feed: (*a*) symbol; (*b*) wiring diagram.

connector and all the white wires together in another. The black jumper is wired to the brass terminal, and the white jumper is wired to the white terminal.

When the cable from the panelboard runs to a lighting fixture instead of to a switch, the wiring is slightly more complicated. The symbols for such a circuit are shown in Figure 10–15(a), and the wiring is shown in Figure 10–15(b). One cable and two jumpers have been added to the wiring used for one lamp holder. Again the black wire from the panelboard is spliced to the other black wires. The white wire from the panelboard is spliced to the white jumper and to the white wire of the cable feeding the second lighting fixture.

In these instructions, one cable is always specified as coming from the panelboard. In practice, the cable does not usually run directly from the panelboard but instead runs from another outlet box where power is always available. For example, several lamp holders are usually wired on the same circuit. The cable runs from lamp holder to lamp holder, and only the first lamp holder is wired directly to the panelboard. An example

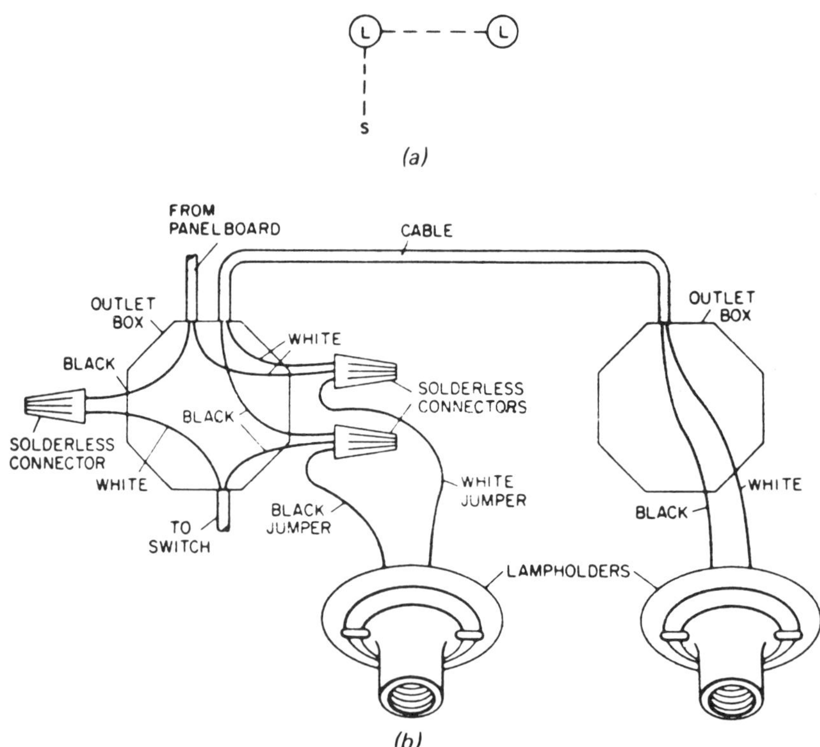

FIGURE 10–15. Wiring for two lamps controlled by one switch and using lamp-holder feed: (*a*) symbol; (*b*) wiring diagram.

of this is shown in Figure 10–16, where a pull-switch lamp holder is fed from the switch that controls a switched lamp holder. This diagram also shows one place where three-wire cable can be used to reduce the amount of wire in a residence.

Although the symbols in Figure 10–16(a) show no connection between the two lamp holders, some of the wiring may be shared as in Figure 10–16(b). The device box is fed by a two-wire cable, but a three-wire cable runs to the outlet box for the switched lamp holder. Another two-wire cable runs to the pull-switch lamp holder. At the device box the two white wires are spliced and run to the switched lamp holder, where a white jumper is spliced to the white cable wires. Although it is spliced, the white wire is continuous to the pull-switch lamp holder. A black jumper is used on the device box to connect the switch. The black wires are spliced here and in the first outlet box to make a continuous wire to the pull-switch lamp holder. The third (red) wire of the three-wire cable is connected to a switch terminal and to the switched lamp holder. The red wire is, of course, wired to the brass terminals on the switched lamp holder.

FIGURE 10–16. Lamp-holder wiring using three-wire cable: (*a*) symbols; (*b*) wiring.

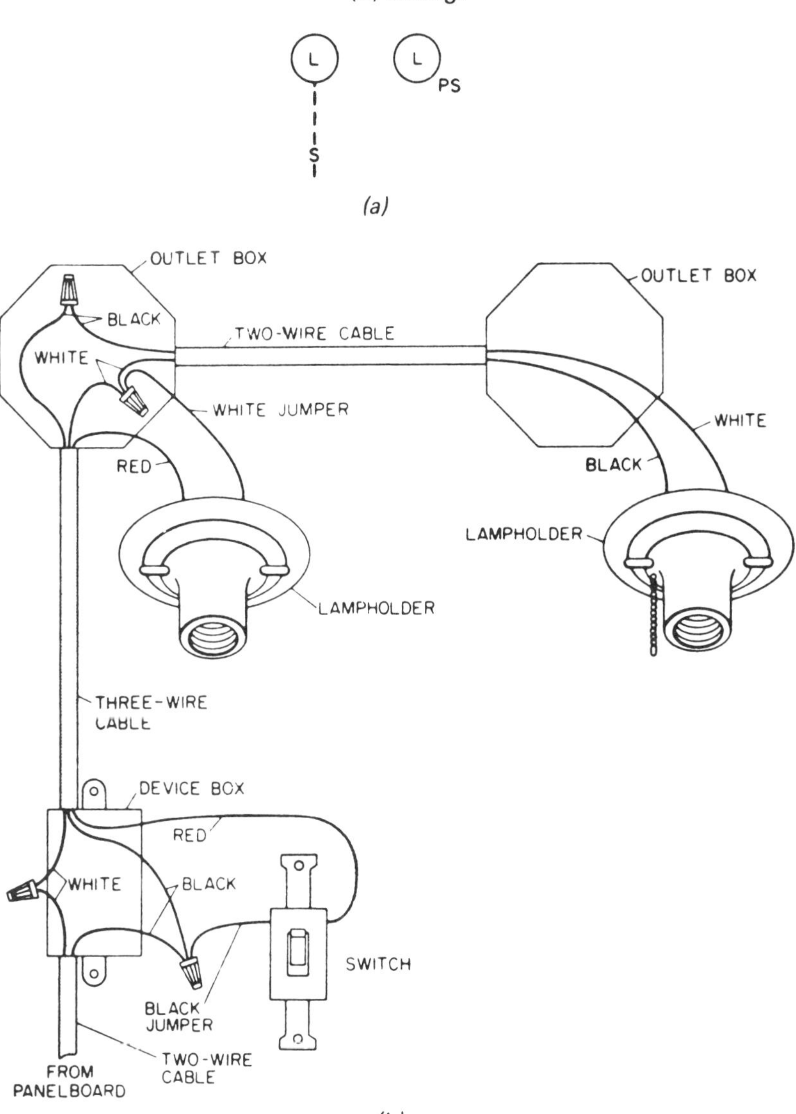

10–9 SWITCHES WITH PILOT LIGHTS

Wiring a switch pilot light is similar to wiring a switch to control two lamps at the same time. The difference is that one lamp is mounted at the switch instead of both lamps at a distance from the switch. The symbols

FIGURE 10–17. Wiring of switch with pilot light: (*a*) symbol for switch feed; (*b*) symbol for lamp-holder feed; (*c*) wiring for switch feed; (*d*) wiring for lamp-holder feed.

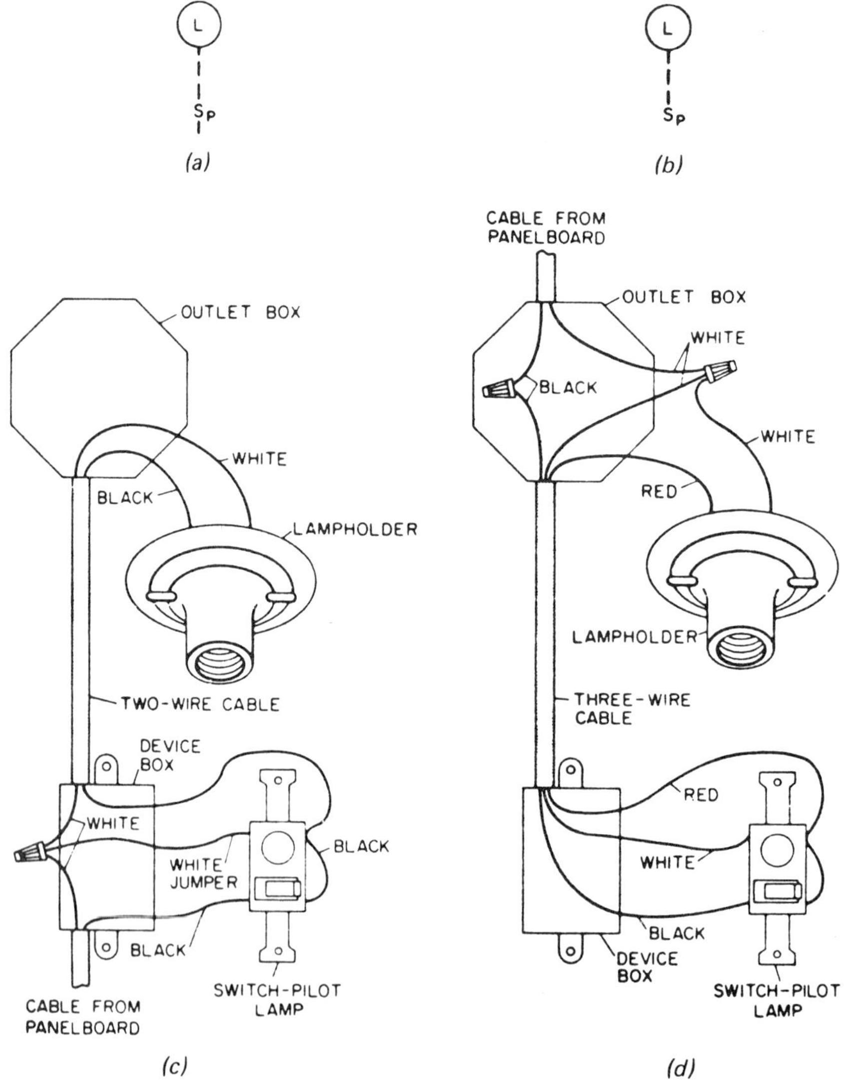

FIGURE 10–18. Wire stripped for combination switch–pilot lamp.

for the wiring are shown in Figure 10–17(a), where the cable feeds the switch, and in Figure 10–17(b), where the cable feeds the lamp holder. In Figures 10–17(c) and (d), the combinations of switch and pilot lamp are shown, with the switch and pilot lamp each having independent terminals. Some manufacturers' products may have an internal lead between the switch and the pilot lamp and will require slightly different wiring.

When the cable from the panelboard feeds the switch, as in Figure 10–17(c), only two-wire cables are needed. The white wires are spliced in the device box with a white jumper, and the jumper is connected to the white terminal on the pilot lamp. The black wire from the panelboard is connected to one side of the switch, and the black wire running to the lampholder is connected to the unused switch and pilot lamp terminals. Because there usually will be a barrier between the terminals, strip this wire as shown in Figure 10–18 to eliminate making a splice. The bare wire between the insulation may be wrapped around one terminal screw, and the bare wire on the other end wrapped around the other screw terminal.

When the cable from the panelboard feeds the lamp holder, as in Figure 10–17(d), it is convenient to use a three-wire cable between the outlet box and the device box. The white wire connects to the white terminals of the lamp holder and the pilot lamp. The black wires from the two cables are spliced in the outlet box. At the device box the black wire is connected to a switch terminal, and at the outlet box the red wire is connected to the lamp holder brass terminal. At the device box the red wire is stripped as shown in Figure 10–18, and connected to the brass terminal of the pilot lamp and a switch terminal as shown in Figure 10–17(d).

10–10 THREE-WAY SWITCHES

When a pair of three-way switches controls a single outlet, there are several combinations in which the cable from the panelboard, the switches, and the outlet may be arranged. The outlet may be either a lamp holder or a receptacle. The most common connections are shown symbolically in Figures 10–19(a), (b), and (c). In Figures 10–19(a) and (b) the cable from the panelboard feeds the lamp holder. In Figure 10–19(c) the cable from the panelboard feeds a three-way switch.

In three-way switches the common terminal is a different color from the other two, and its location varies with the make of switch. In the

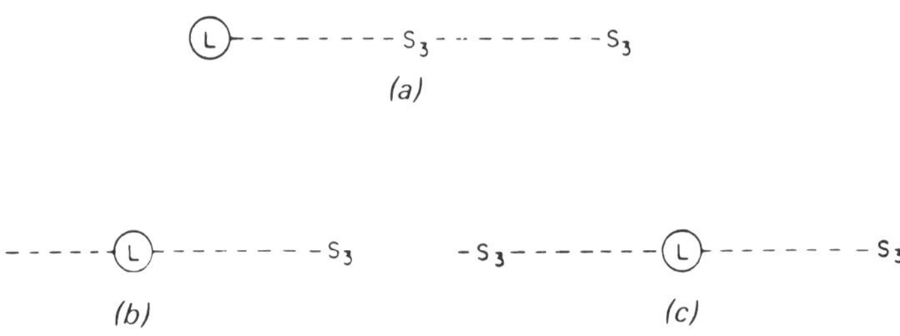

FIGURE 10–19. Symbols showing combinations for three-way switch connections: (*a*) and (*b*) lamp-holder feed; (*c*) switch feed.

following examples the common terminal is shown alone on one side. The white wire from the panelboard will always run to the white terminal of the lamp holder without being switched, although it may be spliced. The black wire from the panelboard will be connected to the common terminal of one three-way switch with either a black wire or a white wire, and the wire from the common terminal of the other three-way switch will be connected to the lamp holder brass terminal. At the lamp holder there will always be a white wire and one of another color. When three-wire cable is used, the other wire will be colored either black or red.

The wiring for the three-way switch circuit of Figure 10–19(a) is shown in Figure 10–20(a). This is the easiest of the circuits to wire because the fewest splices are made. The white wire from the panelboard is wired directly to the lamp holder, but the black wire is spliced to a white wire from the common terminal of switch S_2. In the device box for switch S_1 the white wires from the two-wire and three-wire cables are spliced. At both switches S_1 and S_2 the red wires are connected to the upper terminals and the black wires to the lower terminals on the same side of the switches. Either or both of these two wires from the three-wire cable could be reversed and the circuit would still function properly. The black wire on the outlet box at the lamp holder is part of the two-wire cable and is connected to the common terminal of S_1.

The wiring for the circuit of Figure 10–19(b) is shown in Figure 10–20(b). A two-wire cable from the panelboard supplies power to the lamp holder outlet box, and two three-wire cables connect to two device boxes for switches. Again the white wire from the panelboard is connected to the white terminal of the lamp holder. This time the black wire from the panelboard cable is spliced to the black wire of the cable running to S_1, where it is connected to the common terminal of S_1. At both S_1 and S_2, red wires are connected to the upper terminals and white wires to the lower terminals on one side of each switch. At the outlet box for the lamp holder, both white wires from the three-wire cables are spliced. The red

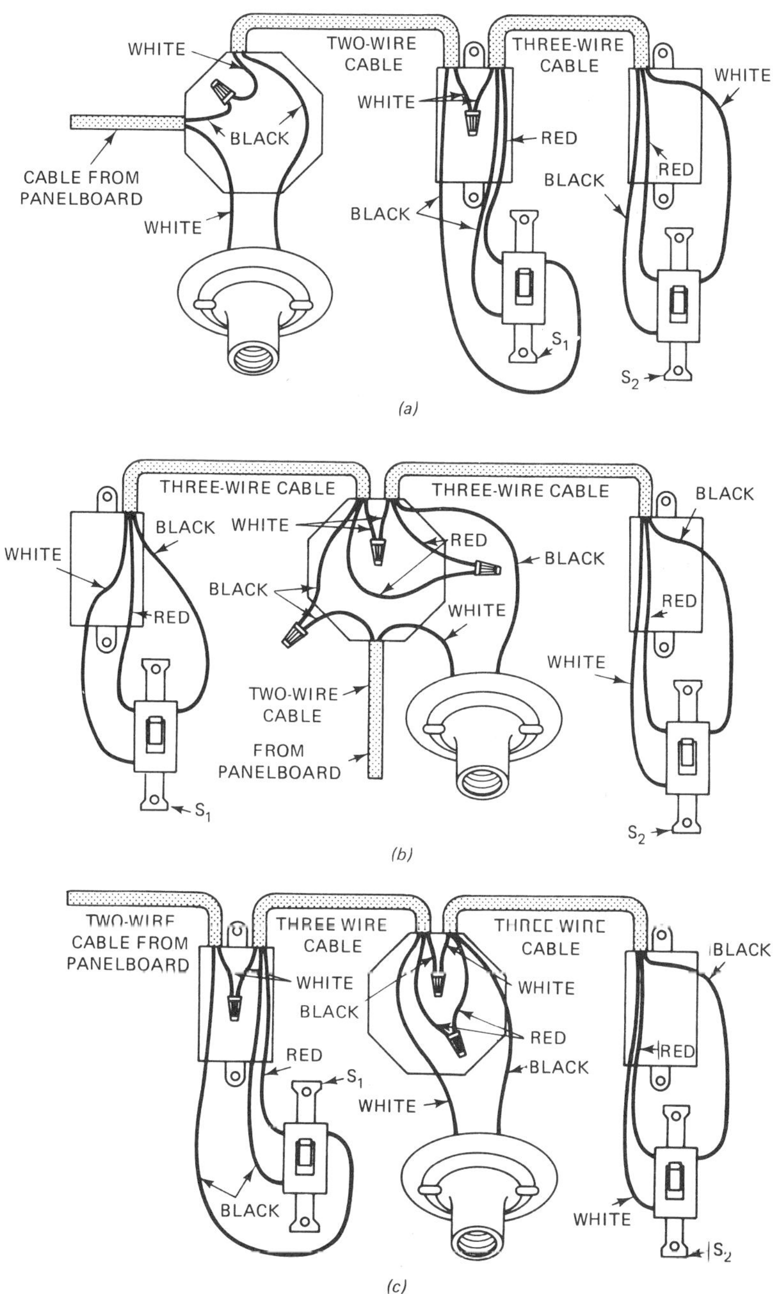

FIGURE 10–20. Wiring of three-way switches (S_1,S_2): (*a*) lamp-holder feed with switches on one side; (*b*) lamp-holder feed with lamp holder in the middle; (*c*) switch feed.

wires are also spliced in this outlet box. The common terminal of S_2 is connected to the lamp holder by a black wire.

The wiring for the circuit of Figure 10–19(c) is shown in Figure 10–20(c). A two-wire cable from the panelboard supplies the power, and three-wire cables connect the boxes together. This time the white wire from the panelboard cable is spliced in the device box for S_1 to a white wire from the three-wire cable. At the lamp holder outlet box this white wire is connected to the lamp holder white terminal. The black wire from the panelboard cable is connected to the common terminal of switch S_1. In the outlet box in the center the red wires from S_1 and S_2 are spliced together. In the same outlet box the black wire from S_1 in spliced to a white wire from S_2. The brass terminal of the lamp holder is connected to the common terminal S_2 by a black wire. Note that S_1 is connected by a red wire and two black wires, whereas S_2 is connected by a red wire, a black wire, and a white wire.

10–11 DOOR CHIMES AND DOORBELLS

Transformers for use with door chimes and doorbells are made with fixed outputs of either 16 or 10 V. The usual mounting methods are through a ½-in. knockout of an outlet box or on a flat plate that covers an outlet box. With these mountings the transformer is outside the outlet box, the primary wires are inside the outlet box, and the connecting screws or wires for the low-voltage (secondary) winding are outside the outlet box. Transformers are also made for surface mounting inside an outlet box or panelboard, but they take up space that could be used for making other connections.

The primary of the transformer is connected to the wires of a 120-V lighting or special-purpose circuit by splicing within an outlet box. One primary wire is connected to the neutral wire, and the other wire is connected to an ungrounded wire. The secondary connections are made with bell wire. The secondary connections for door chimes are shown in block diagram form in Figures 10–21(a) and (b). In Figure 10–21(a) the door chime can be wired to sound a double note for the front door and a single note for the back door. When wiring the two-door chime, run one wire from one transformer secondary terminal to the terminal labeled TRANS of the chime. Run another wire from the transformer secondary terminal to one terminal on the front-door switch and one terminal on the back-door switch. The wire may be spliced at a convenient place rather than run from the front-door switch to the back-door switch. Run a wire from the unused terminal on the front-door switch to the FRONT terminal

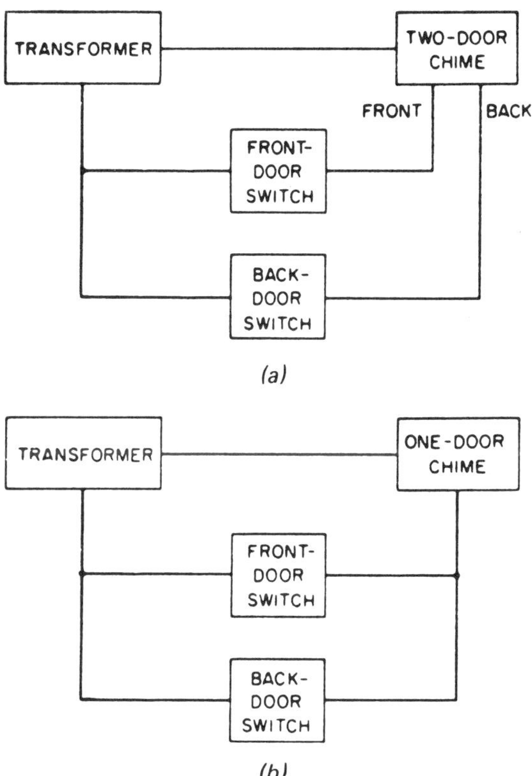

FIGURE 10–21. Door-chime connections: (*a*) two-door and (*b*) one-door.

of the chime. Run a wire from the unused terminal of the back-door switch to the REAR (BACK) terminal of the chime. When the two-door chime is tested, it should sound the proper notes for both doors.

A one-door chime is often used with only one door switch, but the connections for using a one-door chime with two door switches are shown in Figure 10–21(b). The wiring is very similar to that for the two-door chime except that the one-door chime has only two terminals; so the wires from the front- and back-door switches are spliced together and a single wire is run to the chime. If only one switch is to be used, the connections to the second switch are not installed. Then, for example, one wire will run from the front-door switch to the transformer, and another wire from the same switch to the chime terminal. Additional switches can be wired in parallel to the switches shown, or the one-door chime can be replaced by a bell or buzzer.

10–12 APPLIANCE CIRCUITS

The NEC classifies appliances in three groups: fixed, portable, and stationary. Fixed appliances are permanently fastened in one place, for example, water pumps and heating-system motors. Portable appliances are those that can be easily moved around in normal use. Stationary appliances usually remain where they are installed, although they are not fastened into place. For instance, a stationary appliance like a refrigerator can easily be moved from one house to another.

Every appliance must be provided with a disconnect means and overcurrent protection. With portable appliances the plug and receptacle arrangement is satisfactory if the plug and receptacle ratings are equal to or greater than the appliance ratings.

Fixed and stationary appliances rated below 300 W or ⅛ hp require no special disconnecting means. The branch circuit fuse or circuit breaker provides adequate protection.

When larger fixed and stationary appliances are connected to a circuit serving other loads, no special disconnect means is needed if the circuit is protected by a circuit breaker or fuses mounted on a pull-out block. If the circuit is protected by plug fuses, however, you must install a separate enclosed switch for each appliance. Although the switch does not have to be fused, the usual practice is to install a fused switch. Enclosed switches contain one fuse for 120-V circuits and two fuses for 240-V circuits. Install fuses that have an ampere rating not greater than that of the fuse protecting the branch circuit.

When a fixed or stationary appliance is supplied by its own individual circuit, follow the above recommendations. However, the fuse or circuit breaker ratings should not exceed 150 percent of the appliance ampere rating unless the appliance is rated at less than 10 A. In that case, the 15-A fuse or circuit breaker may be used. If the appliance has an automatically started motor, the overcurrent device should not be rated over 125 percent of the ampere rating of the motor being protected. Although the NEC does not require a separate circuit for each automatically started motor, such as a water pump or oil burner, it is wise to provide separate circuits.

10–13 WIRING FOR APPLIANCES

Smaller appliances can be connected in the standard receptacle, which is rated at 15 A. Heavy appliances often need larger receptacles with special contact arrangements, such as those shown in Figures 10–22(a) and (b).

The receptacle in Figure 10–22(a) is rated at 30 A and 250 V and is commonly used for electric clothes dryers. The receptacle in Figure 10–22(b) is rated at 50 A and 250 V and is commonly used for electric ranges. Note the difference in the contact arrangement, which prevents the wrong plug from being inserted. These receptacles may be used in either three-wire 120/240 V circuits or 240-V circuits, with the third contact used for grounding.

Some 240-V appliances, such as water heaters, do not need connections to the neutral and may be wired by a two-wire cable. However, the standard two-wire cable is made up only with black and white wires. The white wire should be used only for the neutral, which creates a recognition conflict. However, if both ends of the white wire are painted black, the wire can be considered black and then be used as an ungrounded wire. Some other appliances, such as electric clothes dryers, may operate on a 120/240 V combination and do require a neutral wire.

Heavy appliances may be wired with either cable or conduit. Ranges and clothes dryers may be wired with service-entrance cable, which has a bare neutral. When the appliance is to be connected by a cord and plug, run the cable up to the receptacle, which may be either flush or surface mounted. Flush-mounted receptacles are used with square outlet boxes and wallplates. Follow the receptacle manufacturer's recommendations when deciding which size of outlet box to install. Locate the receptacle within 6 ft of the intended appliance location. Good planning makes it possible to locate the receptacle closer to the appliance and thus simplify the final installation. The cords, or pigtails, for electric ranges and clothes dryers are made in lengths up to 6 ft, but these appliances are more likely to be delivered with cords about 2 ft long.

Clothes dryers are basically 240-V loads, although some have 120-V motors. Clothes dryer frames must be grounded. You can do the grounding with the neutral wire if it is no. 10 or heavier. Thus a three-wire cable must be used even if the entire clothes dryer operates at 240 V. The 30-A receptacle in Figure 10–22(a) and the no. 10 wires are heavy enough for

FIGURE 10–22. Contact arrangements for common power receptacles: (*a*) clothes-dryer receptacle; (*b*) range receptacle.

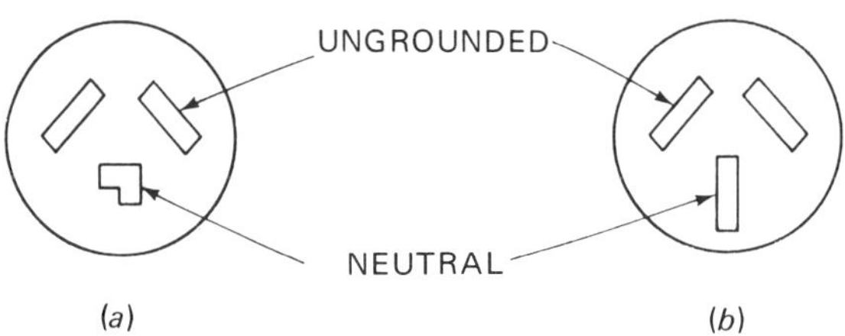

the typical home clothes dryer. Run the white wire to the neutral terminal and each ungrounded wire to one of the ungrounded terminals. Tighten the connections securely, because high currents will be carried. Although the NEC permits the use of the plug and receptacle device as the disconnecting means, a separate switch may be installed when the circuit is protected by plug fuses.

A water heater rated at less than 4600 W may be supplied by a two-wire no. 12 cable. Paint both ends of the white wire black, and protect the installation with a 20-A, two-pole circuit breaker, or fuses on a pull-out block, or a separate fused switch. Although grounding is not required, connecting a ground wire to a water pipe is recommended. Sometimes a plastic pipe is used for plumbing, or the metal water pipes may be disconnected; therefore, a grounding wire makes the water heater safer. If the branch circuit is protected by plug fuses, install a separate switch. If the branch circuit is protected by a two-pole circuit breaker or fuses on a pull-out block, a separate switch is not needed.

10–14 RANGES

The wiring for self-contained ranges is similar to that for electric clothes dryers. Because of the manner in which the oven and individual burners are connected within the range, the neutral wire carries less current than the ungrounded wires. So you may use a neutral wire one size smaller than the ungrounded wires. For most ranges, the ungrounded wires are no. 8 and the neutral is no. 6. If the range is connected by a pigtail cord, use a receptacle that has the contact configuration shown in Figure 10–22(b). Connect the neutral wire to the neutral terminal and the ungrounded wires to the ungrounded terminals. The pigtail cord serves as the disconnecting means, and serves to ground the range through the neutral wire.

In sectional ranges the oven is a separate unit installed in the wall or between kitchen cabinets, and the burners are installed on the kitchen counter at a convenient location. The NEC calls the units wall-mounted ovens and counter-mounted cooking units. Self-contained ranges are considered stationary appliances, whereas such sectional ranges are considered fixed appliances.

The two basic methods of wiring sectional ranges are either to supply each section by its individual branch circuit or to supply both sections from one 50-A circuit. Any wiring method may be used, including the use of service-entrance cable with a bare neutral. The frames of the sectional units may be grounded to the neutral wire if the wire is no. 10 or larger.

If the sections being installed do not have the frames connected to the neutral terminals, make sure that they are grounded.

Wall-mounted ovens are usually rated for about 4600 W. At 240 V the current is nearly 20 A, so no. 12 would be suitable. Use no. 10 for grounding. The wires may run directly to the oven, but it is more convenient to install the same kind of 30-A pigtail cord and receptacle that is recommended for clothes dryers. A counter-mounted cooking unit is installed in the same manner as the wall-mounted oven. Use no. 10 wire unless the section is rated above 6900 W; then it requires heavier wires.

The wall-mounted oven and the counter-mounted cooking unit may be supplied by a single 50-A cable containing two no. 6 wires and one no. 8 wire. The circuit for this wiring is shown in block-diagram form in Figure 10–23. A 50-A cable is run from the panelboard to the junction box, where it is tapped to make the two cables connecting the 50-A receptacles. Below the receptacles the 50-A pigtail cords are used to connect the wall-mounted oven and the counter-mounted cooking unit. If two junction boxes are mounted as close as possible to the section and only enough wire to permit serivicing is used, no. 10 wire can be run from each junction box to each section. Receptacles and pigtail cords are not required, but they will make the installation more convenient. Alternatively, the sections may be wired directly to the junction box shown on the branch circuit.

Pigtail cords are not considered as disconnecting means for fixed appliances. The disconnecting means may be a two-pole circuit breaker, cartridge fuses on a pull-out block, or a separate enclosed switch with cartridge fuses.

FIGURE 10–23. Sectional range supplied by a 50-A cable.

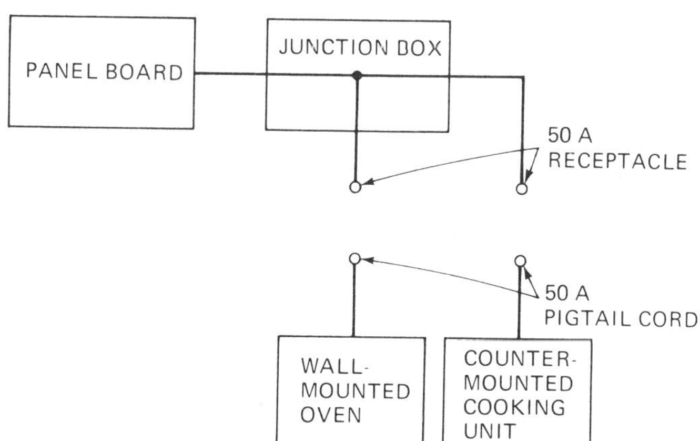

10–15 ELECTRIC SPACE HEATERS

Fixed electric space heaters are supplied from individual branch circuits rated at 15, 20, or 30 A. Because the heaters are considered to be continuous loads, the circuit current cannot be over 80 percent of the wire ampacity. For example, a 20-A branch circuit should not carry over 0.80 × 20 = 16 A. When a long cable supplies the heater, a larger-sized cable may be needed to reduce the voltage drop and maintain the heater output.

A baseboard-type heater controlled by a two-pole line-voltage thermostat is shown in Figure 10–24(a). The heater is supplied by a two-wire, 240-V circuit with an ampacity of 20 A. The typical two-pole line-voltage thermostat is rated at 5000 W, but the load should be limited to 80 percent of the rating, or 4000 W. Because a no. 12 copper wire rated at 20 A is used at a reduced ampacity of 16 A, the two-pole line-voltage thermostat

FIGURE 10–24. Wiring of electric space heaters: (*a*) line-voltage control; (*b*) low-voltage control.

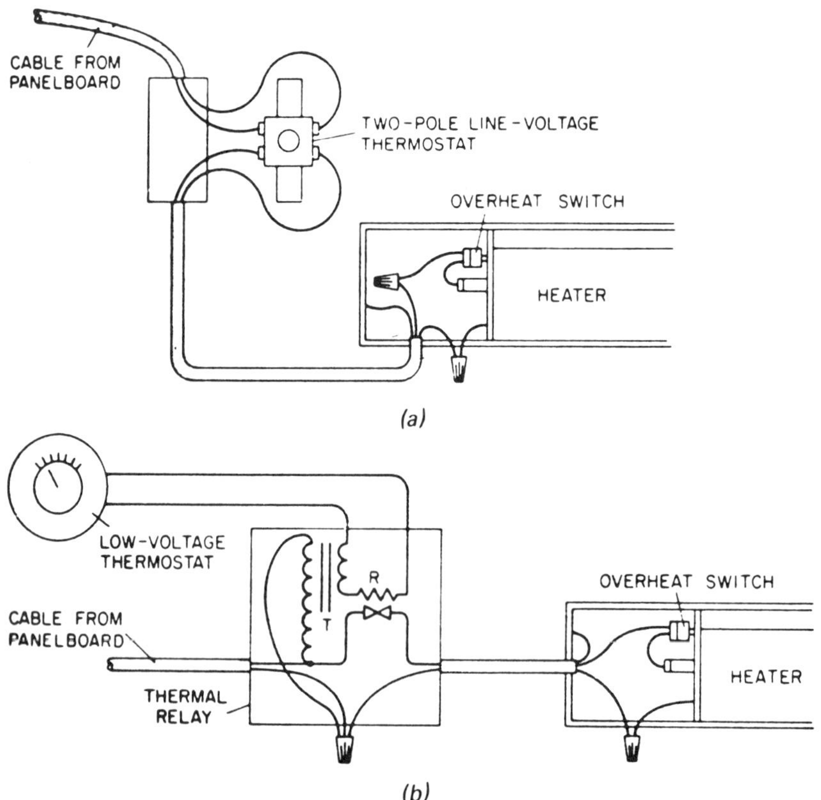

will not normally control more than 16 × 230 = 3680 W. Although the standard circuit voltages are 120 and 240 V, all calculations are made with 115 and 230 V to agree with the NEC.

Thermostats are mounted on 2- by 3-in. device boxes. In Figure 10–24(a), a two-wire cable from the panelboard runs to the two-pole line-voltage thermostat. Another two-wire cable runs from the two-pole line-voltage thermostat to the overheat switch on the heater. The overheat switch is a thermostat that runs the full length of the heater; it shuts the heater off if its temperature rises too high. This could happen, for example, if the heater were completely blocked off by draperies. The overheat switch is made part of the heater to increase the safety of the installation. From the overheat switch a wire runs to one connection of the heating element. The other wire from the two-pole line-voltage thermostat connects to the other terminal wire of the heating element within the heater. When the two-wire cable contains a grounding wire, the grounding wire is fastened to the heater frame.

The low-voltage control system in Figure 10–24(b) uses a thermal relay to control the power to the heater. The thermal relay includes a transformer *T*, which steps 240 V down to 24 V. The low-voltage thermostat controls the flow of current through a resistance heater *R*. When the thermostat closes, calling for heat, a current flows through resistance heater *R*, which is wound on a bimetallic strip made from two suitable different metals. The resistance heater heats the bimetallic strip and causes it to bend and close the switch contacts of the thermal relay. When the room temperature rises, the thermostat opens and stops the flow of current through the resistance heater so that the bimetallic strip cools. After the bimetallic strip cools, the switch contacts of the thermal relay open. In this way the thermal relay silently controls the power flowing to the heater.

In Figure 10–24(b), the cable from the panelboard runs to the thermal relay, where the transformer primary is connected between the two wires. After the splice, one wire runs directly to the heating element, but the other is connected to a terminal on the switch of the thermal relay. From the thermal relay terminal a wire runs to the overheat switch of the heater and then to the heating element.

In Figure 10–24(b), the diagram shows how the thermal relay directly controls the power flow to the heater. The thermal relay could also be used to control an electrically operated switch such as a relay or contactor. This flexibility permits you to place the low-voltage thermostat and thermal relay in the most convenient places while running short cables to the heater.

Fixed electric space heaters are rated as fixed appliances and should be protected as explained previously.

10–16 TESTING

Continuity Tests

Branch circuits are sometimes tested for short circuits and continuity with a tester made from a doorbell or a buzzer in series with a battery consisting of two 1.5-V cells. The test is made after the wiring is completed but before the power is connected. Doorbells, water heaters, and appliances with thermostats and automatic switches must be disconnected.

Branch circuits may be quickly checked for short circuits by connecting the free lead of the tester to the neutral bus of the panelboard. When the test lead from the battery is touched to any ungrounded wire, the tester should not ring. If the tester rings, it indicates a short circuit either between the ungrounded wire and the neutral or between the ungrounded wire and the metallic outlet box.

If the test does not indicate any short circuits, the continuity test can be made. Turn all circuit breakers on and install all fuses. Connect temporary jumpers between the neutral bus and both input terminals of the panelboard. Be sure to remove the jumpers after the test is completed.

Then take the tester to each outlet where a fixture or receptacle is to be installed. Touch it across the black and white wires; it should ring to indicate continuity. The circuit should be completed through the tester just as it will be completed later through a lamp or an appliance. To make the test, you must short together the wires that will run to switches in order to duplicate conditions when switches are turned on. Because cables are looped from outlet box to outlet box, you will need temporary connections with duplicate splices in outlet boxes and at receptacles. After this test, connect the tester between the black wire and the metal box. If the grounding wire or armored cable is well grounded at each box, the bell will ring. Because the grounding circuit has a higher resistance than the neutral wire, the bell may not ring loudly. The grounding should be well done as a safety precaution.

Insulation Test

The NEC requires that all wiring shall be so installed that the system will be free from short circuits and unspecified grounds. Most such faults are caused by faulty insulation. Determine whether the insulation of the wiring has been damaged by crushing, by excessively sharp bends, or by penetration by nails. Sometimes switches, receptacles, and other devices may be poorly insulated because of in-transit breakage, defective materials, or improper installation. The wiring testing can be best done after all switches, receptacles, and circuit breakers are installed, but before lamps are installed and appliances and other loads are connected.

The test doorbell will detect short circuits but will not detect damaged insulation when the resistance is high, 100 Ω, for example. The resistance of insulation can be better measured with either a VOM (volt-ohm-milliammeter) or a Megger. (A VOM can also be used, of course, to measure voltages, resistances, and currents.) Insulation is measured by placing the instrument switches in the proper positions to measure the highest resistance. The test leads are then touched together for a moment. The resulting short circuit should be indicated on the instrument scale as 0 Ω. If the pointer does not indicate exactly zero, correct the pointer position with the 0-Ω adjusting knob. This procedure is called zero adjustment and should be carried out before short-circuit tests. The test leads are then connected to the circuit under test, and the resistance is read on the proper scale of the instrument.

A Megger is a measuring instrument that is made specifically to measure high resistances; it is superior to the VOM because it tests with a high voltage, which makes it possible for it to find defects missed by the VOM. Meggers usually operate at voltages between 150 and 500 V, whereas the VOM is usually limited to about 7.5 V. The leads of a Megger are connected to the circuit to be tested, and the resistance is read directly in megohms.

Common Wiring Faults

Three common faults of electrical wiring are open circuits, short circuits, and grounds. An open circuit, as the name implies, is simply a circuit that is no longer complete because the continuous path for the current has been broken accidentally. As might be expected, the failure occurs largely at splices and at terminals where the mechanical fasteners may loosen and open the circuit. The danger from this type of fault is arcing at the point where the circuit opens. This arcing can cause considerable damage without tripping the circuit breaker or blowing the fuse unless a ground or short circuit results from the original fault.

A short circuit results when there is a contact between the ungrounded and neutral wires that bypasses the load. Because the resistance of the circuit is considerably reduced during the short circuit, a high current flows in that part of the circuit. The high current trips the circuit breaker or blows the fuse. The direct result is a dangerously high current flow that overheats the insulation and may even cause fire.

A ground, or a ground fault, in a system occurs when an ungrounded wire accidentally comes into contact with the ground or the grounding wire. Since the resistance of the ground can be quite high, 20 Ω or more, a ground fault may not trip the breaker or blow the fuse. The first indication of the ground fault may come as a high electric bill or some difficulty in operating the appliances on the grounded circuit. A grounded circuit always presents a possible shock hazard. This is illustrated in Figure 10–25. The post light is mounted on the upper end of a metal pipe that has its lower end buried in concrete. Because the post light is fed by a two-wire cable, it is not connected directly by a wire to the ground of the electrical system. There is a ground path through the earth, but its resistance is high compared with the resistance of a copper grounding wire. Therefore, when the ungrounded wire accidentally touches the metal pipe, not enough current flows to trip the circuit breaker or blow the fuse. Consequently, 120 V exists between the post light-support pipe and the electrical system ground. A person using a garden hose near the post light might become part of the ground path and receive a shock.

Faults in Appliances

To remedy an electrical fault in a building, first determine whether it is an equipment fault or a wiring fault. An equipment fault is an electrical failure in a piece of the electrical equipment, such as an appliance. A system

FIGURE 10–25. Ground fault in post light.

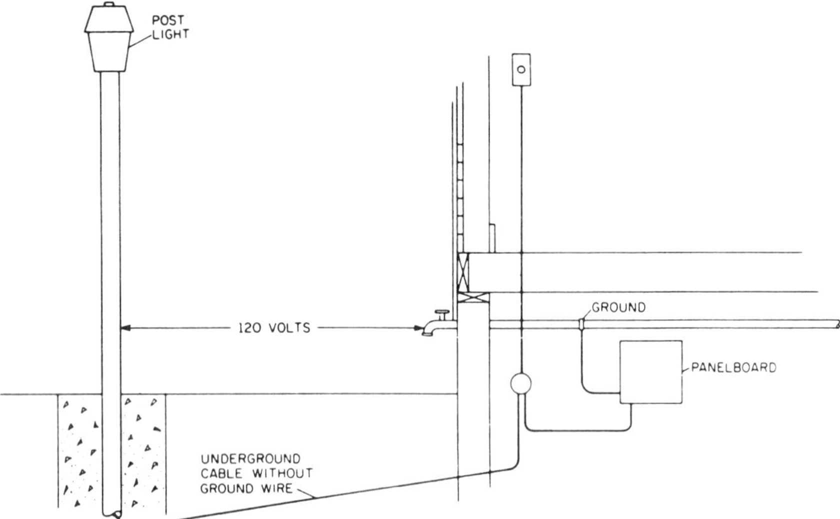

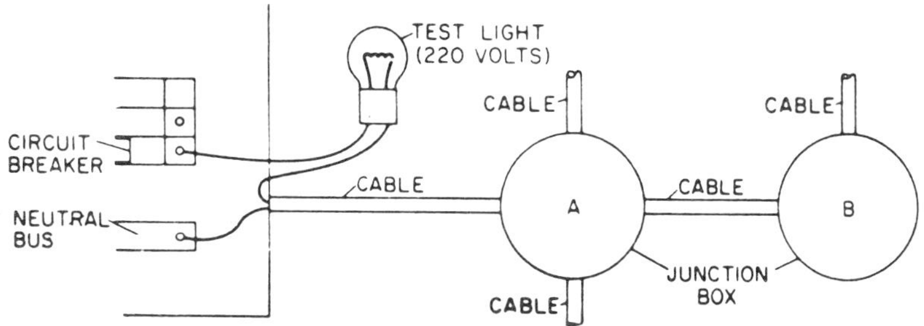

FIGURE 10–26. Use of test light in troubleshooting.

fault is an electrical failure within the building wiring system. Since more faults occur in appliances than in wiring, check first for an appliance fault. Disconnect all appliances to see whether that eliminates the fault. It is easy to disconnect appliances since most of them are of the plug-in type. If the circuit remains normal and circuit breakers do not trip or fuses do not blow, the fault is obviously in one of the appliances. Then examine the appliances one by one to locate the fault.

Faults in Wiring Systems

If the fault is in the wiring system, the breaker will trip when the circuit is tested with the appliances disconnected. The next logical step is to open the outlets at lights, at switches, and at receptacles to try to locate the trouble. If nothing can be found, break the circuit by opening one of the splices, then retest. If the circuit holds, the fault is in the part of the system that was disconnected. To keep from tripping the circuit breaker or blowing the fuse under fault conditions, some resistance can be added to the circuit, as shown in Figure 10–26. Adding such a test light does not remove the fault, but it does add enough resistance to the circuit to keep the breaker from tripping. When there is a fault the lamp will light up. Assume that there is a fault at one of the outlets fed from junction box *A*, for instance. When the circuit is opened at junction box *B*, the test light will remain lit, indicating that the fault is on the line side (circuit-breaker side) of line *B*. When the circuit is opened at junction box *A*, the test light will go out, which will indicate that the fault is on the load side of *A*. With this troubleshooting method, you have narrowed down the fault to the cables feeding from *A*, and can now presume that the cables feeding from *B* are good.

If the system uses plug fuses, screw in the test lamp in place of the blown fuse. Remember, this test is not valid if there is any load on the circuit. Disconnect all lamps, appliances, and motors first. Also, as a safety measure always shut off the power while disconnecting circuits, and be sure that the ungrounded wires are not shorted to ground or a metal outlet box.

Ground-Fault Detection

A ground fault may be detected easily by using a clamp-on meter, which can usually measure ac voltage, ac current, and resistance. Test leads are used to make ac voltage and resistance measurements. When current is being measured, the jaws of the clamp-on meter are opened and placed around the wire. Hold the clamp-on meter so the jaws will close by themselves to ensure an accurate circuit measurement. The circuit does not have to be opened to measure current as it does when a regular ammeter is used.

In Figure 10–27(a), we study a 240-V single-phase motor that is assumed to have a ground fault between the motor winding and the motor frame (case). Because the conduit grounds the motor frame to the panelboard, the ground current flows in the conduit. In other situations

FIGURE 10–27. Method of detecting ground fault in motor: (*a*) layout diagram; (*b*) schematic diagram.

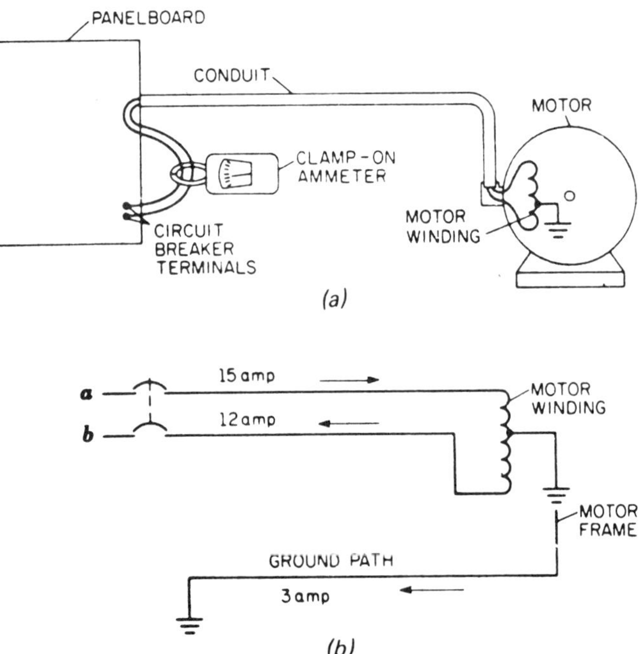

the ground path might be through the earth, water pipes, building steel, or a grounding wire.

The jaws of the clamp-on meter are placed around both current-carrying wires. If there is no ground fault, the meter will register at zero. The current would be zero if the current flowing to the motor were equal to the current returning from the motor. But if a ground fault occurs, part of the current returns to the panelboard through the conduit, and so the currents flowing in the wires are unequal. Accordingly, the clamp-on meter would indicate a current that is the difference between the currents flowing in the two wires.

A schematic diagram of that circuit is shown in Figure 10–27(b). Let's assume that the motor usually draws 12 A, but that a ground fault has occurred and 3 A now flows through the ground path. The current in line *a* has increased from 12 to 15 A, while the current in line *b* remains at 12 A. The difference between the currents is 3 A, which is the same as the current in the ground path. If the jaws of a clamp-on ammeter could be placed around the two lines, it would indicate 3 A. In this example the ground fault occurred near the line *a* end of the motor winding. If the ground fault occurred exactly at the midpoint of the winding, the currents would remain balanced and both currents would increase. For example, both currents might increase to 14 A. The resistance of the ground path would determine how much current flows in the ground path.

When this test is made, enclose all current-carrying wires inside the jaws of the clamp-on ammeter. If the circuit includes a neutral wire, remember to enclose it, too, because it carries current. Do not, however, include grounding wires, because they do not normally carry current.

If the circuit has a grounding wire, measure the unbalanced current with the grounding wire connected and then with it disconnected. There may be a change in unbalanced current due to the difference in resistance of the ground paths. When nonmetallic-sheathed cable is used, the ground should disappear when the grounding wire is disconnected. In circuits that are either made with armored cable or are placed in conduit, there should not be any change in current.

10–17 FINISHING THE INSTALLATION AND FINAL INSPECTION

When the rough wiring is installed, more wire than necessary is left in the outlet boxes. Those wires should be long enough to make connections easily, but not so long as to crowd the outlet box. Do not try to cut off the last little bit when the device is replaced; the new one may require slightly longer wires because of different terminal placement. Remember to strip off only enough insulation to make the connection. There must be no bare wire between the end of the insulation and the terminal screw.

After the device is wired, it is fastened to the device box by machine screws that pass through elongated holes in the strap of the device into the threaded holes in the ears of the device box. The enlarged ends, called plaster ears, may be cut off easily if they are not needed. Because device boxes usually are recessed slightly in the walls, the plaster ears lie on the wall surface and automatically hold the device flush with the wall surface. Elongated holes are provided in device straps to permit devices to be mounted straight up and down when the device box is at an angle.

Faceplates are furnished with screws to fasten them to the devices. When plastic faceplates are used, tighten the screw very carefully to avoid cracking the plate. When faceplates are used with two- or three-gang boxes, the devices have to be carefully lined up so that all screws will mate with their respective holes in the device straps.

Lighting Fixtures

The lamp holders shown in Figures 10–11 and 10–13 are mounted directly on the outlet box by machine screws that turn into the threaded ears of the outlet box. If the lighting fixture is too large to fasten directly to the outlet box, fasten a flat strap to the box instead. You may pass machine screws through holes in the lighting fixture and enter threaded holes in the strap. Other fixtures are made with one large hole that is of the correct size for passing over ⅛-in. running-thread pipe, or nipple. The nipple may be screwed into the center hole of the strap shown in Figure 10–28.

FIGURE 10–28. Lighting fixture mounting parts: (*a*) fixture strap; (*b*) fixture stud; (*c*) adapter.

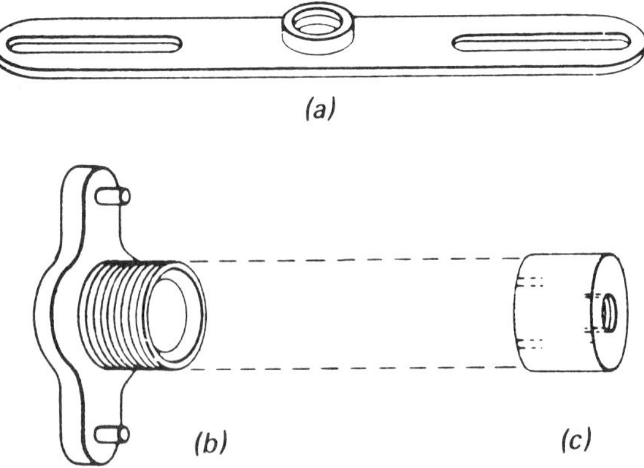

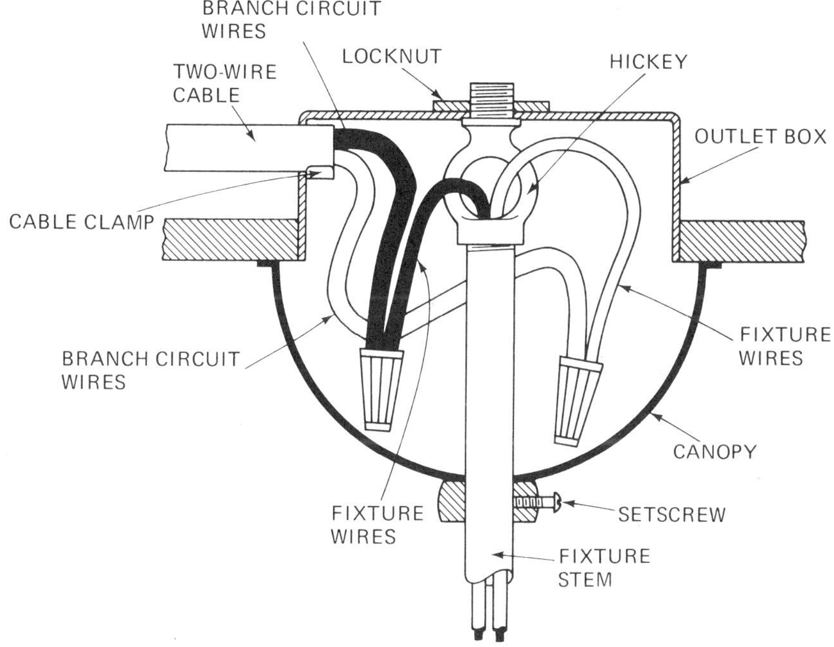

FIGURE 10–29. Installation of ceiling fixture.

Machine screws pass through the elongated holes in the strap to fasten the strap to the outlet box. The fixture is secured to the nipple by a decorative nut (finial). Another method of fastening such a fixture is to bolt the fixture study as in Figure 10–28(b) to the bottom of the box. An adapter is screwed into the stud. See Figure 10–28(c). The small hole in the adapter is threaded to accept a nipple. The nipple-type assembly may be used with either ceiling or wall fixtures.

Fixtures that are larger but weigh less than 50 lb may be fastened as shown in Figure 10–29. The outlet box is supplied by two branch-circuit wires, which are part of a two-wire cable secured by a cable clamp. The fixture stem screws into the hickey, which is fastened by a locknut. Another type of hickey screws onto the fixture study shown in Figure 10–28(b). The fixture wires are spliced to the branch-circuit wires. The canopy is then placed against the ceiling by a setscrew or some other device. Any combustible ceiling finish between the edge of the canopy and the outlet box must be covered with a noncombustible material, such as a piece of sheet steel.

A fixture weighing more than 50 lb must be supported independently of the outlet box. In new work, you may use bar hangers that include a fixture stud.

Completing the Panelboard Wirings

After all the outlets on a circuit are wired, the branch-circuit wires may be connected to the panelboard. Panelboards are not supplied with cable clamps, so it is necessary to use a bushing on each cable. Cables are often wired through knockouts on the side of the box. Select a knockout near the circuit breaker or fuse holder for the particular circuit to reduce the amount of wire within the panelboard. You must leave a longer piece of cable at the panelboard than at the outlet box because all neutral wires must be connected to the neutral bus. When you install nonmetallic-sheathed cable, you also connect the grounding wire to the neutral bus. However, there may not be enough neutral terminals to connect each grounding wire individually. So splice the grounding wires together and connect only one wire to the neutral bus. When you install armored cable, remove the paint where the bushing will enter. Also, install the nut on the bushing so that it makes it dig into the metal and so ensures a low-resistance grounding connection.

Plug-in circuit breakers are normally used for residential installations. Install a single-pole circuit breaker of the correct rating into the panelboard for each 120-V branch circuit. For 240-V branch circuits, plug in two-pole circuit breakers. A single-pole breaker has one terminal for connecting the ungrounded wire of the branch circuit. A two-pole breaker has two terminals, one for each ungrounded wire. After all the circuits have been tested, lock the circuit breakers into position by the hardware provided.

When a fused panelboard has been installed, treat the neutral and the grounding wires in the same way as those of circuit breakers. There is a terminal near each fuseholder for plug fuses to connect the ungrounded wires of a branch circuit. Before installing the plug fuses, insert the correct adapters, because type S fuses must be installed. The 240-V circuits are often protected by cartridge fuses on pull-out blocks. Wire the ungrounded wires as shown in the panelboard wiring diagram. You may have to run some short wires from the output side of the main fuses to the input side of the branch-circuit fuses to feed the 240-V branch circuits. If the pull-out for the main fuses has been plugged in, be sure to remove it before installing these wires. Make sure that the branch-circuit current will flow through the fuses, and that the branch-circuit wires and the wires from the main fuses are not connected to the same pair of terminals. Two cartridge fuses of the correct rating may then be placed in each pull-out block.

Circuit Index

When all the circuits have been completed, the appliances and heating system are connected and tested, and all breakers are installed (both active breakers and spares), you can install the panel trim. You should now mark up the panel index clearly and legibly. The index contains space for you to label each circuit. This task may seem unimportant, but its real value becomes apparent when the system develops a fault that has to be located by someone unfamiliar with the system.

Final Inspection

At this point you should obtain the final inspection certificate. Usually, the inspection is made automatically by an inspector sometime after the rough-wiring inspection is approved. Remember, it is important to obtain that certificate because it assures all persons concerned that the wiring has been installed in a safe and secure manner. That protects the owner because it attests to the fact that the building is electrically safe and insurable without unusual risk. As a disinterested but qualified third party, it also protects you, the electrician, when you can tell the owner that you have wired a system that is electrically safe for his house.

EXERCISES

10–1. Define the terms *new work* and *old work* as they are understood in the electrical trade.

10–2. What is meant in the electrical trade by *roughing-in*?

10–3. How is a grounding clip installed?

10–4. What is the color and shape of the grounding terminal on a receptacle?

10–5. When both wires of a lighting fixture have the same basic color, but one has a tracer, which wire is neutral?

10–6. At what heights would you install receptacles?

10–7. What factors should be considered in installing outdoor receptacles?

10–8. What factors should be kept in mind when selecting switch positions?

10–9. What type of lighting arrangement is recommended near a bathroom mirror? Why?

10–10. Is the use of pendant-type lamps permissible in closets?

10–11. What is the range of operating voltages for door chimes? For buzzers and doorbells?

10–12. In a noncombustible wall, how must the front edge of an outlet box be positioned with respect to the final front surface of the wall?

10–13. Describe two typical arrangements for mounting outlet boxes.

10–14. Describe how armored cable is installed in an outlet box.

10–15. Describe how nonmetallic cable is installed in an outlet box.

10–16. Describe how the terminals of a standard duplex receptacle are identified.

10–17. What is a two-circuit duplex receptacle? What arrangement is made on many standard duplex receptacles to permit easy conversion to the two-circuit type?

10–18. In Figure 10–9, the white wire is attached to a black wire. Under what circumstances, if any, is this permitted by the NEC?

10–19. Describe a *three-wire* circuit. What are the advantages of this type of circuit?

10–20. In a three-wire circuit, suppose that the two ungrounded wires carry currents of 18 and 22 A, respectively. What is the neutral current? Why?

10–21. With what type of cable is a three-wire circuit wired?

10–22. Draw the wiring-diagram symbols for a ceiling-mounted, pull-chain lamp holder. If three such lamps are used on a circuit, how is the last one indicated?

10–23. Draw the wiring-layout symbol for a lighting fixture controlled by a single-pole switch. How would you indicate that the cable from the panelboard runs to the switch? To the lamp holder?

10–24. Describe how you would install two lamp holders controlled by one switch.

10–25. On a wiring diagram, how would you show the arrangement identified in Exercise 10–24?

10–26. Describe how you would wire a pull-chain lamp holder fed from a switch that controls a switched lamp holder.

10–27. Draw the wiring-diagram symbols for a switch with pilot light where (*a*) the cable feeds the switch, and (*b*) the cable feeds the lamp holder.

10–28. Describe two arrangements for installing a pair of three-way switches to control a single outlet.

10–29. Describe the way you would connect a two-door chime.

10–30. Define what is meant by fixed, portable, and stationary appliances.

10–31. If a stationary appliance is rated at 400 W, does it require a special disconnecting means? Explain your answer.

10–32. Does the NEC require a separate circuit for each automatically started motor, such as a water pump or oil burner?

10–33. What are the voltage and current ratings of a receptacle commonly used for an electric clothes dryer? For an electric range?

10–34. In using a standard two-wire cable to wire a water heater, what precaution must be taken to have the cable conform to NEC specifications?

10–35. What size wire may be used for grounding a clothes dryer, and how is the grounding accomplished?

10–36. When a water-heater branch circuit is protected by a plug fuse, is any additional safety element required?

10–37. What wire sizes are used to wire most electric ranges?

10–38. Using a single 50-A cable, describe how you would wire a wall-mounted oven and a counter-mounted cooking unit.

10–39. When electric space heaters are supplied by a 30-A branch circuit, what is the current limitation? Why?

10–40. Describe how a line-voltage thermostat operates to control an electric space heater.

10–41. Describe how the overheat switch of an electric space heater is wired.

10–42. Describe the thermal relay and how it operates in a low-voltage control system for electric space heating.

10–43. Describe one method of checking individual branch circuits for a possible short circuit.

10–44. How is a continuity test made after the roughing-in is completed?

10–45. What instrument is best suited to testing insulation? Why?

10–46. How is it possible for a ground fault to exist without blowing a fuse or tripping a circuit breaker?

10–47. What step would you take first to determine where an electric fault might exist in a wiring system? Why?

10–48. What is the easiest way to locate a ground fault?

10–49. How would you install a fixture weighing more than 50 lb?

10–50. What types of plug-in circuit breakers would you use to complete the panelboard installation in a residence?

11

Old Work

11–1 GENERAL

As you know, old work is the wiring of buildings that were completed before the wiring is started. Such work presents few electrical problems that have not already been discussed in connection with new work. Most of the difficulties will be problems of carpentry or mechanics—how to run wires from one point to another with the least effort and with minimum damage to walls, floors, and ceilings. Old work requires more cable than new work, because it is often better and easier to run the cable the long way around through spaces that can be easily used.

In old work, every building presents a special set of problems because of construction variations. At times it may seem that every carpenter has his own construction methods. If you study buildings that are presently being constructed, you will better understand what is hidden behind the plaster of older buildings. For example, the differences are especially noticeable in the construction around doors, windows, and corners and in the spacing between the different members of the structure.

11–2 MAKING WALL AND CEILING OPENINGS

Cutting good openings for outlet boxes requires a certain skill and lots of common sense. The openings must not be too large, and they must be made neatly. After the building structure has been studied and the wiring has been planned, mark the approximate place where the outlet box will be located. If possible, allow yourself some leeway so that the opening may be shifted slightly in any direction. To determine whether there is a stud or a joist in the way, thump on the wall and ceiling. Thumping will produce a different sound at a timber than in the hollow between timbers. In older housing the plaster is supported by narrow wooden strips called *laths*. The space between the laths is filled with plaster. A newer house may have been plastered over metal lath or dry wall, such as gypsum board, hence, the difference in sounds.

When the house is plastered with wooden lath, dig through the plaster at the approximate outlet location until you determine the space between two laths. Then bore a small hole all the way through and probe with a stiff wire to determine whether a timber is nearby. If the space is clear, hold the outlet box against the wall and make a mark around it. If you make a proper layout for a device box, only one lath will be completely cut off, as shown in Figure 11–1(a). A side view of the installation is seen in Figure 11–1(b). You must locate the center of the box on the center of a lath, not in the space between laths. Drill holes carefully at diagonally

FIGURE 11–1. Proper location of device box with respect to wood laths: (*a*) front view; (*b*) side view.

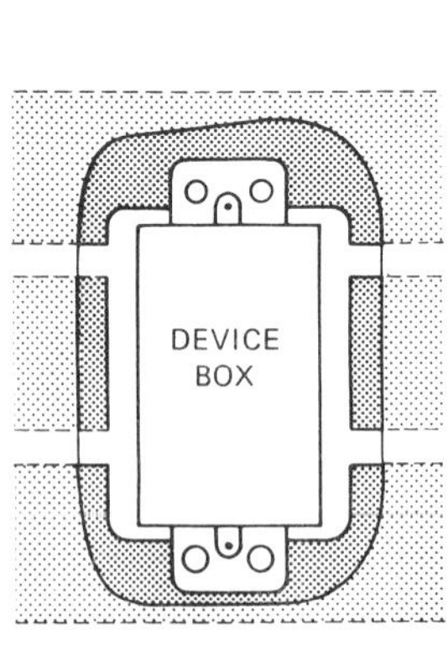

(a)

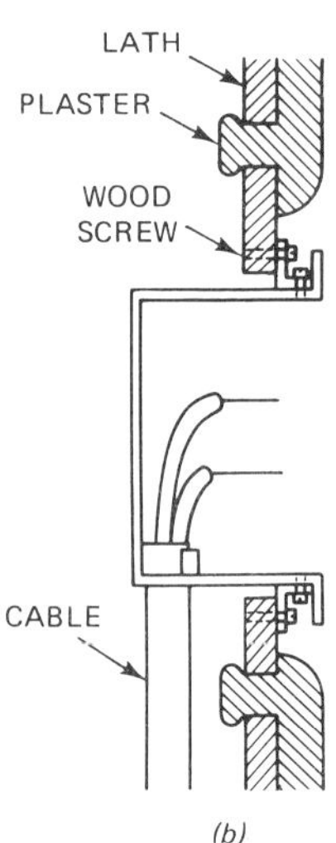

(b)

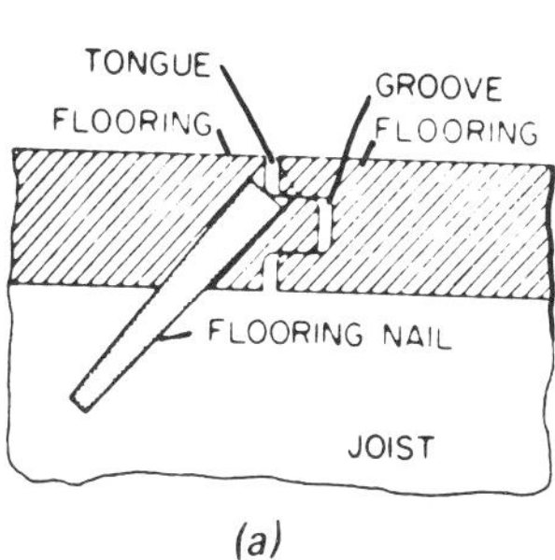

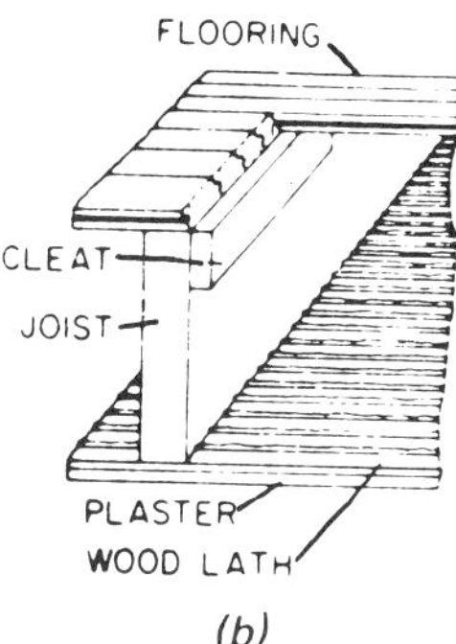

FIGURE 11–2. Typical floor construction: (*a*) tongue-and-groove flooring; (*b*) cleat to support cut end of flooring.

opposite corners large enough for a hacksaw blade to enter. Use the hacksaw blade with the teeth pointing toward you—the opposite of the usual method. You cut when you pull the hacksaw blade toward you. If you push the blade in the usual way, there is a danger that the lath will pull away from the plaster and leave some plaster unsupported. While you saw, hold your hand against the plaster to support it and to reduce the tendency for the plaster to pull off the wall.

Where dry-wall construction has been used, a portable jigsaw, or a saber saw, may be used to make the openings. Equip the saber saw with a metal-cutting blade and use a low cutting speed.

Be careful not to mar the walls or get them dirty. If possible, confine the cutting to the areas that will later be covered by the lighting fixture canopy or the wallplate. If the walls are papered, a repairable hole can be made by first cutting a cross in the paper with a sharp razor blade and then folding back the triangular pieces thus formed. If water will not damage the wallpaper, use water to loosen the sections to be folded back.

11–3 FLOOR OPENINGS

Sometimes outlet and switch boxes may be located so near obstructions that it is necessary to lift the boards of the floor above. Attic flooring made from plain boards is easily lifted because the location of the floor joists can be determined from the nailheads on the surface. Other floors usually are made from tongue-and-groove boards with concealed nailing, as shown in Figure 11–2(a). To lift these boards, it is necessary to cut off the tongue of the board with a very thin chisel, such as a putty knife with the blade cut off short and then sharpened. Drive this chisel down between the two boards to cut off the tongue. The floor joists will be located where the chisel will not penetrate more than an inch. Cut off the tongue for the entire distance between three joists, because the floor will then be stronger when the flooring is replaced. Drive small holes at two points next to the joists. You may use a carpenter's square and a pencil to draw lines at right angles to the flooring and parallel to the joists. Cut off three boards with a saber saw or a keyhole saw beginning at the small holes. A keyhole saw has a small blade that tapers to a point. The electrically operated saber saw will cut faster, but it tends to splinter the surface of flooring. After the tongue is removed, insert a chisel at a number of points along the board to pry it up. Pound on the adjacent board with a hammer to help remove the first board. Use a block of wood under your hammer to protect the floor from being damaged.

After you remove the first board, the other boards may be easily lifted to provide access for wiring. Before you replace the floorboards, nail cleats on the sides of the joists to support the ends of the floorboards, as shown in Figure 11–2(b).

Remove the nails from the floorboards and joists before replacing the floorboards. Replace each piece in its original position and nail it to the joists and cleats with finish nails. Finish nails have a small head about twice the diameter of the main part of the nail, whereas common nails have a larger flat head. Especially in hardwood floorboards, the nail will be easier to drive if you first drill nail holes through the floorboards. Use a drill about the same diameter as the main stem of the nail. After the head of the finish nail is flush with the flooring, you may countersink it by using a carpenter's nail set and a hammer. You may blend in the nail holes and saw cuts with the floorboards if you fill them with one of the soft, colored, putty-like compounds that are made in many shades and sold at most hardware stores and lumberyards.

11–4 MOUNTING OUTLET BOXES

For new work, all outlet boxes must be at least 1½ in. deep; for old work, shallower boxes may be used if a deeper box would seriously weaken a member of the building. For example, recessing a 1½-in. deep box into a floor joist would seriously weaken that joist. The outlet box should be large enough to accept the wires without their being jammed together. Never use outlet boxes less than ½ in. deep. A typical shallow box for use with nonmetallic sheathed cable is shown in Figure 11–3. Shallow boxes are also available with a built-in fixture stud.

Although a shallow outlet box may be fastened directly to wood lath

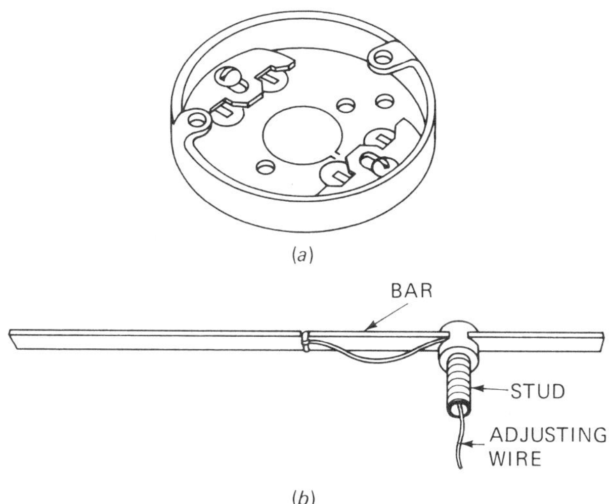

FIGURE 11–3. Shallow outlet box (*a*) and old work hanger (*b*).

with screws, it is better to fasten it to a joist. If the flooring is removed above the place where you want to mount the outlet box, you can nail a wooden block with a nominal thickness of at least 1 in. between the joists to support the lighting fixture. Outlet boxes may be supported between joists by using the old-work hanger shown in Figure 11–3(b). Slip the hanger through a hole in the ceiling large enough to clear the bar and the stud. Then pull the wire until the stud is centered in the hole. The bar should be at right angles to the lath for the best support. The stud passes through a hole in the bottom of the outlet box where you have removed a knockout. Use a locknut to secure the outlet box to the stud.

11–5 MOUNTING DEVICE BOXES

The device box shown in Figure 11–1 may be fastened to the wood lath by four small wood screws passing through the holes in the mounting ears. The mounting ears, which are adjustable so that the front edge of the outlet box may be made flush with the wall, are shown more clearly in Figure 11–4. When the device box is mounted on wallboard or similar materials, you cannot use screws, because the material is too weak to hold them. Install the device box instead with the device-box supports shown in Figure 11–4(a). These supports consist of a pair of metal straps with ears. Note that the projection of the strap beyond one ear is longer than that beyond the other. Slip the device box into the opening until it is

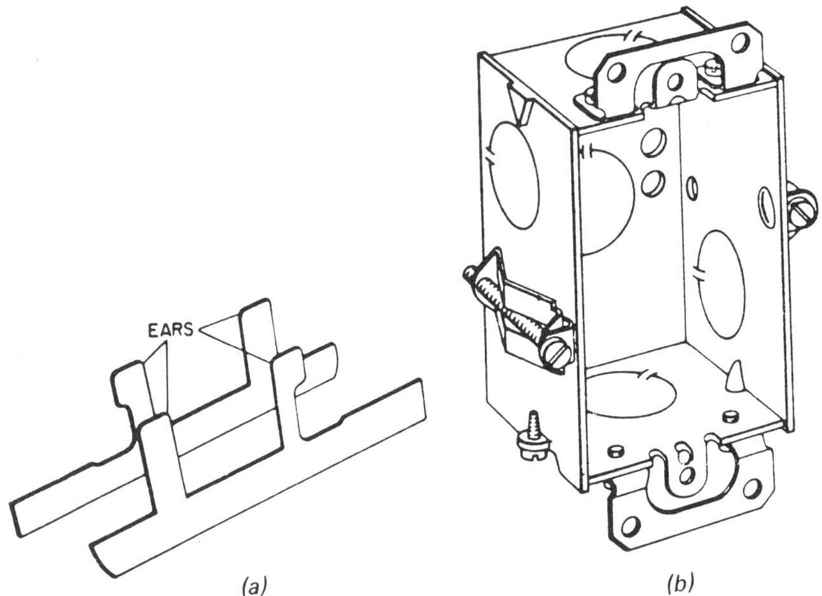

FIGURE 11–4. Device-box supports (*a*) and screw-clamp device box (*b*).

restrained by its mounting ears. Next insert the long leg of a strap until the ear rests against the edge of the opening. Then it is possible to insert the short end. Move the strap until both ends are hooked behind the wallboard; pull the strap up firmly by one ear; bend the ear down inside the box. Repeat the process with the other ear and strap. Be sure that the ears remain close to the inside surface of the box or grounds may result.

The device box in Figure 11–4(b) is easy to install, but it requires a larger hole to clear the metal strips that are on the sides of the box. After you slip the device into the wall opening, tighten the screws on the sides since they cause the metal strips to expand outward and grip the wall. Be careful. Because the expanding metal strips are not directly opposite each other, the device box tends to twist in the wall opening as you tighten the screws.

11–6 RUNNING CABLE

In old work, you install the cable after the outlet box openings are cut and before the outlet boxes are installed. Use nonmetallic-sheathed and armored cable normally because they are flexible and easily fished into the empty spaces. In general, you will find that it is easier to run cables from the cellar or the attic into a wall than through other parts of the building.

Removing the baseboards in the room above will often make it possible to reach a ceiling outlet. Ceiling outlets generally do not line up with switch outlets, so removing the baseboards makes it possible to run a cable across the wall without disturbing the finished part of the wall. Cut a groove in the plaster where it will be covered by the baseboard. If two wall outlets in the same room are to be connected, remove the baseboard to help in running the cable.

The exterior walls of newer homes can present a problem not encountered in older homes. For instance, fishing cables is difficult in residences in which insulation has been installed to make heating and air conditioning more economical. Fiberglass insulation should be handled carefully; it can be irritating to the skin and dangerous to the eyes.

Fish tapes are useful in installing cable, because the fish tapes can often be pulled from one opening to the other where it would be difficult to push a cable. Another efficient method is to feed a fish tape in from each opening and hook the ends together. One fish tape is used to pull the other fish tape and the cable through the openings. A small weight on a string is also useful for fishing between the walls of partitions.

11–7 MAKING CONNECTIONS

At the new outlet, the general procedure is to prepare the end of the cable as it would be prepared for new work and then secure the cable in the outlet box. Next work the outlet box with the cable back into the opening and then fasten it to the wall. An oversize opening makes installation of the box easier, but it may also make it difficult to fasten the box properly. A smaller opening is a sign of good workmanship. The opening around the outlet box should later be filled with patching plaster or plaster of paris.

Make connections to the new device in the same way as for new work. Remember to include a grounding wire and install grounding receptacles where it is practical to obtain a good ground. When you must connect new wires to the old wires, the neutral and ungrounded wires may not be easy to identify. Be careful; because the neutral wire may appear black, it is best to determine which wire is actually the neutral.

One method of testing is shown in Figure 11–5, where a neon circuit tester is connected between the brass terminal on a cleat lamp holder and a cold-water pipe. The lamp in the circuit tester will be illuminated with a pink glow if there is normal voltage between the brass terminal and the cold-water pipe. Use caution when making this test so that you will not become part of the circuit and receive a shock. The neon circuit tester may be used if no cold-water pipe or ground is available nearby. Because a high resistance is included in the neon circuit tester, one lead can be

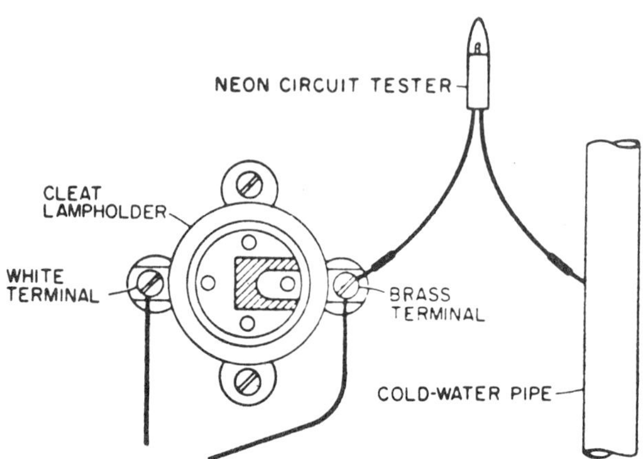

FIGURE 11–5. Use of neon circuit tester.

grasped safely in your hand while the other is touched to the terminals. The terminal at which the neon lamp glows is the ungrounded one. However, the neon bulb will not glow as brightly as when it is connected to a good ground.

Remove the fuse or turn the circuit breaker off before you connect the new wires to the old ones. Make any splices or connections well and insulate them properly.

When you add outlets to older homes, you may find a type of wiring that has not been discussed because it is not ordinarily used in new installations. This wiring is called *knob-and-tube work;* it uses individual wires rather than cables. When these old wires pass through a timber, such as a joist, they are protected by porcelain tubes; when they run along a surface, they are supported by cylindrical porcelain insulators called knobs. Consult the NEC, which lists the requirements for knob-and-tube wiring.

11–8 SURFACE WIRING

Surface wiring is another method of adding outlets to old wiring. In this method you use surface-mounted outlets and cables. Use is made of a special plastic two-wire cable, which is fastened to the walls by driving small round-head nails through holes that have been prepunched in the cable. The cable is flexible enough that it can be bent in either direction to go around corners. Surface-mounting switches, lamp holders, receptacles, and junction boxes are all made for use with the cable. A pull-chain lamp holder used with this cable is shown in Figure 11–6.

11–9 MODERNIZING THE INSTALLATION

Many residential wiring systems are obsolete because they do not have enough circuits to supply all the appliances presently in common use. The general-purpose, or lighting, circuits usually remain adequate for the intended loads. It is the additional kitchen and laundry appliances and air conditioners that are most likely to overload the circuits. Some appliances even operate from 240-V circuits, which may not be available in a given building. The service-entrance equipment is also likely to be too small for today's load.

Before deciding that all the old wiring needs replacement, analyze the load. If all the appliances were disconnected, the general-purpose circuits for lighting and receptacles for loads such as radios, television sets, and vacuum cleaners are probably adequate. It is easier and cheaper to modernize the wiring system if the existing branch circuits can be used. The work is likely to require adding a new service entrance, kitchen circuits, laundry circuits, and other appliance circuits.

Plan the installation to determine how many additional branch circuits are needed and the size of service-entrance equipment. Allow two 20-A branch circuits for the kitchen receptacles. Also, plan for individual branch circuits for other appliances, like furnace motors, air conditioners, and ranges. The service-entrance equipment should be large enough so that additional circuits may be added later.

FIGURE 11–6. Surface-mounting pull-chain lamp holder.

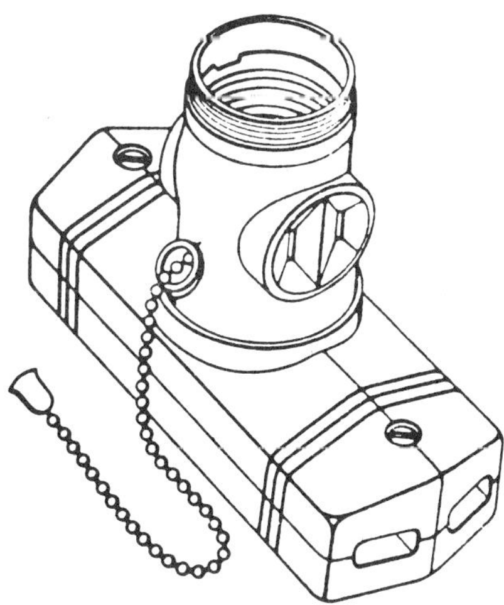

11–10 SERVICE-ENTRANCE CHANGES

Check with the power supplier for the recommended procedure for changing over to a new service entrance. Knowing power company practices will help you to improve the wiring with the least inconvenience to your customer. The general practice is to install the new branch circuits without connecting them. The new service-entrance equipment with a current rating of at least 100 A can be installed near the present service-entrance equipment unless a better location is available. You will have to have the new service entrance inspected before the power company will connect the service drop to it.

One method of connecting the new panelboard to old wiring is to completely remove the old panelboard and replace it with a junction box, as shown in Figure 11–7(a). Because the old wires or cables will usually not be long enough, it is necessary to lengthen them by splicing cables to them. The splices are made in the junction box. All circuit breakers or fuses are in the new panelboard.

Another method is to use the old panelboard as a distribution center. See Figure 11–7(b). A three-wire cable protected by a two-pole circuit breaker connects the two panelboards. If the old panelboard contained main fuses, or service overcurrent devices, you may leave them in position.

FIGURE 11–7. Basic connection methods in rewiring installations: (*a*) junction box replacing old panelboard; (*b*) using old panelboard as a distribution center.

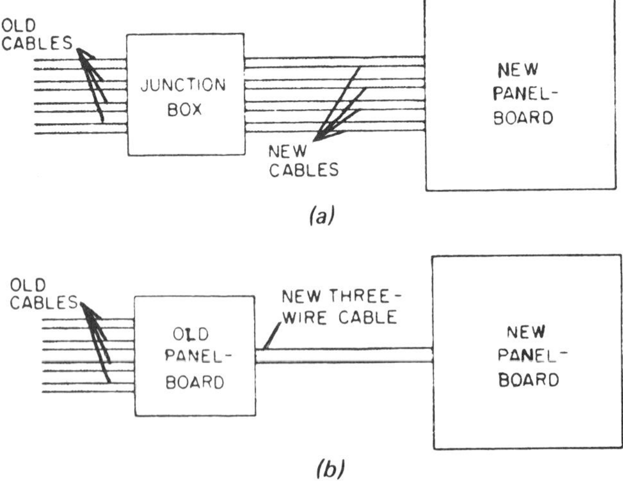

Connect the new three-wire cable below the main fuses but ahead of the branch-circuit fuses.

Make sure that the three-wire cable is heavy enough to carry the load current. If the existing service-entrance equipment is rated at 30 A, a three-wire no. 10 cable protected by a two-pole, 30-A circuit breaker is adequate. If the existing service-entrance equipment is rated at 60 A, a three-wire no. 6 cable protected by a two-pole, 50-A circuit breaker is adequate. An alternative method is to replace the old panelboard with a new distribution center that contains circuit breakers with the proper ratings.

EXERCISES

11–1. In a house plastered with wooden lath, describe how you would go about making a wall opening.

11–2. In cutting through a wall plastered with wooden lath, how is the hacksaw blade held? Why?

11–3. With respect to the laths, how should a box be located in a plastered wall? Why?

11–4. How can a repairable hole be made in wallpaper?

11–5. Describe how you would lift floorboards when the floor is made from tongue-and-groove boards.

11–6. What should you do before attempting to drive finish nails in a hardwood floor?

11–7. What should you do after driving finish nails in a hardwood floor?

11–8. What is the minimum depth of an outlet box for new work? For old work?

11–9. Describe how a device box should be mounted in old work.

11–10. What type of cable is normally used in old work? Why?

11–11. When insulation has been blown into the walls of a newer home, what precautions should be taken? Why?

11–12. Describe a test for determining which wire is neutral.

11–13. How is it possible to use a neon circuit tester when no suitable ground is available?

11–14. Describe tube-and-knob work.

11–15. What is *surface wiring* and what type of components are required for work of this type?

11–16. In modernizing a residential electrical system, where are modifications usually required?

11–17. If a modification of an old service entrance is needed, what is the minimum capacity the NEC permits for the new service entrance?

11–18. Describe two methods for connecting a new panelboard to old wiring.

11–19. In doing old work, what allowance should be made for the kitchen?

11–20. If, in connecting a new panelboard to old wiring, you find that splices are needed, where must they be made?

V

ELECTRIC MOTORS AND FARM INSTALLATIONS

12

Motors

12–1 GENERAL

An electric motor is a device that converts electrical energy into mechanical energy. All electric motors operate on the same basic principle—the interaction of two magnetic fields to produce mechanical rotation.

12–2 BASIC MOTOR ACTION

Figure 12–1 illustrates the action of a current-carrying conductor in a magnetic field. Assuming that the current through the conductor is directed into this page, as you look at it, the magnetic field surrounding the conductor has its lines of force running counterclockwise. The magnetic field provided by the poles of the permanent magnet are, of course, directed from north to south.

Lines of force resemble elastic bands in that they always try to contract to their shortest length. In Figure 12–1, the lines of force of both

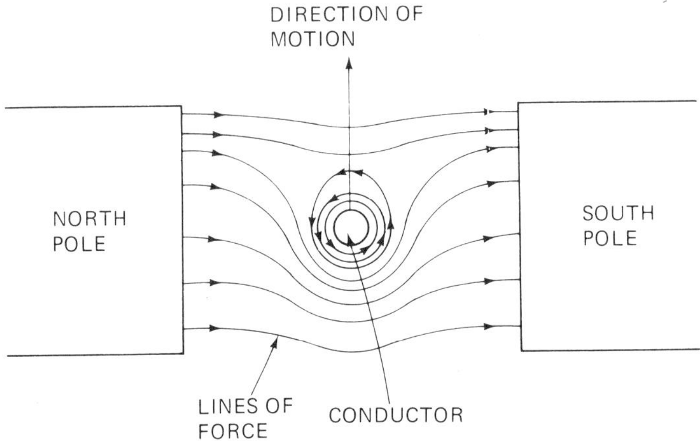

FIGURE 12–1. Action of a current-carrying conductor in a magnetic field.

FIGURE 12–2. A simple dc motor.

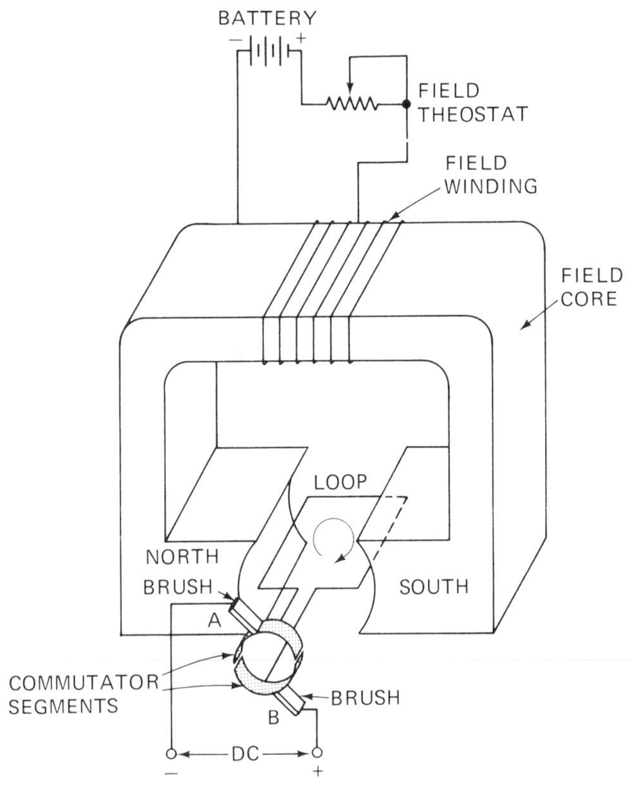

fields are in the same direction *under* the conductor, so they add. Above the conductor, however, the lines of force are in opposition, so the total field in that vicinity is weakened. The crowded lines of force below the conductor try to straighten out and, in so doing, exert an *upward* push on the conductor.

Now let us apply this principle to a dc motor to see how rotation of a motor shaft is produced. A simple dc motor is illustrated in Figure 12–2. Here a single loop is located between the poles of a permanent magnet. The ends of the loop connect to metallic commutator *segments*, which are separated by a small air space. Connection to the external circuit is made through carbon *brushes* that seat on the commutator segments. Current flows from the negative terminal of the dc supply, through commutator segment and brush *A*, the loop, brush *B* and its associated commutator segment, and then back to the positive terminal of the supply.

Refer now to Figure 12–3. The + mark in the center of the conductor represents the tail of an imagined arrow and shows current *flow into* this book page. The dot in the center of the opposite conductor represents the

FIGURE 12–3. Loop rotation as a result of field interaction.

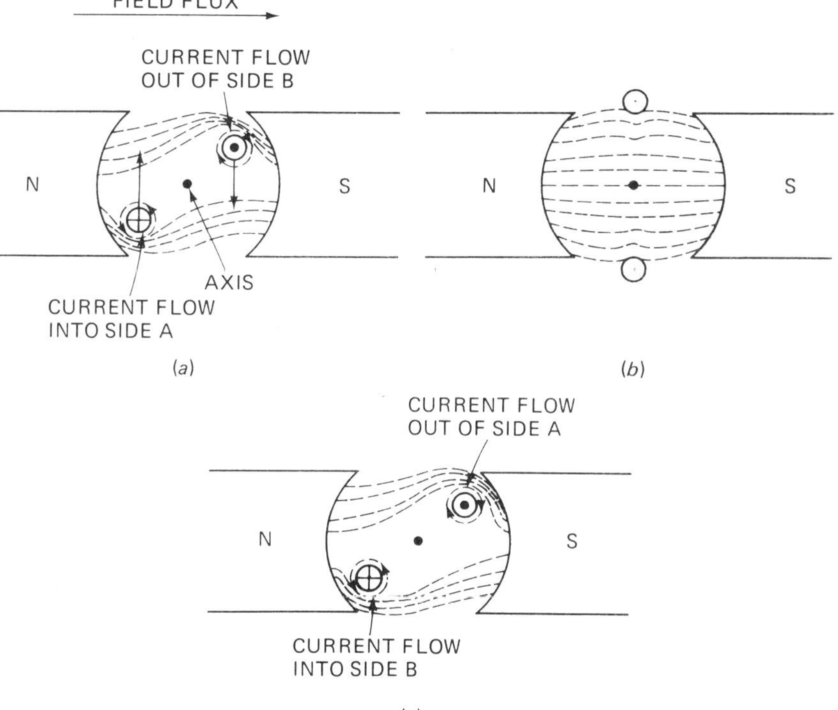

arrow's head and shows current *flow out* of our "demonstration" page. With the sides of the loop positioned as shown in Figure 12–3(a), there is an upward thrust on one side of the loop and a downward thrust on the other side. The resulting twisting force exerted on the loop is called *torque*.

When the sides of the loop are perpendicular to the lines of force created by the permanent magnet, the two segments of the commutator are short-circuited by the brushes. As a result, essentially no current flows through the loop. The point at which this occurs is the *point of commutation*. Seemingly, at this point, our single-loop generator would come to rest. The inertia of the loop is great enough, however, to carry it past the point of commutation, and the situation depicted in Figure 12–3(c) then exists. Although the sides of the loop have reversed position, the direction of current through the loop has not. Thus the loop continues to rotate in a clockwise direction.

In an actual working motor, unlike the theoretical motor we have just examined, the single loop is replaced by a group of *multiple-loop* coils, called *armature coils*, and each armature coil is connected to a set of commutator segments that are electrically insulated from one another by strips of mica. The coils are mounted in slots in an iron core called the *armature*. In place of the permanent magnet, all but very small motors employ *electromagnets* to provide the magnetic field in which the current-carrying armature loops rotate. A metal shaft, which passes through the center of the armature, is supported at both ends of the motor housing in such a manner as to permit easy rotation.

The torque that would be produced by a single-loop motor would, of course, be very small. Moreover, that torque falls to zero twice during each rotation of the loop, that is, when the loop is passing through the point of commutation. Not only does the practical motor produce the torque needed to do a meaningful amount of work, but the armature coils are actually arranged in such a way that torque never drops to zero.

Some large motors use multiple electromagnets to increase the torque. These electromagnets are called *poles*, and there is always an *even* number of poles. The poles are attached to motor housing, and, because they do not rotate, the poles and housing are called the motor *stator*. The armature, accordingly, is called the *rotor*. In all multipole motors, the armature coils are spaced so that opposite sides of a given armature coil pass beneath opposite-polarity poles at all times.

Direction of Rotation

When the direction of the current flow through either the armature or field of a dc motor is reversed, the direction of motor rotation is also reversed. The reversal of armature current changes the direction of the lines of force around the current-carrying conductors. Careful inspection of Figure 12–3 indicates that, when this occurs, the direction of rotation must reverse.

When current through the field windings is reversed, the poles change polarity, and the resulting interaction between the field and armature again causes a reversal in the direction of rotation.

Counter EMF

When a conductor cuts lines of force, a voltage is generated in the conductor. You will recall from Section 1–20 that, because this voltage opposes the voltage applied to the motor, it is called *counter electromotive force*, or CEMF.

The magnitude (strength) of the CEMF depends on the *number* of armature conductors, the *speed* of rotation, and the *strength* of the magnetic field. In any given motor, of course, the number of conductors is fixed, but the speed of the motor and the strength of the magnetic field may be varied at any time. Thus the magnitude of the CEMF is controllable. Either an increase in speed or an increase in the number of magnetic lines provided by the field acts to increase the CEMF.

The amount of CEMF generated by a motor regulates the armature current. Because the resistance of an armature is very low, CEMF is the main factor limiting the flow of armature current in a dc motor. When the armature is turning slowly and generating only a small CEMF, there is a large armature current, which causes the motor to have strong torque. This makes the motor speed up as it overcomes the opposition of the mechanical load. As the motor speeds up, it generates more CEMF. This then decreases the armature current, decreases torque, and decreases the speed. The motor will turn at a speed that will produce just enough torque to balance the mechanical load on the motor.

Neutral Plane

A plane perpendicular to the lines of force of the field is called the *neutral plane*. For proper commutation, each armature coil must be in the neutral plane when it is shorted by the brushes. This seemingly simple condition is not quite so simple to achieve in practice.

Armature Reaction

When current passes through the armature, the interaction between the lines of force of the armature and the field is such as to distort the field. This distortion causes a shifting of the neutral plane; it is called *armature reaction*. Since current through the armature varies with the load, the neutral plane also varies with the load. If each coil of the armature is not in the neutral plane when it is shorted by the brushes, excessive sparking occurs at the brushes; this sparking causes pitting of both the brushes and the commutator segments.

It is not practical, of course, to shift the position of the brushes each time a change in the position of the neutral plane occurs. It is practical, however, to use the small interpoles (also called commutating poles) between the main poles. The windings of the interpoles are in series with the armature winding. With this arrangement, the strength of the interpole field changes with changes in armature current and cancels the distortion caused by armature reaction. Thus the interpoles improve commutation under varying loads.

Although the interpoles correct for field distortion due to armature reaction, the field is weakened as a result of armature reaction. To keep the field strength constant under varying-load conditions, some motors have compensator windings connected in series with the armature and interpole windings; they are placed in slots on the faces of the pole pieces.

Torque and Horsepower

As you know, the twisting force produced by a motor is called torque. The amount of torque produced by a motor is proportional to the armature current and the strength of the field. Increasing either the armature current or field strength increases torque. Maximum torque is usually produced when the motor stalls as a result of excessive load because the rotor is standing still, no counter electromotive force is produced, and armature current is maximum.

While the amount of torque that a motor produces is an indication of how powerful the motor may be, a torque rating does *not* tell us how much *work* a motor can do. To do work, a motor must move a load. A stalled motor may be producing a large torque, but no work gets done unless the motor turns. For this reason, motors are rated according to their capacity to do work—in horsepower.

Horsepower (hp) is the unit rating that expresses the mechanical power of the motor. For example, a motor is rated at 1 hp when it can lift the equivalent of 33,000 lb 1 ft in 1 min. It might also lift 16,500 lb 2 ft in 1 min, or 11,000 lb 3 ft in 1 min, or any other proportionate equivalent.

As indicated on the motor nameplate, a given motor will produce a given horsepower at a *rated* speed.

As noted in Section 1–10, the electrical equivalent of mechanical horsepower is expressed in watts; 746 W is equal to 1 hp. A motor with 746-W electrical power input would not, however, deliver 1 hp to the load at the output shaft. Mechanical and electrical losses in the form of heat and friction use up a portion of the input power. The actual output will be the input power multiplied by the efficiency rating of the motor (a factor that depends on the design of the motor).

Speed Regulation

The ability of a motor to maintain its rated speed under load is called its *speed regulation* and is expressed as a percentage. Mathematically,

$$\% \text{ regulation} = \frac{\text{no load speed} - \text{rated load speed}}{\text{rated load speed}} \times 100 \qquad (12\text{–}1)$$

A small percentage indicates good speed regulation. If, for example, a given motor runs at 1500 rpm with no load and slows to 1450 rpm with its rated load, its speed regulation is

$$\% \text{ regulation} = \frac{1500 - 1450}{1500} \times 100 = 3.33\%$$

12–3 DC MOTOR TYPES

There are three common types of dc motors. Each type has different torque and speed characteristics.

Series Motor

As shown in Figure 12–4, a series motor has its field and armature connected in series. Thus any current through the armature must also flow through the field winding. The field winding consists of a few turns of heavy wire of low resistance.

The series dc motor is noted for its high starting torque, which it develops as a result of the heavy current through both the field and armature windings when the motor is stalled. When the motor starts, current is limited only by the low resistance of the armature and field

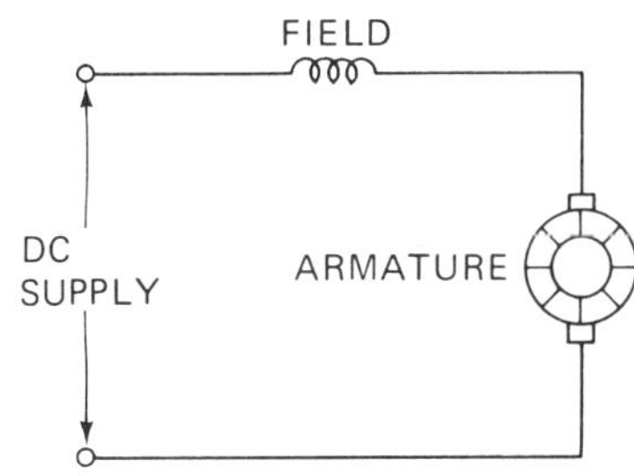

FIGURE 12–4. Schematic diagram of a series motor.

windings. Since torque is proportional to armature current and field strength, a heavy armature current increases both factors that make up torque.

Whenever the load on a series motor is increased, the motor slows down and develops more torque. The reason it slows down so much is that the increased load increases the armature current, which also flows through the field winding and strengthens the field. With a strong field, the armature develops more CEMF. The increased CEMF then limits the armature current. If regulated to turn more slowly, the armature can draw enough current to develop the torque needed to balance the mechanical load.

The series motor has very poor speed regulation and will run away if it is operated without a load. Under no-load conditions, the motor develops a high starting torque and picks up speed rapidly. As the armature rotates, CEMF limits armature current, but this also weakens the field strength. Now the armature must turn even faster to generate the same CEMF, but each increase in speed further weakens the field. The process is cumulative and will lead to destruction of the motor. For this reason, series dc motors are coupled directly to their loads.

Series motors are used in heavy industrial applications; they are not encountered in residential wiring systems.

Shunt Motors

As shown in Figure 12–5(a), the field winding of a *shunt* motor is connected in parallel with the armature. Thus the full supply voltage is applied to

FIGURE 12–5. Shunt motor: (*a*) schematic diagram; (*b*) speed–torque characteristics.

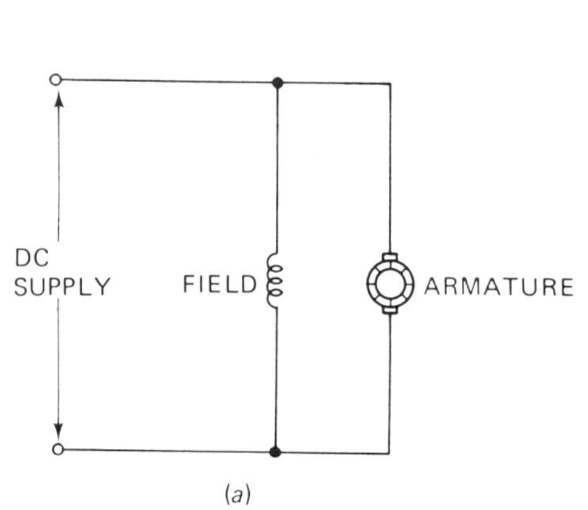

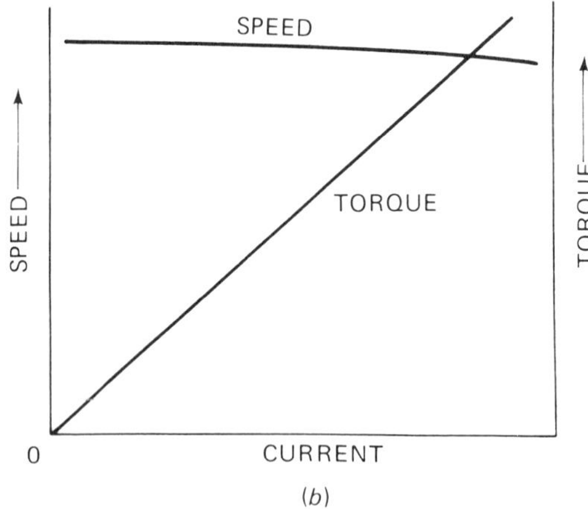

both windings. The field consists of many turns of wire because the supply voltage is applied to it. The resistance of the armature is low, and current through it is limited mainly by CEMF.

The shunt motor has nearly constant speed from no load to full load, and is often called a *constant-speed* motor. Under no-load conditions the motor speeds up enough to produce sufficient CEMF to limit armature current to a safe value. Because the field coil and armature are in parallel, the field strength is almost independent of armature current. At a particular no-load speed, the CEMF almost equals the applied voltage, and there is no further increase in speed. As load is applied, the motor slows down slightly and generates less CEMF. This allows more armature current to flow, and the motor produces more torque to balance the increased load. Since a small change in CEMF produces a large change in armature current, the motor does not slow down appreciably.

Another factor that helps the shunt motor maintain its speed under load is the weakening of the field as a result of armature reaction. As armature current increases, armature reaction increases, and the field becomes weaker. A weaker field causes less CEMF for a given speed, and the motor does not have to slow down very much to produce the necessary torque.

A rheostat connected in series with the field coil can control the speed of a shunt motor. Any increase in resistance causes a decrease in current through the field winding and makes the field weaker. With a weaker field the armature generates less CEMF. This allows more current to flow, which, in turn, tends to produce more torque. The increased torque causes the armature to speed up until the torque produced by the motor balances the mechanical load.

When the resistance is decreased, more field current flows and the field becomes stronger. Now the CEMF is increased and less armature current flows. With less armature current, the motor does not produce enough torque to balance the load, and the motor slows down. Now the CEMF decreases, allowing enough current to flow to produce the necessary torque to balance the load at reduced speed. We can sum up these two conditions by saying that decreasing the field strength causes the motor to speed up, and increasing the field strength causes the motor to slow down.

The field winding of a shunt motor should never be opened while the motor is running, or the motor may run away. As you know, the motor speeds up as the field is made weaker. If the field winding is opened, there may be enough residual magnetism in the pole pieces to give the motor a very weak field. The motor will accelerate and may reach a speed that will cause its destruction.

The shunt motor has good starting torque, though not as good as a series motor. When the motor is started, armature current is limited only by the resistance of the armature. Thus a large armature current flows and produces a large torque. Because the field strength is almost constant, the torque produced is almost directly proportional to armature current.

The speed and torque characteristics of a shunt motor are shown in the graph in Figure 12–5(b). The speed curve shows that the speed remains almost constant over a wide range of current. As current increases with an increasing load, the speed falls off slightly. Notice that the torque curve rises steadily as the current increases.

The shunt motor is used where constant speed under varying load is important. Also, since it is easy to adjust the speed by changing the field strength, the shunt motor is used where the motor speed must be adjusted accurately.

Compound Motors

The compound motor has both a series-field coil and a shunt-field coil. These two fields give compound motors characteristics that are a combination of the series- and shunt-type motors. The schematic diagram of a compound motor is shown in Figure 12–6. Notice that a rheostat is used to adjust the strength of the shunt field. Because the series field carries the total current of both the armature and shunt field, it is made up of only a few turns of heavy wire. If the series field is connected so that its magnetic field aids that of the shunt winding, the motor is called a *cumulative* compound motor. If the series field is connected so that it opposes the shunt field, the motor is called a *differential* compound motor.

The cumulative compound motor resembles the series motor because it has a high starting torque, and it also resembles a shunt motor because it has a maximum speed. When it is loaded, the series field increases the

FIGURE 12–6. Schematic diagram of a compound motor.

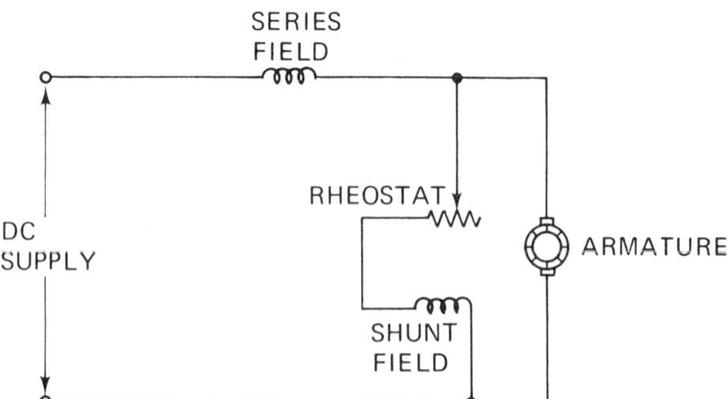

total field strength and causes the torque to increase. At the same time, the stronger field causes the motor to slow down. Under no-load conditions, the presence of the shunt field ensures that the motor will have a maximum speed that it will not exceed, just as in the shunt motor. The cumulative compound motor is used where high starting torque is needed, but the runaway characteristic of a series motor cannot be tolerated.

The differential compound motor tends to speed up when a load is applied to it. The load causes increased current through the series field coil. Since the series field opposes the shunt field, the total field is weakened, and the motor tends to speed up. The differential compound motor is seldom used because its speed is unstable. When starting under load, it is possible for the series field to overcome the shunt field, and the motor can run away, or even reverse direction. For this reason, the series field coil is usually shorted out while the motor is being started.

Of the various types of motors described, the only one that you are likely to encounter in residential or farm wiring is the shunt motor.

12–4 MOTOR STARTERS

Small motors (1 hp or less) are usually started by placing the motor directly across the power-supply line. The motor draws several times the normal running current, but the resistance of the winding is sufficient to limit the current to a safe value until the motor gets up to speed. When the motor is running, CEMF opposes the applied voltage and limits the armature current to its normal value.

Motors larger than 1 hp generally employ a starting device that limits starting current to a safe value. Most starting devices are simply variable resistances that can be inserted in series with the motor winding. The resistors limit the initial current to a safe value and are cut out of the motor circuit, one by one, until the motor is finally connected directly across the line. Starters may be designed for either manual or automatic operation.

A *starting box* intended for manual operation with a shunt-type dc motor is shown in Figure 12–7. The diagram is drawn so as to show the mechanics of operation as well as the circuitry. One side of the supply is connected directly to the armature and the field coil through switch S. The other side of the supply is connected to the other side of the armature and the field coil through switch S and the starting box. The armature is connected to point 5 on the starting resistance R, and the field coil is connected through a *holding coil* to point 1 on resistance R. A *wiper arm, W*, is connected to switch S.

When the wiper arm is in the position shown in Figure 12–7, no

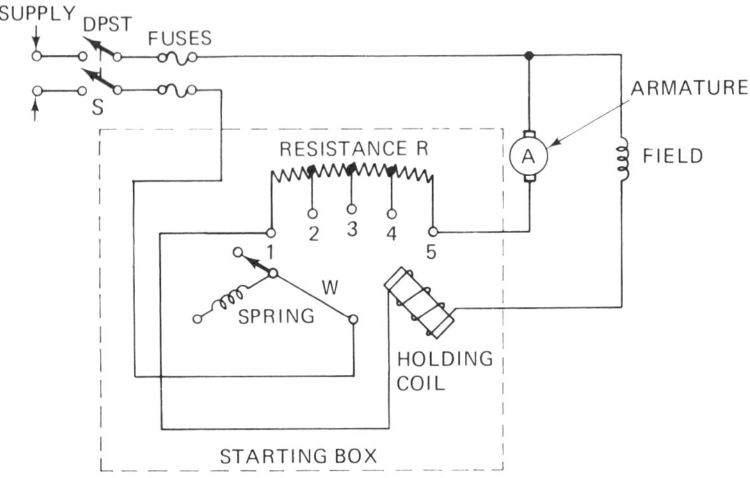

FIGURE 12–7. Schematic diagram of a starting box intended for manual operation with a shunt-type dc motor.

current can flow either to the field or to the armature even with switch *S* closed. Thus closing switch *S* has no immediate effect on motor operation. If we close switch *S* and then move the wiper arm to position 1, however, current will flow to the motor through two paths. Current will flow from position 1, through the holding coil to the field winding, then to the other side of the line. This places full line voltage on the field winding, allowing maximum field strength for starting.

The other current path is from position 1 through resistance *R* and the motor armature to the other side of the line. Thus the resistance limits the armature current but has no effect on the field current. The armature current produces a torque, and the motor starts to turn. As it turns, the CEMF builds up and reduces the armature current until the motor arrives at the speed determined by resistance *R* and the CEMF. At this point the motor speed cannot increase unless the armature current is increased.

Moving wiper *W* to position 2 removes the resistance between positions 1 and 2 from the armature circuit and allows more current to reach the armature. This increased armature current allows the motor to turn faster, but at the same time builds up a stronger CEMF, which eventually limits the speed once again. By moving the wiper arm to position 3, we again get an increase in armature current, speed, and CEMF. Thus, as the wiper arm is moved from position 1 through position 5, both speed and CEMF are built up gradually until the normal operating speed is obtained.

The progressive movement of the wiper arm places the resistance of the starting box in series with the field coil, and a decrease in field current occurs. Because the resistance of the starting box is very small compared to the resistance of the field winding, the decrease in field current is slight.

At position 5, none of the resistance is in the armature circuit, and the motor is running at full speed with a normal CEMF. At this point, the holding coil, which is in series with the field winding, acts to hold the wiper arm in position 5, hence its name. The holding coil may work in one of two ways: it may either hold the wiper arm directly by electromagnetic attraction between the core of the coil and the wiper arm itself, or the starting box may have a mechanical latch that actually holds the wiper. In the latter case, the latch is held in position by the magnetism of the coil.

In either case, if the coil is suddenly deenergized, the wiper arm is released. When it is released, the spring tension on the arm returns it to the off position, removing all current from the motor. This gives us an automatic undervoltage protection, and also ensures that, if the field circuit opens, the armature circuit will also open so that the motor cannot run away. To stop the motor, we merely open switch S, which removes all current from the motor circuit. The wiper arm then moves to its off position, resetting the starting box.

In operating a starter of the type described, the wiper arm must not be moved either too quickly or too slowly. On the one hand, if the resistance is bypassed too quickly, the CEMF will not build up fast enough to limit current to a safe value. On the other hand, if we move the arm too slowly, a heavy starting current may flow long enough to damage the starter resistance. Consequently, the operator of a motor starting box must judge the movement of the wiper arm carefully.

The sound of a motor is usually the key to the proper operation of the starting box. As a motor starts and comes up to speed, it produces a low-pitched rumble, which gradually increases to a high-pitched, whining noise. At full speed, the noise usually disappears, because the pitch rises until it is above the range of the human ear. When a starting box is used, the noise will increase in pitch until the speed limit for a particular starting position is reached. Then it becomes almost constant as long as the motor remains at that speed. Thus the operator holds the arm at each position until the pitch of the motor noise is no longer increasing; then he moves the arm to the next position.

All resistance starters operate on the same basic principle. They may vary greatly in mechanical design and physical appearance, but electrically they are the same. They simply provide a method of inserting, and then gradually decreasing, a resistance in the armature circuit of the motor.

12–5 AC MOTORS: GENERAL

Due to the widespread distribution of ac power, most motors that you are likely to encounter will be of the ac type. Single-phase motors are generally used for applications requiring 1 hp or less. Many types of single-phase motors are made with different characteristics. Some have high starting torque, some are made for intermittent operation only, and still others are made with one major thought in mind, economy.

Above 1 hp, *three-phase* ac motors are much used. The most common type of motor in this general class is the *induction* motor.

12–6 THREE-PHASE INDUCTION MOTORS

Induction motors are so called because current in the rotor is induced from the stator. The rotor has neither slip rings nor a commutator, so it has no direct electrical connection to the supply. The stator windings are connected to the supply line and act like the primary of a transformer. The rotor windings act like the secondary windings of a transformer. When current flows in the stator windings, a voltage is induced in the rotor windings, causing current to flow there, too.

In a typical three-phase induction motor, the frame supports the stator core. The core is made of laminated steel to cut down on eddy current and hysteresis losses. The stator windings are embedded in slots in the core.

The rotor core is also made of laminated steel. Heavy copper bars are embedded in slots on the core, and the bar ends are welded to copper shorting rings. Because of its appearance, a rotor of this type is usually called a *squirrel-cage* rotor. The bars and shorting rings form short-circuited windings. When a voltage is induced in these windings, a large current flows through them.

Operation

The three-phase voltage applied to the stator of an induction motor sets up a rotating magnetic field. To see how this rotating magnetic field is produced, let us look at the connections to the stator.

Figure 12–8 shows a delta-connected, two-pole, three-phase stator connected to a three-phase ac supply. (Two-pole means two poles for each phase.) The windings on poles 1 and 4 are supplied by phase A, the windings on poles 3 and 6 are supplied by phase B, and the windings on poles 2 and 5 are supplied by phase C. They are wound so that, when one pole of the pair is north, the other pole of the pair is south.

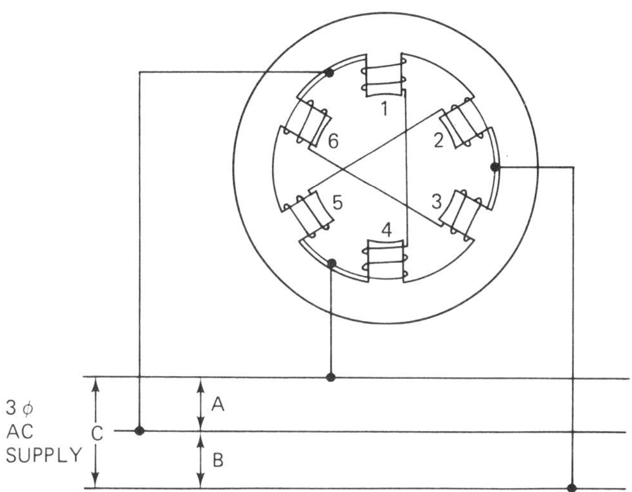

FIGURE 12–8. Delta-connected, two-pole, three-phase stator connected to a three-phase ac motor.

Recall that in a three-phase voltage each phase is displaced 120° from the other two phases. As these voltages rise and fall, they cause currents to flow in the stator windings. To see why the magnetic field will rotate, let us examine the magnetic fields at different times in one cycle.

Figure 12–9(a) shows the current waveform I_A through windings 1 and 4 (phase A of Figure 12–8) for one cycle, and the magnitude and polarity of the field this current sets up between poles 1 and 4 at different times in the cycle.

Figure 12–9(b) shows the same thing for current waveform I_B, which lags waveform I_A by 120°, and Figure 12–9(c) shows the same thing for current I_C, which lags current I_B by 120°.

Figure 12–9(d) shows the combined field strength and direction for the three phases. The heavy black arrows represent the resultant magnetic field. As shown, the field rotates clockwise, and its strength is unchanged as it rotates.

Initially, current waveform I_A in Figure 12–9(a) is maximum positive, and the field between poles 1 and 4 is maximum, with 1 as the south pole and 4 as the north pole. At the same time, current I_B is a small negative value, setting up a small magnetic field with 6 as the south pole and 3 as the north pole. Current I_C is also a small negative value, setting up a small magnetic field with 2 as the south pole, and 5 as the north pole.

These three magnetic fields combine, as shown in Figure 12–9(d), to form a single strong magnetic field with the south pole at 1 and the north pole at 4.

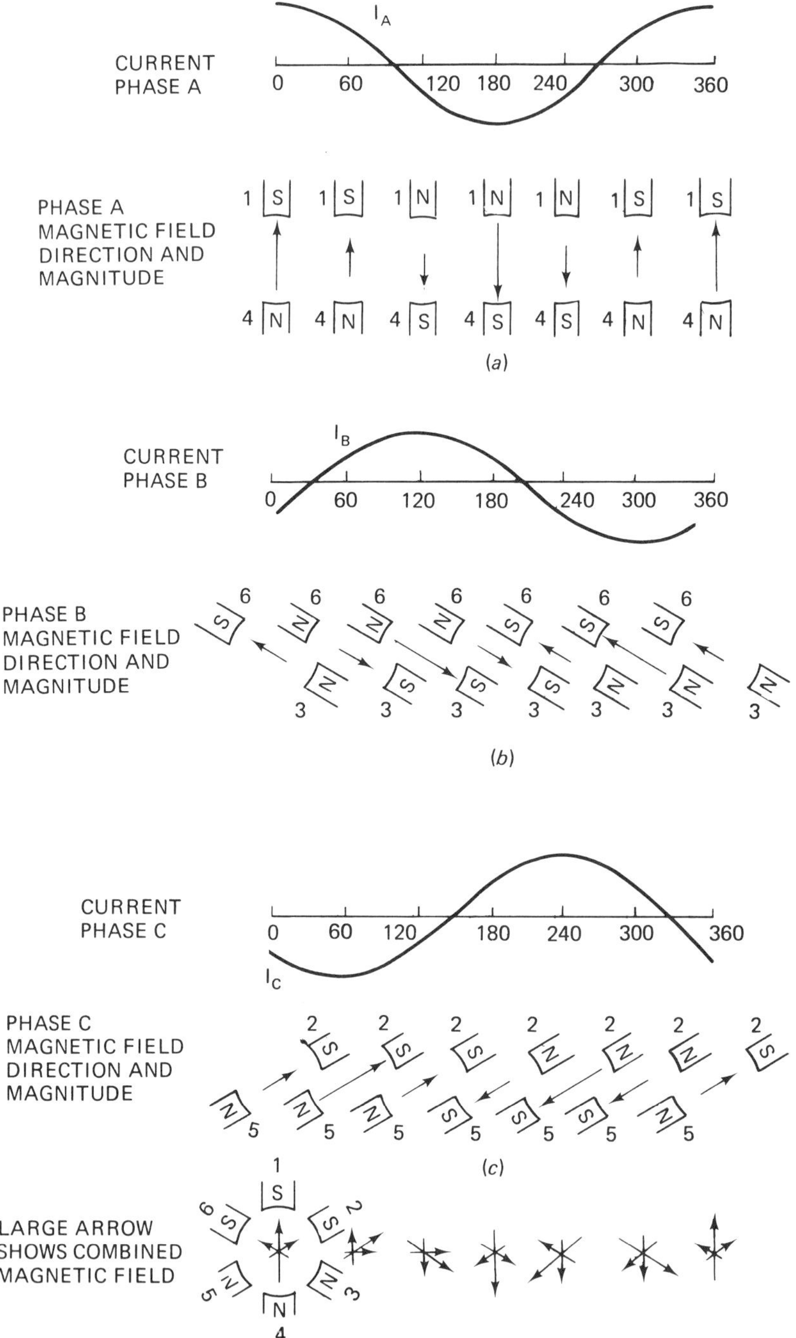

FIGURE 12–9. How a rotating magnetic field is produced by the stationary armature of a three-phase ac motor.

By examining the currents and fields through one complete cycle, you can see that the strength of the resultant magnetic field remains constant, and that the field rotates one complete revolution for each cycle of the applied voltage.

Now let us see how this rotating magnetic field can be used to cause rotation of a squirrel-cage rotor that has many shorted conductors. When the rotor is placed in the rotating field established by the stator, the lines of force cut the shorted conductors. This induces a voltage in the conductors and causes a large current to flow. The resulting magnetic field set up around the rotor is opposite to that surrounding the stator. Thus the south pole is induced in the rotor opposite a north pole in the stator. As the stator field rotates, it attracts the rotor field, causing the rotor to turn in the same direction as the rotating stator field.

Because the rotor turns in the same direction as the rotating magnetic field of the stator, the direction of rotation can be changed by reversing the direction of field rotation, which depends upon the phase relationships of the currents in the stator windings. Interchanging any *two* of the supply leads will change the phase relationship of two of the phases and cause the field to rotate in the opposite direction.

The rotor does *not* turn as fast as the rotating field of the stator. Let us see why. You know that the torque produced by a motor is proportional to the strength of the two fields that are interacting to cause rotation.

When the rotor is not turning, the stator field is cutting the rotor conductors at a maximum rate and, therefore, inducing the maximum voltage in them. When the rotor follows the rotating field and turns, less voltage is induced in the rotor conductors because the field cuts the conductors at a slower rate. Remember, the rotor and stator are turning in the same direction. If the rotor speed were to become equal to that of the stator field, no voltage would be induced in the rotor. With no voltage induced in the rotor, there would be no current flow, no field around the rotor, and no turning force, so the rotor would slow down. As soon as the rotor slowed down, of course, the lines of force of the stator field would once again cut the rotor conductors and create a new rotor field, which would react with the stator field to produce rotation.

Synchronous Speed

The speed at which the rotating stator field turns is called the *synchronous* speed of the motor. With a 60-Hz supply voltage, the synchronous speed is 3600 rpm. The synchronous speed can be decreased by adding poles. Looking back at Figure 12–8, suppose that the six poles are moved around so that they occupy only half the circumference of the stator. One cycle of supply voltage will still cause the magnetic field to rotate past the six poles, but now this will be only a half-turn. To get even torque we must, of course, fill in the other half of the stator with poles. This gives us two pairs of poles per phase, and the field will rotate at half-speed. Using three pairs of poles per phase will cut the speed to one third.

The synchronous speed for any ac induction motor is expressed by the formula

$$N = \frac{f \times 120}{P} \qquad (12\text{–}2)$$

where N is the synchronous speed in rpm, f is the frequency of the applied voltage in hertz, and P is the number of poles per phase. In an ac motor with four poles per phase and a 60-Hz supply, for example, the synchronous speed is

$$N = \frac{60 \times 120}{4} = 1800 \text{ rpm}$$

Slip and Torque

As noted earlier, the rotor of an induction motor turns more slowly than the rotating field. The difference between the synchronous speed and the speed of the rotor is called *slip*. With more slip, more current flows in the rotor, and the rotor develops more torque. Also, as the slip increases, the speed of the motor decreases.

Thus motor speed is equal to synchronous speed minus slip. For example, a motor with a full-load speed of 1750 rpm and a synchronous speed of 1800 rpm has a slip of 50 rpm. Slip is expressed as a percentage of the synchronous speed. In this case,

$$\% \text{ slip} = \frac{1800 - 1750}{1800} \times 100$$

$$= 2.77\%$$

Increasing the load on the motor will increase both slip and torque. Induction motors have very good torque at near-synchronous speeds. Figure 12–10 shows the speed–torque curve for a typical three-phase induction motor. The percent of full-load torque is plotted against the percent of synchronous speed.

At no-load, the percent of full-load torque is nearly zero and the motor runs at nearly 100 percent of its synchronous speed. As the load is increased, the percent of full-load torque increases, and the speed falls off slightly. At full load (100 percent full-load torque) the slip is about 4

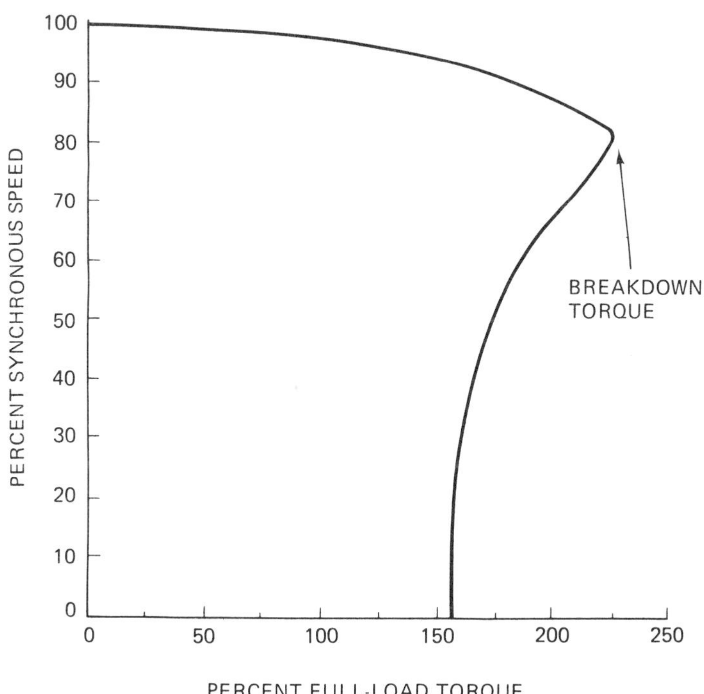

FIGURE 12–10. Speed–torque curve for a three phase squirrel-cage induction motor.

percent, so the motor runs at 96 percent of its synchronous speed. With the load increased so that the torque is 200 percent of full load torque, the speed is still nearly 90 percent of the synchronous speed.

As the load is increased still further, the motor slows down and torque decreases. The point at which torque decreases as speed drops is called the *breakdown torque*. Stalling occurs when the speed drops to zero and torque is about 160 percent of the full-load torque. This is the starting torque of the rotor, also called *locked-rotor torque*.

It would appear that the motor should have its greatest torque when the rotor is not turning, because the motor is then drawing maximum current. With the rotor stopped, however, there is a physical displacement of the rotor field in relation to the stator field. With a locked rotor, this displacement of the rotor field is enough so that part of the rotor field is developing a torque in the opposite direction. This subtracts from the total torque.

To see why this loss of torque takes place, we must examine the *inductive reactance* of the rotor at different speeds. By definition, inductive reactance is the opposition of a coil to an alternating current. Mathematically,

$$X_L = 2\pi fL \tag{12–3}$$

where X_L is the inductive reactance in ohms, π is a constant having the approximate value of 3.14, f is the frequency of the alternating current in hertz, and L is the inductance of the coil in henrys.

Recall that the field set up in the rotor is *in phase* with the rotor current. The rotor current *lags* the induced voltage by an amount dependent upon the inductive reactance of the rotor. As indicated by Equation 12–3, inductive reactance varies with frequency; the higher the frequency, the greater the reactance.

When the rotor is stopped, the voltage induced in the rotor is the same frequency as the line voltage. At the line frequency (60 Hz) the inductive reactance of the rotor is high enough to cause rotor current to lag the induced voltage by a large angle. The field set up around the rotor is in phase with this current. Thus the rotor field is physically lagging the stator field far enough so that a part of the rotor conductors is displaced into a field of the opposite polarity. These conductors produce a torque in the opposite direction and subtract from the total torque.

As the rotor speeds up, the frequency of the induced voltage, which is equal to the line frequency minus the rpm of the rotor, decreases.

If, for example, the synchronous speed of a motor with a 60-Hz supply is 3600 rpm (60 rps), and the slip is 4 percent, the rotor will be turning at 96 percent of the synchronous speed: $0.96 \times 60 = 57.6$ rps. The frequency of the induced voltage under these conditions is $60 - 57.6 = 2.4$ Hz. The inductive reactance of the rotor winding at this frequency is negligible. The rotor field lags the stator field by a small angle, and essentially no opposing torque is developed. The total torque of the motor is, therefore, increased.

12–7 TWO-PHASE INDUCTION MOTORS

Although two-phase motors find very limited application, we shall take a brief look at this type of induction motor for the sake of completeness.

The schematic diagram of a two-phase induction motor is shown in Figure 12–11. The separate windings for each phase are displaced from one another by 90°. The rotor is the squirrel-cage type.

As shown in Figure 12–11(b), poles 1 and 3 are energized by phase *A*, and poles 2 and 4 by phase *B*. The voltages for the two phases are 90° out of phase.

FIGURE 12–11. Schematic diagram of a two-phase motor.

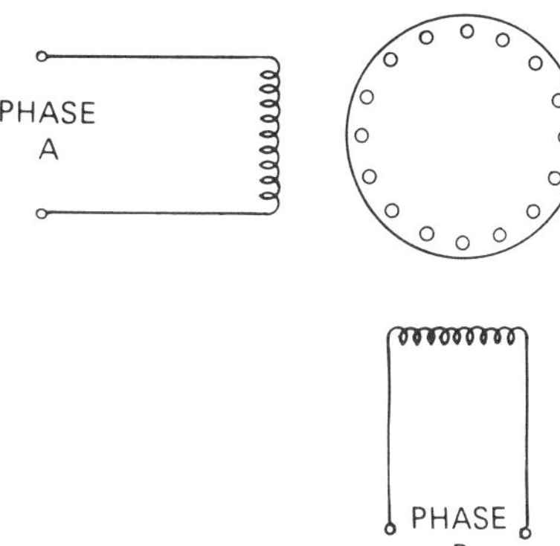

The magnetic fields formed by the supply voltage combine to form a single resultant field. Just as in the three-phase motor, the resultant magnetic field has a constant strength and rotates one complete revolution for each cycle of the supply voltage. Thus the synchronous speed for a two-pole stator is 3600 rmp, per Equation 12–2. With more poles per phase, the synchronous speed decreases.

The direction of rotation depends on the phase relation of the supply voltages. If phase A leads phase B by 90°, rotation is in a clockwise direction. If either phase A or phase B is shifted 180°, phase B will lead phase A by 90°, and rotation will be counterclockwise.

The explanation on slip and torque given for the three-phase motor applies also to the two-phase motor.

12–8 OTHER POLYPHASE MOTORS

The two-and three-phase induction motors are the most common polyphase types. Other types are made, however, for special purposes. Some large three-phase motors have a wound rotor with slip rings and brushed connected to the external circuit. The wound-rotor construction makes it possible to control the speed and torque of the motor by use of an external resistance.

Another special-purpose three-phase motor is the *synchronous* motor. In this motor, the rotor windings are supplied with direct current. The dc-produced field surrounds the rotor and follows the rotating magnetic field around the stator. The rotor locks in step with the rotating field, and since there is no slip, the rotor turns at a constant synchronous speed. The rotors often have shorted squirrel-cage windings that allow the motor to operate as an induction motor until it comes up to speed.

12–9 SINGLE-PHASE MOTORS

By itself, a single-phase supply will not produce a rotating field in a motor. Instead, the flux field pulsates; that is, it builds up in one direction, falls to zero, and then builds up in the opposite direction. Under these conditions, the inertia of the rotor prevents rotation. If the rotor is mechanically turned until it is brought up to speed, however, it will then continue to rotate at near synchronous speed. The rotor will rotate in either direction, depending on how it is started.

To use single-phase alternating current for an induction motor, therefore, some means must be used to start the motor. This is accomplished by *phase splitting,* which gives the *effect* of two phases and makes it possible to have a rotating field. Let us examine several methods of phase splitting.

Resistance-Start Motor

One means of splitting the phase of the supply voltage is to use a separate additional field winding during starting. Figure 12–12 shows a schematic diagram of the motor. There are two field windings and a squirrel-cage rotor. Notice that the main-field winding and the starting winding are spaced 90° apart like the windings in a two-phase motor.

The main winding has very low resistance and high inductance; as a result, the current lags the line voltage by nearly 90°. The starting winding, on the other hand, has high resistance and low inductance so that its current is nearly in phase with the line voltage. Thus there is a phase difference between the currents in the two windings. The difference is not a full 90° as in a two-phase motor, but the effect is the same, and a rotating field is generated.

FIGURE 12–12. Split-phase motor.

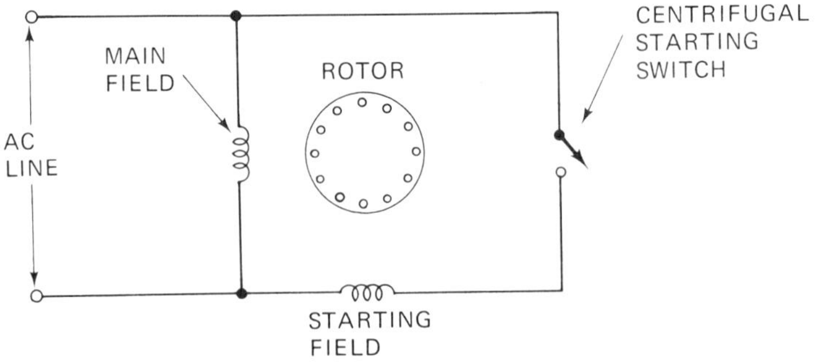

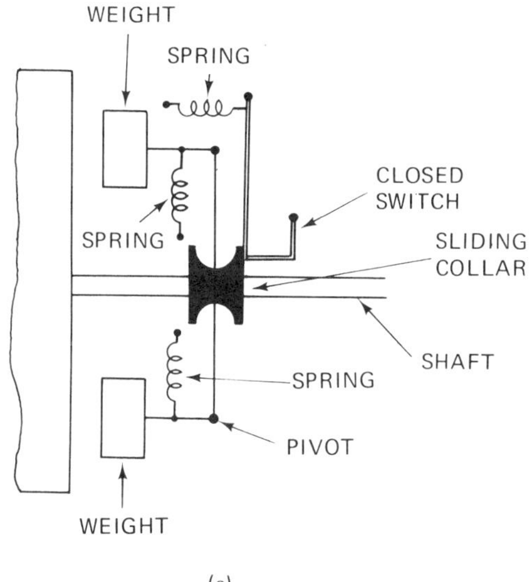

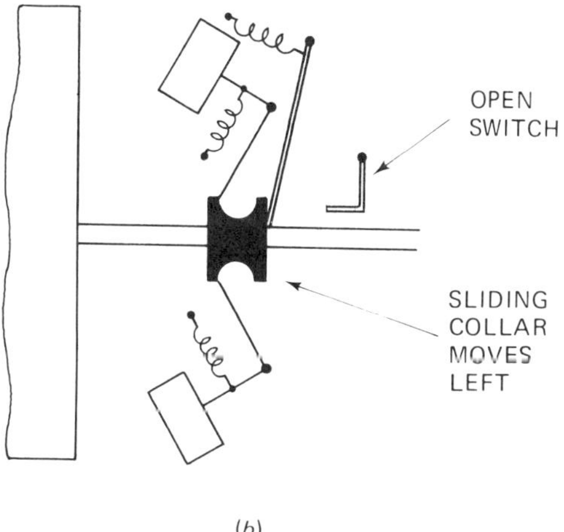

FIGURE 12–13. Centrifugal starting switch: (*a*) armature just starting to turn; (*b*) armature rotating near full speed.

As the motor builds up speed, a centrifugally operated switch opens the starting winding, and the motor continues to run as a single-phase induction motor. Figure 12–13 shows one type of centrifugally operated switch. In Figure 12–13(a), the motor is stopped and the switch is closed. The switch is held closed by a pair of movable weights connected by pivoted levers to a movable collar on the shaft. As the shaft turns, centrifugal force causes the weights to move outward against spring

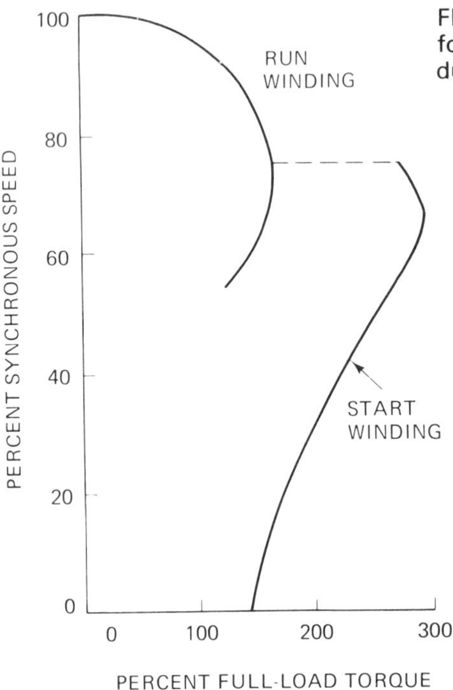

FIGURE 12–14. Speed–torque curve for a split-phase, resistance-start induction motor.

tension. The pivoted levers push the collar to the left on the shaft, and, as shown in Figure 12–13(b), the switch opens and so disconnects the starting winding.

Figure 12–14 shows the effect on motor torque and speed when the switch opens. As the motor starts, the torque is about 140 percent of full-load torque. As the motor speed increases, the torque builds up. At the dashed horizontal line in Figure 12–14, the switch opens and the motor operates as a single-phase induction motor. Notice the drop in torque.

The slip for a given load is greater than for polyphase motors. With no load, the speed approaches synchronous speed.

Capacitor-Start Motor

In the resistance-start motor, the phase difference between the currents in the two windings is not a full 90°. We can make the phase angle nearer 90° by using a capacitor to shift the phase in the starting winding.

The schematic diagram of a capacitor-start motor is shown in Figure 12–15(a). This motor also has two windings spaced 90° apart and uses a squirrel-cage rotor. The centrifugal switch connects the starting winding to the line through a capacitor during starting. Current in the main winding lags the applied voltage. The inductive reactance of the starting winding is canceled by the opposite reactance of the capacitor to keep the current through the starting winding in phase with the applied voltage. Thus the current in the main winding lags the current in the starting winding by 90°, and the motor acts like a two-phase motor.

The speed–torque curve in Figure 12–15(b) shows much higher starting torque than for the resistance-start motor. The dashed horizontal line shows that the torque changes when the starting winding is cut out. Now the motor comes up to speed as a single-phase induction motor. A

FIGURE 12–15. Capacitor-start motor: (*a*) schematic diagram; (*b*) speed–torque curve.

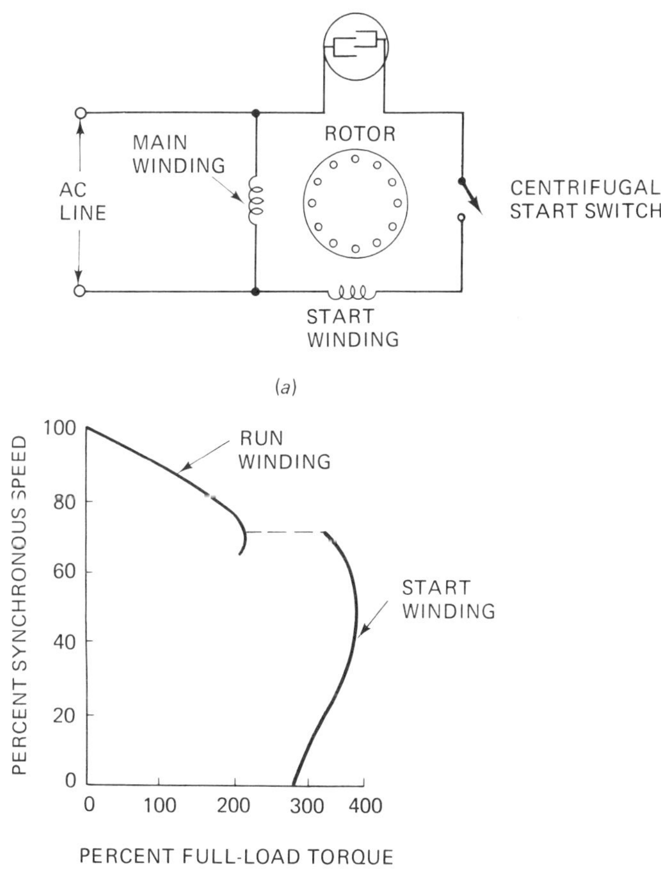

large capacitor and a low-resistance starting winding are used so that the starting winding draws heavy current to obtain a high starting torque. The motor would overheat if the large capacitor were left in the circuit at all times. If a smaller capacitor and a higher-resistance winding are used, however, the centrifugal switch can be omitted. The capacitor is then left in the circuit at all times, and overheating does not occur. Under these conditions, the motor will not produce as much torque, but it will still be self-starting.

The capacitor-start motor is probably the most frequently employed of the single-phase motors. They are made with many variations, but all employ the basic principle described here.

Split-Phase, Reversible Motor

Like the capacitor-start motor, the split-phase, reversible motor uses a capacitor to shift the phase. Refer to the schematic diagram of Figure 12–16. Two leads are brought out to a reversing switch. With the switch up, field *A* is the main winding, and field *B* is supplied through the capacitor. Then the current in field *A* lags that in field *B* by a large angle. With the reversing switch moved down, field *B* becomes the main winding, and field *A* is supplied through the capacitor. Then field *B* lags field *A*, and the motor turns counterclockwise.

Shaded-Pole Motor

Another type of small ac induction motor that is widely used is the shaded-pole motor. See the schematic diagram of Figure 12–17(a). The motor has a squirrel-cage rotor and a single-field winding. The two-pole pieces are split, and a copper ring is fastened around one part of each pole piece.

FIGURE 12–16. Split-phase reversible induction motor.

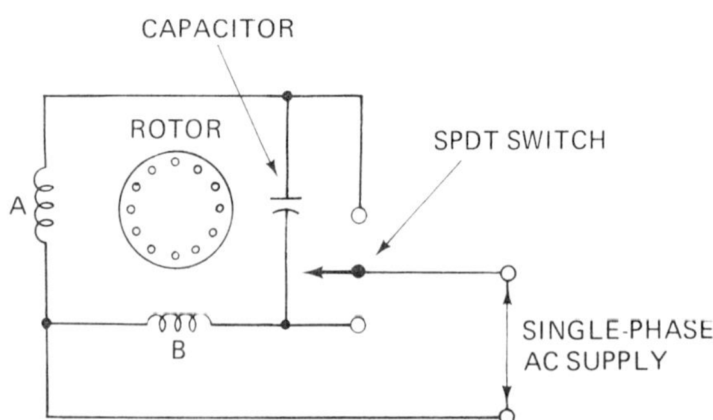

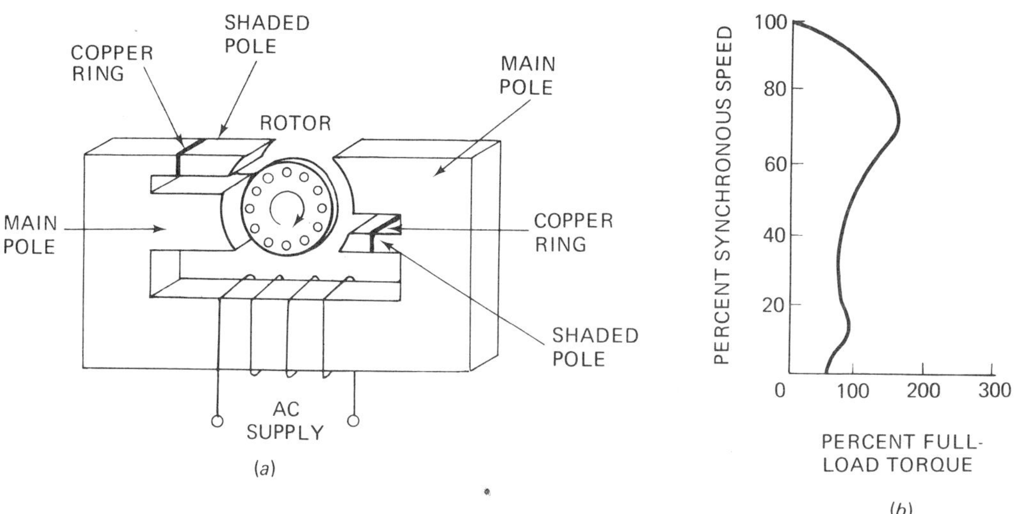

FIGURE 12–17. Shaded-pole motor: (*a*) schematic diagram; (*b*) speed–torque curve.

This part is called the *shaded pole*. The copper ring acts like the secondary winding of a transformer. As the current in the field winding builds up on one half-cycle, a flux builds up in the poles. This changing flux cuts the copper ring and induces a current in it. The current in the copper ring then sets up a flux that opposes the flux producing it. The main pole has a strong field, while that of the shaded pole is relatively weak.

When current in the field winding decreases, the main-pole field also decreases, but current continues to flow in the copper ring. The shaded pole now has a strong field, and the main-pole field is decreasing. When the main-pole field is zero, the shaded-pole field is maximum. Now the main-pole field reverses direction as current through the field winding reverses. The shaded-pole field changes later than the main-pole field.

In effect, the field in the shaded-pole section lags behind that of the main pole by nearly 90°. This gives the effect of a rotating field and produces rotation of the rotor. The rotor turns toward the shaded-pole section. In Figure 12–17 the direction of rotation is clockwise.

As shown in the speed–torque curve of Figure 12–17, the shaded-pole motor develops very low starting torque, has high losses, and cannot be reversed. Its top speed with no load is less than any of the other single-phase motors. This indicates high losses. Low initial cost makes the shaded-pole motor popular, however, in applications where no-load starting is permissible. Fan motors are one very popular application.

12–10 SINGLE-PHASE MOTORS WITH BRUSHES

General

Several types of ac motors use commutators and brushes like those found on dc motors. These ac motors all have higher torque characteristics than single-phase induction motors. They are made for many special purposes where small size and high starting torque are needed and the only power supply available is alternating current.

Universal Series Motor

This motor has a commutator and brushes and operates from either an ac or dc supply. A schematic diagram of the motor is shown in Figure 12–18. Note that it is exactly like the series dc motor. Its principle of operation on alternating current is the same as on direct current. Recall that reversing both the field and armature currents reverses the direction of rotation of a dc motor. With an ac supply, the current through both the field and armature reverses 120 times a second. While the torque drops to zero 120 times a second, the torque is always in the same direction.

The big advantage of the universal motor is *not* that it can be operated on either alternating or direct current. The universal series motor has about the highest starting torque for its size of any ac motor. Its speed–torque characteristics are similar to those of the series dc motor. This makes it extremely useful for applications in saws, portable drills, and similar devices.

With a large bit on a power drill offering a heavy load, for example, the motor will slow down and produce large torque. The same drill using a small bit with a light load will speed up. Thus both drill bits are operated close to their best cutting speed.

Universal series motors are made in many specialized ratings, so a wide range of full-load operating speeds is possible.

FIGURE 12–18. Universal series motor.

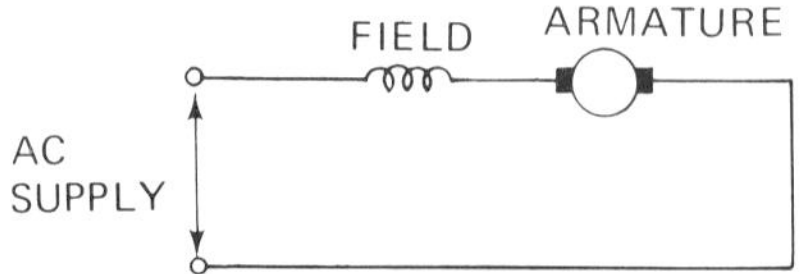

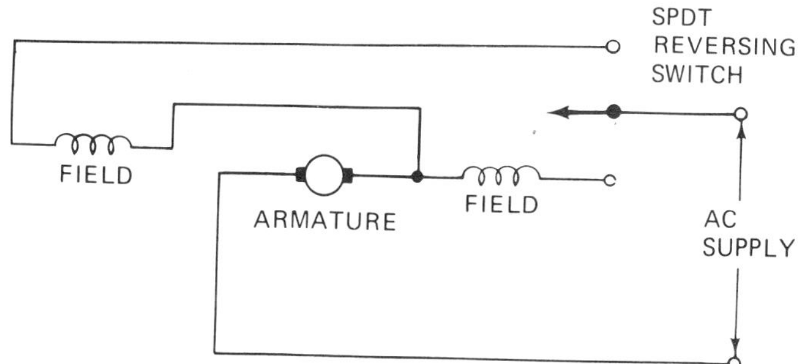

FIGURE 12–19. Two-field reversible motor.

Two-Field Reversible Motor

A variation of the universal series motor is the two-field reversible motor shown in Figure 12–19. Two separate field windings are wound on the stator in opposite directions. When one winding is energized, the motor turns clockwise; when the other winding is energized, the motor turns in a counterclockwise direction. A lead from each field is brought out to a SPDT switch. One field winding is energized at a time to give the desired direction of rotation. In all other ways, this motor is like the universal series motor.

Repulsion Motor

This is another small ac motor having high starting torque. Like the universal series motor, it has a commutator and brushes. Its principle of operation, however, is somewhat different. Refer to the schematic diagram in Figure 12–20. The stator winding is connected directly across the ac line. There is no electrical connection to the rotor or brushes, and the brushes are shorted together. The magnetic field of the stator induces a voltage in the rotor, and the shorted brushes allow a heavy current flow in the rotor windings. This current sets up a field in the rotor that reacts with the stator field to produce rotation.

The position of the brushes is important in determining the direction of rotation and the speed and torque of the motor. When the brushes are in line with the flux of the field, they are said to be in their *hard neutral* position. A high current will then flow in the rotor windings, but the motor will not turn because the fields are in opposite directions.

If the brushes are shifted 90° so that they are at right angles to the field flux, there will be no current flow in the rotor windings. This position

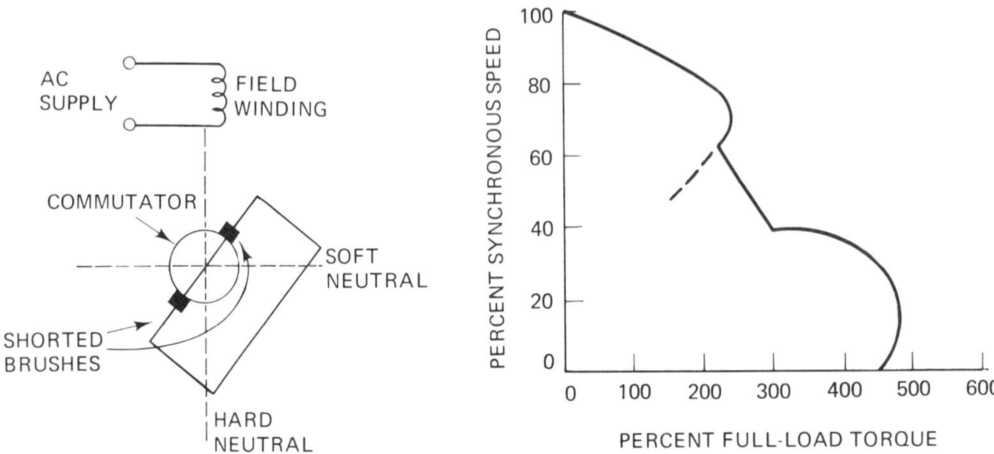

FIGURE 12–20. Repulsion motor: schematic diagram; (*b*) speed–torque curve.

is called *soft neutral,* and, again, the rotor will not turn. To develop torque in the repulsion motor, the brushes must be set between hard neutral and soft neutral. Then there is current flow in the rotor windings and the rotor field is in phase with the stator field. Commercial motors of this type usually operate with the brushes set 15 to 20° from hard neutral. Moving the brushes to the opposite side of hard neutral will reverse the direction of motor rotation.

Some motors of this type start as repulsion motors and, after they are up to speed, operate as induction motors. A centrifugal governor operates a switch to short the commutator segments when the motor is up to speed. The rotor then acts like a squirrel-cage rotor.

The speed–torque curves for a typical repulsion-start motor are shown in Figure 12–20. Note the extremely high starting torque for this motor—over 400 percent of full-load torque. In the figure, note also at the dashed line on the curve that the commutator segments are shorted and the motor comes up to speed as an induction motor. These motors are used where very high starting torque is needed.

Single-Phase Synchronous Motors

Some synchronous motors operate from a single-phase supply. Like the three-phase synchronous motor, their speed is constant. These motors are widely used in electric clocks and other types of timing devices.

Many different kinds of small synchronous motors are designed to operate on single-phase ac. The exact mechanical arrangement varies widely, but they all depend on the same basic principle for operation at synchronous speeds.

The shaded-pole motor of Figure 12–17(a) will operate at synchronous speed if its squirrel-cage induction rotor is replaced by a rotor of hardened steel, which tends to retain the magnetism induced in it. The rotating field induces a field in the hardened-steel rotor. The reaction between the rotating field and the field induced in the rotor produces torque that causes the rotor to turn. The motor starts in the same way as an induction motor. As the motor approaches synchronous speed, the magnetic properties of the core tend to produce permanently magnetized poles along some fixed diameter of the rotor. When this happens, the rotor locks in step with the rotating field and rotates at synchronous speed. This motor is self-starting and is synchronous over a limited torque range. Its speed is 3600 rpm when operated on 60-Hz current.

Figure 12–21 shows another type of small, single-phase, synchronous motor. The speed of this motor is locked to the supply frequency, but as it runs at a submultiple of the synchronous speed, it is called a subsynchronous motor. This motor is also self-starting and has the advantage of running at slow speed.

The speed of the motor depends on the number of teeth on the rotor. The rotor moves a pair of teeth past the poles for each cycle of alternating current. A motor with 16 teeth will, therefore, run at 450 rpm with a 60-Hz supply voltage.

FIGURE 12–21. Sixteen-pole subsynchronous motor.

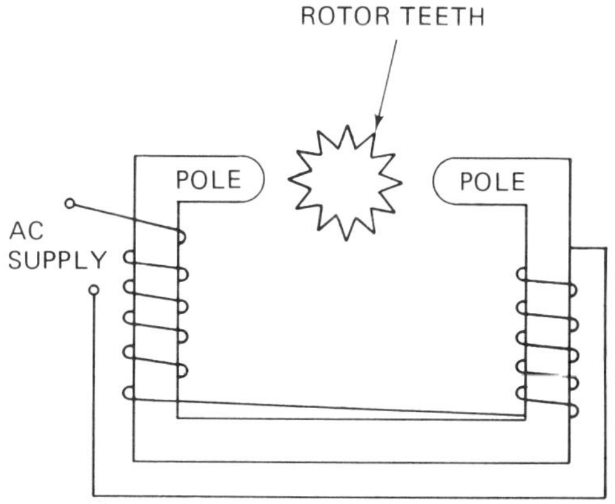

12–11 GENERAL CARE AND MAINTENANCE

Preventive Maintenance

Preventive maintenance includes such jobs as testing, cleaning, drying, oiling, and adjusting motors. For best results, each motor should have its own history card, which may either be attached to the machine or kept in a separate file. The card lists such items as the name of the manufacturer, model number, serial number, dates of purchase and installation, and the dates of inspection, brush changes, or overhaul.

In doing any work on motors, refer to the manufacturer's literature for operating limits, types of lubricants, replacement parts, and so forth.

Service Factors

Formerly, motors were rated on the basis of temperature rise *above* a specified ambient temperature. The amount of temperature rise was generally limited to 40°C (72°F). Today, motors have a *service factor* stamped on their name plates. This factor ranges from 1.00 to 1.35. A rating of 1.00 means that, when the motor is installed where the ambient temperature is not over 40°C (104°F), it can safely deliver its rated horsepower continuously. When the service factor is greater than 1.00, multiply the rated horsepower by the service factor to determine the horsepower that the motor can safely deliver on a continuous basis with the same temperature limitation. If, for example, a 2-hp motor has a service factor of 1.15 (most common for motors over 1 hp), it can safely and continuously deliver $1.15 \times 2 = 2.3$ hp, provided the ambient temperature does not exceed 40°C (104°F). At higher temperatures, continuous operation at the full-rated horsepower will seriously limit the life of the motor.

The service-factor rating of a motor should not be confused with the *overload capacity* of almost any motor. Overload capacity simply means that the motor can deliver, for relatively short periods of time, 1.5 to 2.0 times its normal horsepower without harm. Consistent overload operation is a very poor practice, but when the overload does not occur frequently, the additional expense of securing a motor with a larger horsepower rating may not be justified.

Resistance Readings

The resistance between the field or armature winding and the frame should be a very high value, usually measured with an instrument called a megohmmeter. This measurement should be taken with the power line completely disconnected and should be recorded on the history card. The resistance will vary from time to time owing to variations in temperature, the amount of moisture between the windings, and foreign materials, such as dust, on the windings. The best time to take resistance readings is immediately after the machine has been shut down so that the heat has dried out the moisture. Moisture causes the resistance to read lower than normal.

A large change in a motor's resistance reading may occur over a period of months. This is an indication of trouble, which may be caused by the accumulation of dirt, grease, or moisture in the insulation of the windings. If the windings are not cleaned and dried out, the lowered resistance will cause leakage currents, which will eventually destroy the insulation and cause short circuits.

Cleaning Methods

An air hose, vacuum cleaner, or clean, unused paint brush will usually remove surface dust or dirt. Where dirt is encrusted on the windings, you will have to scrape it off with a wooden or plastic paddle. *Never* use metal or anything with a sharp edge, since either might damage the insulation.

Oil and grease can be removed by wiping the insulation with a piece of cheesecloth dampened with an approved petroleum solvent, such as Stoddard solvent. Carbon tetrachloride should *never* be used because it will cause deterioration of the rubber insulation and commutator film; furthermore, its fumes are poisonous.

A safe method of cleaning electrical machinery is to use fresh water and household detergent. The winding should *not* be soaked in the water. Apply the water with a scrub brush. This should be done as quickly as possible and the windings should then be wiped dry. Before reassembling the machine, it should be baked so that all moisture is removed. If the machine is small, it can be baked by placing it on top of an oven, steam pipes, or a radiator. Larger units should be placed in an oven or heat chamber.

Commutators

Commutators acquire a film after a week or so of operation. This film is made up of particles from the commutator, carbon particles from the brushes, and moisture from the air. The film may range in color from very light tan to almost black. The best color for good contact and minimum wear is about the color of milk chocolate.

The film, which is the result of friction between copper and carbon, has a definite advantage. It reduces wear, acts as a lubricant, and tends to reduce arcing at the brushes. To keep the proper composition of this film, all replacement brushes must be the exact type recommended by the manufacturer.

Commutators should be cleaned often enough to prevent the surface from becoming oily or greasy. Hand-woven duck canvas is a suitable cleaning material. Press the canvas against the rotating commutator to wipe it efficiently. *Never* use solvents or any other item except the canvas to clean the commutator.

The color and uniformity of the film on the commutator indicates its condition. Look also to see if the mica between commutator segments is properly undercut. If the mica is high, it will cause the brushes to ride up, and arcing will take place between the brushes and the commutator segments. Arcing will soon pit the copper and cause excessive wear. In extreme cases, the commutator will have to be resurfaced by a shop specializing in motor overhaul.

Brushes

The brushes are critical parts of any motor. They are fitted into a brush holder and rest against the commutator. The pressure of the brush is governed by a spring that presses against the outside end of the brush. Brushes will give very little trouble when they are the correct type and the correct pressure is applied.

When a new brush is installed, it should be shaped to the commutator. One way to do this is to put a piece of sandpaper between the commutator and the brush, with the sanded surface pressing against the brush. Rotate the commutator back and forth by hand until the face of the brush assumes the contour of the commutator.

Bearings and Lubrication

In small motors the bearing may be a bronze sleeve with grooves to allow oil to reach the shaft. Larger motors have ball bearings. Some motors also have an additional bearing, called a thrust bearing. It may be a single steel ball between the end of the shaft and the housing.

Never lubricate a motor without first referring to the manufacturer's literature. *Much damage can be done by overlubrication.* Moreover, an excess of oil and grease decreases the resistance of the insulating material, collects dirt, and causes fire hazards.

Indications of faulty bearings are noisy operation, overheating of bearings, or freezing of the shaft. Clearance between rotors and stators is very small. A bad bearing can cause a rotor to rub against the field poles.

Unless you have specialized training in motor overhaul, confine your work on motors to their installation and routine maintenance as described here.

12–12 MOTOR WIRING

It is unlikely that, until you have gained considerable experience in wiring, you will be called upon to design a motor-circuit installation. Instead, it is much more likely that you will assist in the installation of a motor circuit designed by an experienced electrical contractor who is familiar with the various ordinances governing such work. It is important, however, that you gain from this text some understanding of the factors involved in such jobs.

Because we are not concerned with industrial applications in this text, our discussion of motors in this section is concerned primarily with overcurrent protection, how a motor is totally disconnected from a circuit, and the selection of wire sizes. Moreover, we shall concern ourselves solely with single-phase ac motors, since this is the type that you will most often encounter. For more detailed information on specific motor installations, refer to Code Article 430, Motors, Motor Circuits, and Controllers.

Motor Circuits: General

When a single motor is connected to a branch circuit, four components are required: (1) an overcurrent protection device, (2) a controller to start and stop the motor, (3) running overcurrent protection, and (4) a disconnecting means to disconnect totally the motor and controller from the circuit. Quite often, more than one of these components are combined in a single device.

Overcurrent Protection

Code Reference 430-42(d) reads as follows: "The branch circuit short-circuit and ground-fault protective devices protecting a branch circuit to which a motor or motor-operated appliance is connected shall have sufficient time delay to permit the motor to start and accelerate its load."* Since the starting current of a motor is much heavier than its running

*Reprinted with permission from the 1978 National Electrical Code, Copyright 1977, National Fire Protection Association, Boston, Mass.

current, this reference indicates that either a thermal-type circuit breaker or a time-delay fuse should be used for overcurrent protection. The amperage of the overcurrent device is, of course, determined by the size of the wire used in the branch circuit.

Controllers

On equipment that starts and stops automatically, such as a refrigerator, the manufacturer provides the proper controller for the equipment. On manually operated controllers, however, the selection of the controller is left to the electrician.

The branch-circuit protective device may act as the controller for a stationary motor rated at ⅛ hp or less that is normally left running and is constructed so that it cannot be damaged by overload or failure to start, for example, a clock motor.

For portable motors rated at ⅓ hp or less, the attachment plug and receptacle act as the controller.

For motors rated at 2 hp or less, and 300 V or less, a general-use switch having an ampere rating of not less than twice the full-load current rating of the motor is satisfactory. Or, on ac circuits, general purpose snap switches that are suitable for use on alternating current only are permitted as long as the motor full-load current does not exceed 80 percent of the ampere rating of the switch.

For motors larger than 2 hp, the switch must be rated in horsepower and the rating must be at least equal to that of the motor. Special motor starters, which combine start–stop control and overload protection, are also available. They are rated in horsepower and are operated either by a toggle switch or a push button. On push-button types, the button may be mounted directly on the device enclosure or at some remote location.

A branch-circuit inverse-time circuit breaker, rated in amperes only, may be used as a controller provided it meets the requirements for motor running-overcurrent protection. These are given below.

Running-Overcurrent Protection

The term *running overcurrent,* as used here, is synonymous with the term *overload.* Notice also that overload is an operating overcurrent that, when it persists for a sufficient length of time, causes damage or dangerous overheating of the apparatus. It does not include short circuits or ground faults. Because the starting current of a motor is generally greater than a running overcurrent, the fuse or circuit breaker used for overcurrent protection on a branch circuit will not protect a motor from overload. Thus a separate device is needed to provide that kind of protection.

Manually controlled, 120-V portable motors of 1 hp or less may be plugged into ordinary 15- or 20-A branch circuits. Running-overcurrent protection is usually, but not always, provided by a *thermal protector,* which is intergral with the motor circuit. The notation "Thermally Protected" will be found on the nameplate of such motors.

As the name implies, a thermal protector is simply a heat-sensitive device that acts to open the circuit when the motor temperature exceeds a safe value. Once tripped, the thermal protector usually requires manual resetting.

Permanently installed motors not equipped with an integral thermal protector require separate running-overcurrent protection. Most often, this device is a *motor starter* containing heater coils that match the amperage of the motor. Such starters carry the normal running current indefinitely and a small overload for some time, but trip on heavy overload or when the motor fails to start. The starter then acts as a combined controller and running-overcurrent protector.

If a separate circuit breaker is used for overload protection, it must be rated at not more than 125 percent of the motor amperage. The circuit breaker then acts as a combined controller, overload protector, and disconnecting means.

When a knife switch combined with a fuse is used for overload protection, it must be rated at not more than 125 percent of the motor amperage. The fuse or fuses must, of course, be of the time-delay variety to prevent *blowing* when the motor starts. The switch–fuse combination also combines the functions of controller, overload protection, and disconnecting means.

Disconnecting Means

Regardless of size, the plug and receptacle serve as the disconnecting means for *portable* motors. The explanations that follow apply only to permanently installed motors.

No separate disconnecting means is required for motors rated at ⅛ hp or less. The branch-circuit overcurrent device then serves as the disconnecting means.

A knife switch–fuse combination may serve as the disconnecting means on larger motors.

Motors having an integral thermal protector, or those employing a combined controller–overload protector, require a separate disconnecting means. An enclosed knife switch–fuse combination, with a horsepower rating not less than that of the motor, is often used for this purpose, even though the code does not require the fuse inasmuch as overload protection is already provided.

Motors rated at 2 hp or less many also use a knife switch–fuse combination rated in amperes *provided* the rating is at least two times the motor amperage. Alternatively, you may use a general-purpose ac toggle switch rated at 125 percent or greater of the motor amperage. For either arrangement, an SPST switch is used for 120 V and a DPST switch for 240 V.

Wire Sizes

The wires used with any motor must have an ampacity rating of no less than 125 percent of the motor nameplate current rating.

EXERCISES

12–1. Explain what happens to a current-carrying conductor in a magnetic field.

12–2. How is rotation of a motor shaft produced?

12–3. What is meant by *torque*?

12–4. What is CEMF and how is it produced?

12–5. What regulates the amount of CEMF produced by a motor?

12–6. What is the *neutral plane* of a motor?

12–7. What is meant by *armature reaction* and how does it affect a motor?

12–8. What measures are taken in practical motors to counteract the effects of armature reaction?

12–9. When does a motor produce its maximum torque?

12–10. Define horsepower.

12–11. What is the electrical equivalent of mechanical horsepower?

12–12. Does a motor with 746-W electrical-power input produce 1 hp at the output shaft? Explain your answer.

12–13. If a motor runs at 1500 rpm at no load, and at 1460 rpm with its rated load, what is its speed regulation?

12–14. Describe a series motor and draw its schematic diagram.

12–15. What are the speed–torque characteristics of a series motor?

12–16. Can a series motor be run without a load? Why?

12–17. Describe a shunt motor and draw its schematic diagram.

12–18. What are the speed–torque characteristics of a shunt motor?

12–19. How may the speed of a shunt motor be controlled?

12–20. Describe a compound motor and draw its schematic diagram.

12–21. What are the speed–torque characteristics of a compound motor?

12–22. What is the difference between a *cumulative compound* and a *differential compound* motor?

12–23. Draw the schematic diagram of a starting box intended for manual operation of a shunt-type dc motor and explain its operation.

12–24. Explain the operation of a three-phase induction motor.

12–25. In a three-phase induction motor, the rotor does not turn as fast as the rotating field of the stator. Why not?

12–26. What is meant by *synchronous* speed of a motor?

12–27. What formula is used to determine the synchronous speed of any ac induction motor?

12–28. What is meant by the *slip* of an induction motor, and what formula may be used to determine the slip of a given motor?

12–29. What are the speed–torque characteristics of a typical three-phase induction motor?

12–30. How does inductive reactance affect the torque of an induction motor?

12–31. Explain the operation of a two-phase induction motor.

12–32. What is a synchronous three-phase motor?

12–33. By itself, a single-phase supply will not produce a rotating field in a motor. What general technique is used, therefore, to start a motor?

12–34. Describe the operation of a resistance-start, single-phase motor.

12–35. Explain the operation of a centrifugal-start switch.

12–36. How does the slip of a single-phase, resistance-start motor compare with that of a polyphase motor?

12–37. Describe the operation of a capacitor-start motor.

12–38. Describe the operation of a split-phase reversible motor.

12–39. Describe the operation of a shaded-pole motor.

12–40. Describe the operation of a universal series motor.

12–41. Describe the operation of a repulsion-type motor.

12–42. In a repulsion-type motor what is meant by the expressions *hard neutral* and *soft neutral*?

12–43. Describe how a shaded-pole motor can be made to operate at synchronous speed.

12–44. What is meant by the *service factor* of a motor?

12–45. What conditions may cause a large change in the resistance readings of motor windings?

12–46. Explain the methods for cleaning electrical machinery.

12–47. Describe the general appearance and maintenance of commutators.

12–48. What precautions should be taken when new brushes are installed?

12–49. Explain the general provisions for overcurrent protection in motor circuits.

12–50. Explain the general provisions for controllers, running-overcurrent protection, and disconnecting means in motor circuits.

13

Farm Installations

13–1 GENERAL

In addition to the electrical devices and appliances found in city residences, farms require such things as milking machines, hay dryers, silo fillers, milk coolers, and special hot-water heaters for the dairy and chicken house. Consequently, the electrical power consumption on a farm is usually much greater than for a city residence. Moreover, the wiring system is spread out over much greater distances and may be a combination of overhead and underground.

Since the need for additional appliances and machines changes as a farm grows, the initial planning should provide for the liberal addition of circuits. The cost of installing additional capacity that is not used immediately is generally less than the cost of any major rewiring job, particularly in these days of sharply rising prices.

To avoid the problems associated with icing, lightning, and the movement of tall machinery commonly used on farms, an underground system should prove more satisfactory than an overhead system. The specifications in Section 9–9 of this text for residential underground services apply equally well to a farm.

On most farms, the service drop terminates on a meter pole in the farmyard. Separate feeders then run to the residence, barn, chicken house, repair shop, storage areas, and so on.

In each building there is a service entrance, but no meter. A ground is required for any building that contains more than one circuit or that houses livestock.

13–2 GROUNDS: GENERAL

Inasmuch as municipal water systems are not generally available to farms, *made electrodes* must be used for grounding.

In all such cases the grounding electrode shall consist of a driven pipe, driven rod, buried plate, or other device approved for the purpose and conforming to the following requirements (NEC Sec. 250-83)*:

- *Concrete-encased electrodes.* Not less than (1) 20 ft of bare copper conductor not smaller than no. 4, or (2) ½-in.-diameter steel reinforcing bar or rod encased by at least 2 in. of concrete and located within and near the bottom of a concrete foundation footing that is in direct contact with the earth.
- *Plate electrodes.* Each plate electrode shall present not less than 2 ft^2 of surface to exterior soil. Electrodes of iron or steel plates shall be at least ¼ in. thick. Electrodes of nonferrous metal shall be at least 0.06 in. thick.
- *Pipe electrodes.* Electrodes of pipe or conduit shall be not smaller than ¾-in. trade size and, where of iron or steel, shall have the outer surface galvanized or otherwise metal coated for corrosion protection.
- *Rod electrodes.* Electrodes of rods of steel or iron shall be at least ⅝ in. in diameter. Nonferrous rods or their equivalent shall be listed and shall be no less than ½ in. in diameter.
- *Installation.* Where practicable, made electrodes should be embedded below permanent moisture level. Where rock bottom is not encountered, the electrode shall be driven to a depth of 8 ft. Where rock bottom is encountered at a depth of less than 4 ft, electrodes not less than 8 ft long shall be buried in a trench. All electrodes shall be free from nonconductive coatings, such as paint or enamel.

* Reprinted with permission from the 1978 National Electrical Code, Copyright 1977, National Fire Protection Association, Boston, Mass.

13–3 GROUND RESISTANCE

Section 250-84 of the NEC specifies that the resistance to ground of a made electrode shall not exceed 25 Ω. Where the resistance to ground of a single made electrode is greater than 25 Ω, two or more made electrodes are connected in parallel.

The easiest method to determine the resistance to ground of a made electrode is by means of a ground-resistance meter. Because these instruments are quite expensive, however, and not always available, let us discuss an alternative method that provides a reasonable approximation.

Take an ac ammeter with a 0–30 A scale and mount it in a suitable box. Connect one meter terminal to a 30-A fuse mounted in the same box. From the opposite end of the fuse and from the remaining ammeter terminal, bring out test leads with insulated clips at their ends. Use insulated AWG no. 12 wire for the test leads. Study this arrangement, which is shown schematically in Figure 13–1.

Temporarily disconnect the grounding electrode conductor from the made electrode. Now connect the ammeter assembly between the made electrode and the hot conductor of the wiring system. This grounds the hot wire, and the resulting fault current passes through the series-connected ammeter and fuse. Neglecting the resistance of the meter, the resistance of the ground can be determined by Ohm's law. If, for example, the line voltage is 120 V and the ammeter reads 6 A, the resistance to ground is

$$R = \frac{E}{I} = \frac{120}{6} = 20\Omega$$

This is an acceptable ground, but you should try to improve it. Of course, the hot wire should be grounded no longer than necessary to obtain the

FIGURE 13–1. Arrangement of ammeter, fuse, and test leads to form a resistance-to-ground tester.

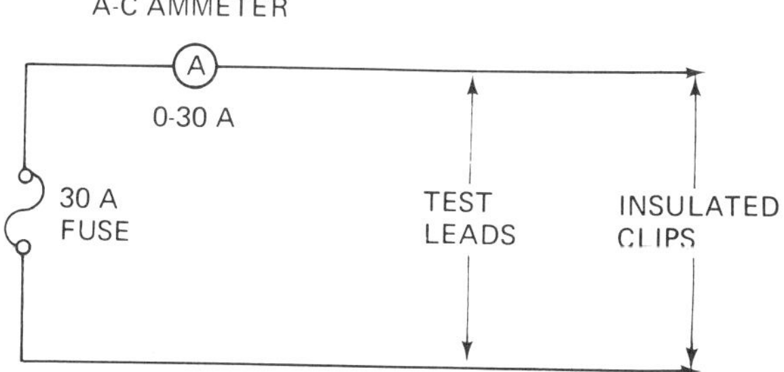

required meter reading. Also, once the meter reading is obtained, make sure that you reconnect the grounding-electrode conductor between the system neutral wire and the made electrode.

Generally, the hot wire will be protected by a 15-A fuse or circuit breaker at the service. In the example given, the fuse will not open, nor will the circuit breaker trip under a brief surge current of 20 A. If the service fuse blows or the breaker trips immediately, you have an excellent low-resistance ground.

Is there any way to improve, that is, reduce, the resistance to ground? Fortunately, there are several methods.

- Install the made electrode where rain falling off the roof will tend to keep the ground around the electrode wet. The reason: the drier the ground, the higher its resistance.
- Drive a second rod 12 to 16 ft into the earth, at a distance of 10 ft or more from the first rod, and connect the two with ground clamps and wire of the same size as the ground wire. In extreme cases of too high resistance, a third rod, spaced 10 ft from, and connected to, the other two may be required.
- Soak the ground around the made electrode with a heavy concentration of saltwater.Repeat this treatment at least once each year.

13–4 METER POLES

As noted earlier, for most farms the service drop ends at a meter pole. To minimize expense, the meter pole should be located near the building with the greatest load. Smaller-sized conductors are required for buildings with smaller loads.

A typical meter-pole installation is shown in Figure 13–2. The service drop terminates at the top of the pole. Usually the neutral wire is the topmost wire, but this should be verified with the electric supplier. Also notice that the neutral is spliced to the neutral lines running to all buildings. In effect, then, the neutral is continuous from the power line to each building. The neutral is also connected to the meter and is grounded at the pole. More is said about grounding later. Right now, let us see what happens to the hot wires.

The ungrounded conductors are run from the top of the meter pole to the meter socket, and from there back to the top of the pole. Including the neutral wire, which grounds the meter socket for safety purposes, three conductors run down the pole and two conductors return to the top of the pole. Depending on local codes, all five conductors may be encased in a

FIGURE 13–2. Typical meter pole.

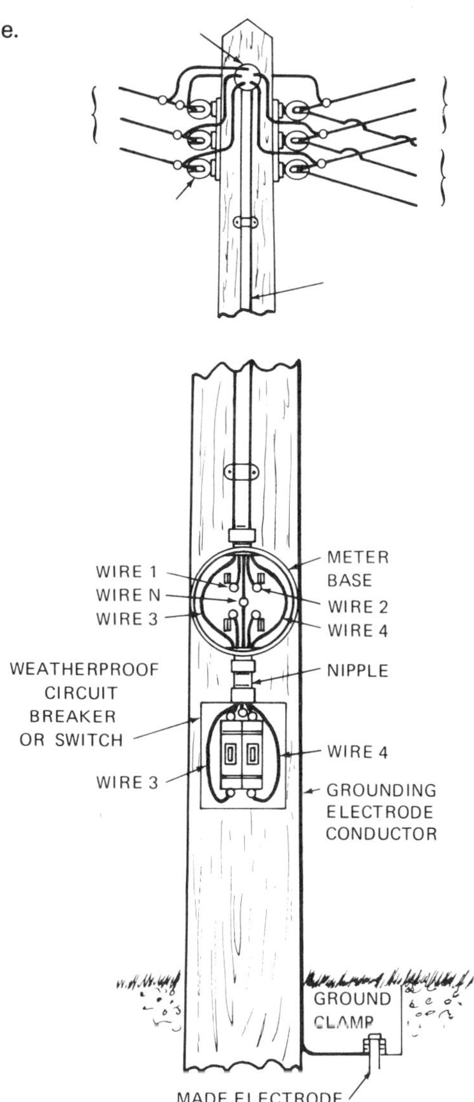

single conduit or, alternatively, separate conduits may be required for the "down" and "up" wires. When all wires are contained in one conduit, the assembly is called a *single stack*. When two conduits are used, it is a *double stack*. With either arrangement, leave at least one third of the pole circumference clear to permit easy climbing.

When a main switch or circuit breaker is used, follow the wiring plan shown in Figure 13–2. The switch or circuit breaker is weatherproof, of course, and is mounted about 5 ft aboveground, with the meter just above it. The two are joined by a nipple.

Insulator racks are installed bear the top of the pole. Use one rack for each set of three wires, and secure the racks to the pole by heavy through bolts that pass completely through the pole.

To solve the water problem, bring the top of the conduit to a point above the topmost insulator and place a service head at its top.

The switch, meter socket,wired conduit, and insulators are preassembled on the pole before it is placed in an upright position.

13–5 DETERMINING FEEDER LOADS

The feeders that supply farm buildings (excluding family dwellings) or loads that consist of two or more branch circuits shall have a minimum capacity computed in accordance with NEC Table 220-40, shown here as Table 13–1.

The total load of the farm for service-entrance conductors and service equipment shall be computed in accordance with the farm dwelling load and demand factors specified in Table 13–2. Where there is equipment in two or more farm equipment buildings or for loads having the same function, such loads shall be computed in accordance with Table 13–1 and

TABLE 13–1. Method for Computing Farm Loads for Other Than Dwellings

Load in Amperes at 230 Volts	*Demand Factor (%)*
Loads expected to operate without diversity, but not less than 125 percent full-load current of the largest motor and not less than first 60 A of load	100
Next 60 A of all other loads	50
Remainder of other loads	25

Reprinted with permission from the 1978 National Electrical Code, Copyright 1977, National Fire Protection Association, Boston, Mass.

TABLE 13–2. Method for Computing Total Farm Load

Individual Loads Computed in Accordance with Table 13–1	*Demand Factor (%)*
Largest load	100
Second largest load	75
Third largest load	65
Remaining loads	50

Reprinted with permission from the 1978 National Electrical Code, Copyright 1977, National Fire Protection Association, Boston, Mass.

may be combined as a single load in Table 13–2 for computing the total load. (See NEC Sec. 230-21 for overhead conductors from a pole to a building or other structure.)

To illustrate the use of the above information, let us compute the load for a farm. Initially, we shall exclude the residence.

Building 1: Dairy barn	
Lighting	2500 W
Water heater	3000 W
Total	5500 W
5500 W/230 V =	24 A
Milker	14 A
Cooler	8 A
Air conditioner	15 A
Total	61 A
Building 2: Chicken house	
Brooder	4000 W
Lighting	600 W
Total	4600 W
4600 W/230 V =	20 A
Building 3: Storage and repair	
5-hp (230-V, single-phase) motor	26 A
2-hp (230-V, single-phase) motor	10 A
Lighting	8 A
Total	44 A

The largest motor employed is 5 hp. From Table 13–1, 125 percent of 26 A is 32.5 A This increases the total load for building 3 from 44 to 50.5 A.

The total load for the three buildings is as follows:

Building 1	61 A
Building 2	20 A
Building 3	51 A
Total	132 A

From Table 13–1:

First 60 A at 100%	60 A
Next 60 A at 50%	30 A
Remainder at 25%	3 A
Total	93 A

Now let us consider the same farm but include the residence. Assume that the house has a floor area of 2000 ft^2, excluding an unfinished attic. The range is rated at 11 kW.

Lighting	6,000 W
Small appliances	3,000 W
Laundry	1,500 W
Total (excluding range)	10,500 W
3000 W at 100%	3,000 W
7500 W at 35%	2,625 W
Net load (excluding range)	5,625 W
Range (NEC Table 220-19)	8,000 W
Net load (including range)	13,625 W
13,625 W/230 V =	59 A
Largest demand, bldg. 1 at 100%	61 A
Second largest demand, residence at 75%	44 A
Third largest demand, bldg. 3 at 65%	29 A
Balance of demand, bldg. 2 at 50%	10 A
Total demand	144 A

13–6 MORE ABOUT THE METER POLE

When service amperage runs 200 A or higher, a current transformer is used to reduce the size of conductors used between the top of the meter pole and the meter proper.

Some current transformers have a single-turn primary in the form of a solid bar. This primary is connected in series with the load to be measured. In farm installations, however, a *doughnut*-type current transformer is more common. The name derives from its resemblance to that familiar pastry.

The doughnut transformer has a single winding—the secondary. The hot conductors whose current is to be measured pass through the center of the doughnut and act as the primary. The secondary is magnetized by the magnetic lines surrounding the primary. The secondary has many turns of small wire and is usually connected to a 5-A ammeter. The ratio of turns between the two circuits depends on the anticipated magnitude of the current to be measured. If, for example, the anticipated current is 200 A, the actual turns ratio would be 40:1. It is customary, however, to refer

FIGURE 13–3. Current transformer.

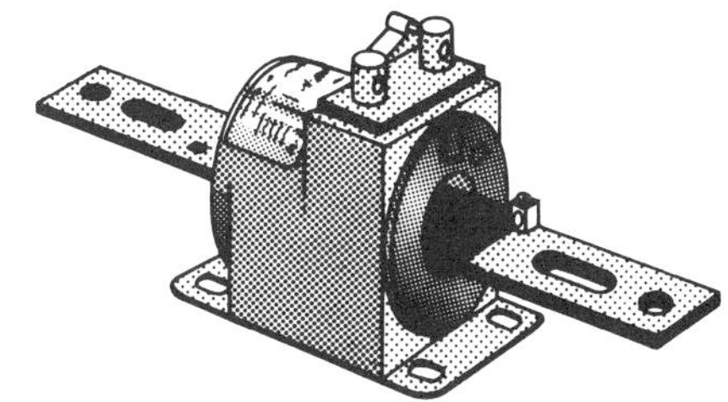

the ratio to the full-scale reading of the ammeter. Thus the ratio is designated 200:5.

Because it consumes very little power, a current transformer is a fairly small device, typically measuring about 5 in. in diameter and just slightly more in length. A mounting bracket is attached to the bottom of the transformer case, and the secondary terminals extend through the top of the case. See Figure 13–3.

The ammeter connected across the secondary effectively forms a short circuit. When the ammeter is not connected, a high voltage is induced in the secondary, because the primary load current is not dependent on the secondary load current. The primary then acts as an inductance, and this creates a voltage drop in the primary. Due to the high secondary-to-primary turns ratio, the resulting open-circuit voltage of the secondary may reach thousands of volts. To remove this safety hazard, a switch is connected in parallel with the secondary and with the ammeter, and is closed when the ammeter is not in use. Appropriately, this switch is called a *shorting* switch. See Figure 13–4.

The current transformer and shorting switch are usually installed in the same cabinet, which is located near the top of the meter pole just under the service-drop insulator bracket. The switch is operated from the

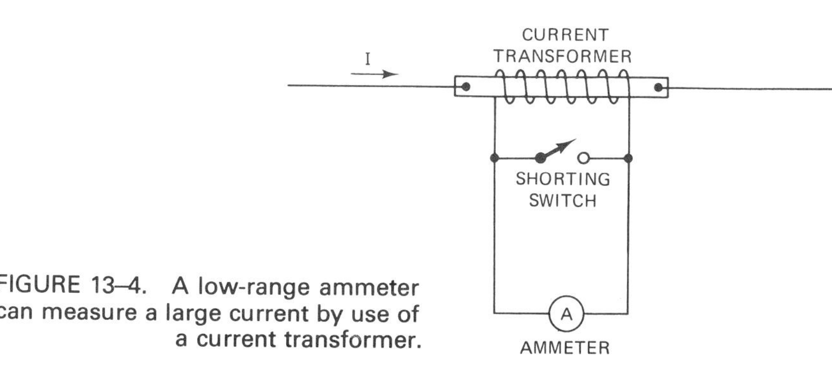

FIGURE 13–4. A low-range ammeter can measure a large current by use of a current transformer.

ground by means of an extension handle secured to the pole and terminating about 5 ft aboveground.

Two AWG no. 14 wires run from the transformer secondary to the current winding of the watthour meter. Two additional AWG no. 14 conductors run from hot wire terminals inside the switch cabinet to the voltage winding of the meter.

For the best protection against lightning, run the neutral wire directly from the top of the pole (near the service drop termination of the neutral) to the ground rod at the bottom of the pole. Depending on local preference, the neutral lead may be stapled to the pole on the side opposite from the conduit, or it may be tucked in alongside the conduit as far as it runs, then run to ground.

The ground rod is typically located about 2 ft from the bottom of the pole with a 1 ft deep trench between the rod and pole. The neutral runs from the base of the pole, through the trench, and is attached to the rod by means of a ground clamp. The entire assembly is then covered.

Avoid driving ground rods in locations saturated by animal excreta, since the chemical breakdown of such degradable matter may destroy all or portions of the ground assembly.

13–7 BUILDING ENTRANCES

A building wired with nonmetallic-sheathed cable and having only one citrcuit need not be grounded. But grounding is mandatory wherever any metal equipment (switch boxes, conduit, etc.) is used.

At its point of entry to the building run the grounding electrode conductor between the neutral and the grounding electrode.

The residence will contain either circuit breakers or a fused-service switch in conjunction with branch-circuit fuses. For convenience, circuit breakers are recommended.

In all buildings excluding the residence, one or more switches must be provided to disconnect all wiring. When buildings are fed current from another building, the switch or circuit breaker that controls the wiring may be located either at the building of origination or at the building of termination. When the fuse at the starting point protects the smallest wire in the entire circuit, the switches in other buildings need not be fused. Conversely, if the fuse at the starting point does not protect the smallest wiring in the circuit, an appropriate fuse is required at the point where the smaller wire is tapped off the larger.

Because good grounds are often difficult to locate on farms, conduit and armored cable are rare; therefore, the NEC recommends nonmetallic-sheathed cable. To make the entire system nonmetallic, and thereby improve safety, use nonmetallic outlet and switch boxes made of various plastic materials. These boxes are installed the same as the metal variety, but the nonmetallic-sheathed cable must be supported within 8 in. of a box. Since the nonmetallic boxes contain no cable clamp, one additional conductor may be installed in the box.

13–8 MISCELLANY

In buildings that house livestock, you must use types NMC or UF cable. Of course, either type is suitable for the entire installation, but there may be occasions when special circumstances dictate some other choice. Always check on the prevailing code.

Locate switches so that they can be operated by a shoulder or elbow. A farmer's hands are often full when he enters a building.

For the same reason that a nonmetallic system is recommended, use only porcelain or Bakelite sockets.

For physical protection, locate lamp outlets between joists so that the lamps do not project too far into closely confined spaces, such as aisles.

For improved lighting, use a reflector with every socket, and clean the reflectors at frequent intervals.

Local codes may make vaporproof receptacles mandatory in the vicinity of haymows.

When cable is located where the possibility of damage exists, make sure that it is provided with some form of protective covering. When the cable passes through a floor, it must be protected by at least a 6 in. conduit or pipe.

Keep exposed cables away from the centers of aisles. Even though more cable may be required, do not run it along the bottoms of joists; instead, run it along the side of a joist.

To ensure operation of a water pump, particularly during periods of emergency such as a fire, connect the pump through an independent service. Simply run wires to the pump house directly from the meter pole, ahead of circuit breakers or main fuses. Mount a weatherproof circuit breaker (or fused-disconnect switch) on the pole.

Make generous use of yard lights and locate one at the meter pole also. Use three-way switches for control between the house and barn. Special lights are available for farmyards, but you must check the local code before connecting the lights to the wiring system.

13–9 LIGHTNING ARRESTERS

In isolated areas, lightning arresters are a very important part of any electrical wiring system. Even though lightning may not strike wires directly, a nearby strike may induce extremely high voltages in the wires. This can cause damage not only to the wires themselves, but to appliances and other equipment attached to the wiring system. When a lightning arrester is installed as part of a properly grounded wiring system, the possibility of lightning damage is reduced significantly.

A lightning arrester has three leads, one white and two black. The white lead is connected to the grounded neutral and the two black wires to the ungrounded conductors. Mount one lightning arrester at the meter pole; for improved safety, mount another at each service switch to a building. The arrester neck is inserted through a knockout, and installation is simple.

EXERCISES

13–1. Why is an underground system recommended on a farm?

13–2. Where does a farm service-drop terminate?

13–3. Describe a meter pole physically.

13–4. Are separate meters required at each building where a service entrance is employed?

13–5. In what buildings is a ground required?

13–6. State the NEC requirements for each of the following types of made electrodes: concrete-encased, plate, pipe, and rod.

13–7. How must made electrodes be installed when rock bottom is encountered at a depth of less than 4 ft?

13–8. What is the highest resistance to ground permitted by the NEC?

13–9. Using an ammeter, describe one method for determining the resistance to ground.

13–10. Give three methods by which the resistance to ground can be decreased.

13–11. In connection with a meter pole, what is meant by the terms *single stack* and *double stack*?

13–12. Where is a main switch or circuit breaker installed on a meter pole? How is it connected to the meter?

13–13. Give the formulas for determining the minimum capacity of service conductors and service equipment at the main point of delivery to farms, including dwellings.

13–14. A typical farm has four major loads: dairy barn, 65 A; chicken house, 26 A; storage and repair, 52 A; and residence, 60 A. What is the total demand?

13–15. The largest motor on a farm draws 30 A. In computing the load, what current is used for the motor?

13–16. Describe a current transformer and how it looks.

13–17. Why is a shorting switch necessary with a current transformer?

13–18. For best protection against lightning, how is the neutral wire routed at the meter pole?

13–19. Describe a typical ground rod installation at the foot of a meter pole.

13–20. What are the switch requirements in all farm buildings, excluding the residence?

13–21. Discuss the overcurrent protection requirements when buildings are fed current from other buildings.

13–22. What type of cable is generally employed in farm electrical systems? Why?

13–23. What types of cable must be used in buildings housing livestock?

13–24. Describe generally how you would install lighting circuits in farm buildings other than the residence.

13–25. What precautions should be taken to ensure operation of the water pump at all times?

13–26. Why is a lightning arrester an essential part of a farm electrical system?

Appendix: Reference Tables

TABLE 1. Coefficients for Three Fluorescent and Three Incandescent Fixtures

Typical Distribution and Maximum Spacing[a]	ρ_{CC}[c] →	80			70			50			30			10			0	Typical Luminaires and Luminaire Maintenance Category[e]
	ρ_W[d] →	50	30	10	50	30	10	50	30	10	50	30	10	50	30	10	0	
	RCR[b] ↓	Coefficients of Utilization for 20 Per Cent Effective Floor Cavity Reflectance, ρ_{FC}																
	1	.66	.64	.62	.64	.62	.60	.58	.57	.55	.53	.52	.51	.49	.48	.47	.45	
	2	.60	.56	.54	.57	.54	.52	.53	.50	.48	.49	.47	.45	.45	.43	.42	.40	
	3	.54	.50	.46	.52	.48	.45	.48	.45	.42	.44	.42	.40	.41	.39	.37	.36	
	4	.49	.44	.41	.47	.43	.40	.44	.40	.38	.40	.38	.35	.38	.35	.34	.32	
	5	.44	.39	.35	.42	.38	.35	.39	.36	.33	.37	.34	.31	.34	.32	.30	.28	
	6	.40	.35	.31	.38	.34	.31	.36	.32	.29	.33	.30	.28	.31	.29	.27	.25	
	7	.36	.31	.28	.35	.30	.27	.33	.29	.26	.30	.27	.25	.28	.26	.24	.22	2-lamp aluminum industrial
	8	.33	.28	.24	.31	.27	.24	.29	.26	.23	.28	.24	.22	.26	.23	.21	.20	with 35° crosswise and
	9	.29	.25	.21	.28	.24	.21	.27	.23	.20	.25	.21	.19	.23	.20	.18	.17	lengthwise shielding
Max. S/MH_{wp} = 1.5	10	.27	.22	.19	.26	.21	.18	.24	.20	.18	.23	.19	.17	.21	.18	.16	.15	LDD Maint. Category II

TABLE 1. *(cont.)*

Typical Distribution and Maximum Spacing[a]	ρCC[c] →	80			70			50			30			10			0	Typical Luminaires and Luminaire Maintenance Category[e]
	ρW[d] →	50	30	10	50	30	10	50	30	10	50	30	10	50	30	10	0	
	RCR[b] ↓	Coefficients of Utilization for 20 Per Cent Effective Floor Cavity Reflectance, ρ_{FC}																
	1	.70	.68	.65	.65	.63	.61	.55	.53	.52	.45	.45	.44	.37	.36	.36	.32	
	2	.62	.58	.55	.58	.54	.51	.49	.47	.44	.41	.39	.38	.34	.33	.32	.28	
	3	.56	.50	.47	.52	.47	.44	.44	.41	.38	.37	.35	.33	.30	.29	.28	.25	
	4	.49	.44	.40	.46	.41	.38	.40	.36	.33	.33	.31	.29	.28	.26	.25	.22	
	5	.44	.39	.35	.41	.36	.33	.36	.32	.29	.30	.27	.25	.25	.23	.22	.20	
	6	.40	.34	.30	.37	.32	.29	.32	.28	.25	.27	.24	.22	.23	.21	.19	.17	
	7	.36	.30	.26	.34	.28	.25	.29	.25	.22	.25	.22	.19	.21	.19	.17	.15	
	8	.33	.27	.23	.30	.25	.22	.26	.22	.20	.22	.20	.17	.19	.16	.15	.13	2-lamp prismatic lens bottom
	9	.29	.24	.20	.27	.22	.19	.24	.20	.17	.20	.17	.15	.17	.15	.13	.11	unit with open top
Max. S/MH$_{wp}$ = 1.2	10	.27	.21	.18	.25	.20	.17	.22	.18	.15	.19	.15	.13	.16	.13	.12	.11	LDD Maint. Category VI
	1	.79	.76	73	.73	.70	.68	.62	.60	.58	.51	.50	.49	.42	.41	.40	.36	
	2	.70	.65	.61	.65	.61	.57	.55	.52	.50	.46	.44	.42	.38	.36	.35	.32	
	3	.62	.56	.52	.58	.53	.49	.49	.46	.43	.41	.39	.36	.34	.33	.31	.28	
	4	.55	.49	.44	.51	.46	.42	.44	.40	.37	.37	.34	.32	.31	.29	.27	.24	
	5	.49	.43	.38	.46	.40	.36	.40	.35	.32	.34	.30	.27	.28	.25	.24	.21	
	6	.44	.38	.33	.41	.35	.31	.36	.31	.28	.30	.27	.24	.25	.23	.21	.19	
	7	.40	.34	.29	.37	.31	.27	.32	.28	.24	.27	.24	.21	.23	.20	.19	.18	Direct-indirect with metal or
	8	.35	.29	.25	.33	.27	.24	.29	.24	.21	.24	.21	.18	.20	.18	.16	.14	dense diffusing sides and
	9	.32	.25	.21	.30	.24	.20	.26	.21	.18	.22	.19	.16	.19	.16	.14	.12	35° x 45° louver shielding
Max. S/MH$_{wp}$ = 1.3	10	.30	.23	.19	.28	.22	.19	.24	.20	.16	.20	.17	.14	.17	.14	.12	.12	LDD Maint. Category II
	1	.78	.77	.74	.76	.75	.73	.74	.72	.71	.71	.70	.68	.68	.67	.66	.65	
	2	.72	.68	.66	.71	.67	.65	.68	.66	.63	.65	.64	.62	.64	.62	.61	.59	
	3	.66	.62	.59	.65	.61	.58	.63	.60	.57	.61	.59	.56	.60	.57	.55	.54	
	4	.60	.56	.52	.59	.55	.52	.58	.54	.51	.56	.53	.51	.55	.53	.50	.49	
	5	.55	.50	.47	.54	.50	.46	.53	.49	.46	.52	.48	.46	.51	.48	.45	.44	
	6	.51	.45	.42	.50	.45	.42	.49	.45	.41	.48	.44	.41	.47	.43	.41	.40	
	7	.46	.41	.37	.46	.41	.37	.45	.40	.37	.44	.40	.37	.43	.39	.36	.35	
	8	.42	.37	.33	.42	.37	.33	.41	.36	.33	.40	.36	.33	.39	.35	.33	.31	Enclosed reflector with in-
	9	.39	.33	.30	.38	.33	.30	.37	.33	.30	.37	.32	.29	.36	.32	.29	.28	candescent lamp
Max. S/MH$_{wp}$ = 1.5	10	.33	.28	.25	.33	.28	.25	.32	.28	.25	.32	.28	.24	.31	.27	.24	.23	LDD Maint. Category V
	1	.66	.65	.63	.64	.63	.62	.62	.61	.60	.60	.59	.58	.58	.57	.57	.55	
	2	.61	.59	.56	.61	.58	.56	.58	.56	.54	.56	.55	.53	.55	.53	.52	.51	
	3	.57	.54	.51	.56	.53	.51	.55	.52	.50	.53	.51	.49	.52	.50	.48	.47	
	4	.53	.49	.46	.52	.49	.46	.51	.48	.45	.49	.47	.45	.48	.47	.44	.43	
	5	.49	.45	.42	.48	.45	.42	.47	.44	.42	.46	.43	.41	.45	.43	.41	.40	
	6	.45	.42	.39	.45	.41	.39	.44	.41	.38	.43	.40	.38	.42	.40	.38	.37	
	7	.42	.38	.36	.42	.38	.35	.41	.38	.35	.40	.37	.35	.40	.37	.35	.34	
	8	.39	.35	.33	.39	.35	.32	.38	.35	.32	.37	.34	.32	.37	.34	.32	.31	Medium distribution reflector
	9	.36	.32	.30	.36	.32	.30	.35	.32	.29	.35	.32	.29	.34	.31	.29	.28	and concave lens
Max. S/MH$_{wp}$ = 1.3	10	.32	.28	.25	.32	.28	.25	.31	.28	.25	.31	.27	.25	.30	.27	.25	.24	LDD Maint. Category V
	1	.75	.73	.71	.73	.71	.70	.70	.69	.68	.68	.67	.66	.66	.65	.64	.63	
	2	.68	.65	.62	.67	.64	.62	.65	.62	.60	.63	.61	.59	.61	.59	.58	.57	
	3	.63	.59	.55	.62	.58	.54	.60	.57	.53	.58	.56	.52	.57	.55	.52	.51	
	4	.58	.53	.50	.57	.53	.50	.55	.52	.49	.54	.51	.48	.53	.50	.48	.47	
	5	.53	.48	.45	.52	.48	.45	.51	.47	.44	.50	.46	.44	.49	.46	.43	.42	
	6	.49	.44	.41	.49	.44	.41	.47	.43	.40	.46	.43	.40	.45	.42	.40	.39	
	7	.45	.40	.37	.45	.40	.37	.44	.40	.37	.43	.39	.37	.42	.39	.36	.35	
	8	.42	.37	.34	.41	.37	.34	.40	.36	.33	.40	.36	.33	.39	.36	.33	.32	Wide distribution reflector
	9	.38	.33	.30	.38	.33	.30	.37	.33	.30	.36	.32	.30	.36	.32	.30	.28	and flat lens
Max. S/MH$_{wp}$ = 1.2	10	.35	.30	.27	.35	.30	.27	.34	.30	.27	.34	.30	.27	.33	.29	.27	.26	LDD Maint. Category V

Reprinted with permission from Chapter 9 of the *Illuminating Engineering Society Handbook*, 5th ed.

[a] Ratio of maximum spacing between luminaire centers to mounting (or ceiling) height above the work plane. See Luminaire Spacing.

[b] RCR = room cavity ratio.

[c] ρCC = percent effective ceiling cavity reflectance.

[d] ρw = percent wall reflectance.

TABLE 2. Lamp Lumen Depreciation Factor

GE Lamp Type		*Watts*	*Mean Lumen Factor* (%)*	*LLD** (%)*
Incandescent	100A	100	93	90
	150A	150	93	90
	150PAR/SP and FL	150	84	78
	150R/SP and FL	150	89	85
	300M/IF	300	91	87
	300R/SP and FL	300	94	92
	500/IF	500	91	88
	1000/IF	1000	92	89
	1500/IF	1500	84	78
Fluorescent	F40CW Mainlighter®	40	87	83
	F40WW Mainlighter®	40	87	83
	F40CWX Mainlighter®	40	83	73
	F40WWX Mainlighter®	40	83	73
	F40CW/S Staybright®	40	90	86
	F40WW/S Staybright®	40	90	86
	F96T12/CW	50	91	87
	F96T12/WW	50	91	87
	F96T12/CWX	50	87	78
	F48T12/CW/HO	60	87	79
	F96T12/CW/HO	110	87	77
	F48PG17/CW	110	77	67
	F96PG17/CW	215	75	65
	F48T12/CW/1500	110	77	67
	F96T12/CW/1500	215	80	71

Reprinted by permission from General Electric Bulletin TPC-42, Nov. 1973.

* Use the Mean Lumen Factor when calculations involve average illumination between relampings (based upon lamp light output at 40–50% of rated life depending upon lamp type).

** Use the LLD Factor when calculations involve minimum illumination between relampings (based upon lamp light output at approximately 70% of rated life).

NOTE 1: Factors for mercury lamps are based upon 24,000 h life.

NOTE 2: Factors shown for fluorescent lamps are based on 3 h/start. Factors for Multi-Vapor and Lucalox lamps are based on 10 h/start. All HID lamp factors are for vertical burning.

TABLE 3. Cavity Ratios

Room Dimensions		Cavity Depth																			
Width	Length	1.0	1.5	2.0	2.5	3.0	3.5	4.0	5.0	6.0	7.0	8	9	10	11	12	14	16	20	25	30
8	8	1.2	1.9	2.5	3.1	3.7	4.4	5.0	6.2	7.5	8.8	10.0	11.2	12.5	—	—	—	—	—	—	—
	10	1.1	1.7	2.2	2.8	3.4	3.9	4.5	5.6	6.7	7.9	9.0	10.1	11.3	12.4	—	—	—	—	—	—
	14	1.0	1.5	2.0	2.5	3.0	3.4	3.9	4.9	5.9	6.9	7.8	8.8	9.7	10.7	11.7	—	—	—	—	—
	20	0.9	1.3	1.7	2.2	2.6	3.1	3.5	4.4	5.2	6.1	7.0	7.9	8.8	9.6	10.5	12.2	—	—	—	—
	30	0.8	1.2	1.6	2.0	2.4	2.8	3.2	4.0	4.7	5.5	6.3	7.1	7.9	8.7	9.5	11.0	—	—	—	—
	40	0.7	1.1	1.5	1.9	2.3	2.6	3.0	3.7	4.5	5.3	5.9	6.5	7.4	8.1	8.8	10.3	11.8	—	—	—
10	10	1.0	1.5	2.0	2.5	3.0	3.5	4.0	5.0	6.0	7.0	8.0	9.0	10.0	11.0	12.0	—	—	—	—	—
	14	0.9	1.3	1.7	2.1	2.6	3.0	3.4	4.3	5.1	6.0	6.9	7.8	8.6	9.5	10.4	12.0	—	—	—	—
	20	0.7	1.1	1.5	1.9	2.3	2.6	3.0	3.7	4.5	5.3	6.0	6.8	7.5	8.3	9.0	10.5	12.0	—	—	—
	30	0.7	1.0	1.3	1.7	2.0	2.3	2.7	3.3	4.0	4.7	5.3	6.0	6.6	7.3	8.0	9.4	10.6	—	—	—
	40	0.6	0.9	1.2	1.6	1.9	2.2	2.5	3.1	3.7	4.4	5.0	5.6	6.2	6.9	7.5	8.7	10.0	12.5	—	—
	60	0.6	0.9	1.2	1.5	1.7	2.0	2.3	2.9	3.5	4.1	4.7	5.3	5.9	6.5	7.1	8.2	9.4	11.7	—	—
12	12	0.8	1.2	1.7	2.1	2.5	2.9	3.3	4.2	5.0	5.8	6.7	7.5	8.4	9.2	10.0	11.7	—	—	—	—
	16	0.7	1.1	1.5	1.8	2.2	2.5	2.9	3.6	4.4	5.1	5.8	6.5	7.2	8.0	8.7	10.2	11.6	—	—	—
	24	0.6	0.9	1.2	1.6	1.9	2.2	2.5	3.1	3.7	4.4	5.0	5.6	6.2	6.9	7.5	8.7	10.0	12.5	—	—
	36	0.6	0.8	1.1	1.4	1.7	1.9	2.2	2.8	3.3	3.9	4.4	5.0	5.5	6.0	6.6	7.8	8.8	11.0	—	—
	50	0.5	0.8	1.0	1.3	1.5	1.8	2.1	2.6	3.1	3.6	4.1	4.6	5.1	5.6	6.2	7.2	8.2	10.2	—	—
	70	0.5	0.7	1.0	1.2	1.5	1.7	2.0	2.4	2.9	3.4	3.9	4.4	4.9	5.4	5.8	6.8	7.8	9.7	12.2	—
14	14	0.7	1.1	1.4	1.8	2.1	2.5	2.9	3.6	4.3	5.0	5.7	6.4	7.1	7.8	8.5	10.0	11.4	—	—	—
	20	0.6	0.9	1.2	1.5	1.8	2.1	2.4	3.0	3.6	4.2	4.9	5.5	6.1	6.7	7.3	8.6	9.8	12.3	—	—
	30	0.5	0.8	1.0	1.3	1.6	1.8	2.1	2.6	3.1	3.7	4.2	4.7	5.2	5.8	6.3	7.3	8.4	10.5	—	—
	42	0.5	0.7	1.0	1.2	1.4	1.7	1.9	2.4	2.9	3.3	3.8	4.3	4.7	5.2	5.7	6.7	7.6	9.5	11.9	—
	60	0.4	0.7	0.9	1.1	1.3	1.5	1.8	2.2	2.6	3.1	3.5	3.9	4.4	4.8	5.2	6.1	7.0	8.8	10.9	—
	90	0.4	0.6	0.8	1.0	1.2	1.4	1.6	2.0	2.5	2.9	3.3	3.7	4.1	4.5	5.0	5.8	6.6	8.3	10.3	12.4
17	17	0.6	0.9	1.2	1.5	1.8	2.1	2.3	2.9	3.5	4.1	4.7	5.3	5.9	6.5	7.0	8.2	9.4	11.7	—	—
	25	0.5	0.7	1.0	1.2	1.5	1.7	2.0	2.5	3.0	3.5	4.0	4.5	5.0	5.5	6.0	7.0	8.0	10.0	12.5	—
	35	0.4	0.7	0.9	1.1	1.3	1.5	1.7	2.2	2.6	3.1	3.5	3.9	4.4	4.8	5.2	6.1	7.0	8.7	10.9	—
	50	0.4	0.6	0.8	1.0	1.2	1.4	1.6	2.0	2.4	2.8	3.1	3.5	3.9	4.3	4.5	5.4	6.2	7.7	9.7	11.6
	80	0.4	0.5	0.7	0.9	1.1	1.2	1.4	1.8	2.1	2.5	2.9	3.3	3.6	4.0	4.3	5.1	5.8	7.2	9.0	10.9
	120	0.3	0.5	0.7	0.8	1.0	1.2	1.3	1.7	2.0	2.3	2.7	3.0	3.4	3.7	4.0	4.7	5.4	6.7	8.4	10.1
20	20	0.5	0.7	1.0	1.2	1.5	1.7	2.0	2.5	3.0	3.5	4.0	4.5	5.0	5.5	6.0	7.0	8.0	10.0	12.5	—
	30	0.4	0.6	0.8	1.0	1.2	1.5	1.7	2.1	2.5	2.9	3.3	3.7	4.1	4.5	4.9	5.8	6.6	8.2	10.3	12.4
	45	0.4	0.5	0.7	0.9	1.1	1.3	1.4	1.8	2.2	2.5	2.9	3.3	3.6	4.0	4.3	5.1	5.8	7.2	9.1	10.9
	60	0.3	0.5	0.7	0.8	1.0	1.2	1.3	1.7	2.0	2.3	2.7	3.0	3.4	3.7	4.0	4.7	5.4	6.7	8.4	10.1
	90	0.3	0.5	0.6	0.8	0.9	1.1	1.2	1.5	1.8	2.1	2.4	2.7	3.0	3.3	3.6	4.2	4.8	6.0	7.5	9.0
	150	0.3	0.4	0.6	0.7	0.8	1.0	1.1	1.4	1.7	2.0	2.3	2.6	2.9	3.2	3.4	4.0	4.6	5.7	7.2	8.6
24	24	0.4	0.6	0.8	1.0	1.2	1.5	1.7	2.1	2.5	2.9	3.3	3.7	4.1	4.5	5.0	5.8	6.7	8.2	10.3	12.4
	32	0.4	0.5	0.7	0.9	1.1	1.3	1.5	1.8	2.2	2.6	2.9	3.3	3.6	4.0	4.3	5.1	5.8	7.2	9.0	11.0
	50	0.3	0.5	0.6	0.8	0.9	1.1	1.2	1.5	1.8	2.2	2.5	2.8	3.1	3.4	3.7	4.4	5.0	6.2	7.8	9.4
	70	0.3	0.4	0.6	0.7	0.8	1.0	1.1	1.4	1.7	2.0	2.2	2.5	2.8	3.0	3.3	3.8	4.4	5.5	6.9	8.2
	100	0.3	0.4	0.5	0.6	0.8	0.9	1.0	1.3	1.6	1.8	2.1	2.4	2.6	2.9	3.1	3.7	4.2	5.2	6.5	7.9
	160	0.2	0.4	0.5	0.6	0.7	0.8	1.0	1.2	1.4	1.7	1.9	2.1	2.4	2.6	2.8	3.3	3.8	4.7	5.9	7.1
30	30	0.3	0.5	0.7	0.8	1.0	1.2	1.3	1.7	2.0	2.3	2.7	3.0	3.3	3.7	4.0	4.7	5.4	6.7	8.4	10.0
	45	0.3	0.4	0.6	0.7	0.8	1.0	1.1	1.4	1.7	1.9	2.2	2.5	2.7	3.0	3.3	3.8	4.4	5.5	6.9	8.2
	60	0.3	0.4	0.5	0.6	0.7	0.9	1.0	1.2	1.5	1.7	2.0	2.2	2.5	2.7	3.0	3.5	4.0	5.0	6.2	7.4
	90	0.2	0.3	0.4	0.6	0.7	0.8	0.9	1.1	1.3	1.6	1.8	2.0	2.2	2.5	2.7	3.1	3.6	4.5	5.6	6.7
	150	0.2	0.3	0.4	0.5	0.6	0.7	0.8	1.0	1.2	1.4	1.6	1.8	2.0	2.2	2.4	2.8	3.2	4.0	5.0	5.9
	200	0.2	0.3	0.4	0.5	0.6	0.7	0.8	1.0	1.1	1.3	1.5	1.7	1.9	2.0	2.2	2.6	3.0	3.7	4.7	5.6
36	36	0.3	0.4	0.6	0.7	0.8	1.0	1.1	1.4	1.7	1.9	2.2	2.5	2.8	3.0	3.3	3.9	4.4	5.5	6.9	8.3
	50	0.2	0.4	0.5	0.6	0.7	0.8	1.0	1.2	1.4	1.7	1.9	2.1	2.5	2.6	2.9	3.3	3.8	4.8	5.9	7.2
	75	0.2	0.3	0.4	0.5	0.6	0.7	0.8	1.0	1.2	1.4	1.6	1.8	2.0	2.3	2.5	2.9	3.3	4.1	5.1	6.1
	100	0.2	0.3	0.4	0.5	0.6	0.7	0.8	0.9	1.1	1.3	1.5	1.7	1.9	2.1	2.3	2.6	3.0	3.8	4.7	5.7
	150	0.2	0.3	0.3	0.4	0.5	0.6	0.7	0.9	1.0	1.2	1.4	1.6	1.7	1.9	2.1	2.4	2.8	3.5	4.3	5.2
	200	0.2	0.2	0.3	0.4	0.5	0.6	0.7	0.8	1.0	1.1	1.3	1.5	1.6	1.8	2.0	2.3	2.6	3.3	4.1	4.9
42	42	0.2	0.4	0.5	0.6	0.7	0.8	1.0	1.2	1.4	1.6	1.9	2.1	2.4	2.6	2.8	3.3	3.8	4.7	5.9	7.1
	60	0.2	0.3	0.4	0.5	0.6	0.7	0.8	1.0	1.2	1.4	1.6	1.8	2.0	2.2	2.4	2.8	3.2	4.0	5.0	6.0
	90	0.2	0.3	0.3	0.4	0.5	0.6	0.7	0.9	1.0	1.2	1.4	1.6	1.7	1.9	2.1	2.4	2.8	3.5	4.4	5.2
	140	0.2	0.2	0.3	0.4	0.5	0.5	0.6	0.8	0.9	1.1	1.2	1.4	1.5	1.7	1.9	2.2	2.5	3.1	3.9	4.6
	200	0.1	0.2	0.3	0.4	0.4	0.5	0.6	0.7	0.9	1.0	1.1	1.3	1.4	1.6	1.7	2.0	2.3	2.9	3.6	4.3
	300	0.1	0.2	0.3	0.3	0.4	0.5	0.5	0.7	0.8	0.9	1.1	1.3	1.4	1.5	1.7	1.9	2.2	2.8	3.5	4.2
50	50	0.2	0.3	0.4	0.5	0.6	0.7	0.8	1.0	1.2	1.4	1.6	1.8	2.0	2.2	2.4	2.8	3.2	4.0	5.0	6.0
	70	0.2	0.3	0.3	0.4	0.5	0.6	0.7	0.9	1.0	1.2	1.4	1.5	1.7	1.9	2.0	2.4	2.7	3.4	4.3	5.1
	100	0.1	0.2	0.3	0.4	0.4	0.5	0.6	0.7	0.9	1.0	1.2	1.3	1.5	1.6	1.8	2.1	2.4	3.0	3.7	4.5
	150	0.1	0.2	0.3	0.3	0.4	0.5	0.5	0.7	0.8	0.9	1.1	1.2	1.3	1.5	1.6	1.9	2.1	2.7	3.3	4.0
	300	0.1	0.2	0.2	0.3	0.3	0.4	0.5	0.6	0.7	0.8	0.9	1.0	1.1	1.3	1.4	1.6	1.9	2.3	2.9	3.5
60	60	0.2	0.2	0.3	0.4	0.5	0.6	0.7	0.8	1.0	1.2	1.3	1.5	1.7	1.8	2.0	2.3	2.7	3.3	4.2	5.0
	100	0.1	0.2	0.3	0.3	0.4	0.5	0.5	0.7	0.8	0.9	1.1	1.2	1.3	1.5	1.6	1.9	2.1	2.7	3.3	4.0
	150	0.1	0.2	0.2	0.3	0.3	0.4	0.5	0.6	0.7	0.8	0.9	1.0	1.2	1.3	1.4	1.6	1.9	2.3	2.9	3.5
	300	0.1	0.1	0.2	0.2	0.3	0.3	0.4	0.5	0.6	0.7	0.8	0.9	1.0	1.1	1.2	1.4	1.6	2.0	2.5	3.0
75	75	0.1	0.2	0.3	0.3	0.4	0.5	0.5	0.7	0.8	0.9	1.1	1.2	1.3	1.5	1.6	1.9	2.1	2.7	3.3	4.0
	120	0.1	0.2	0.2	0.3	0.3	0.4	0.4	0.5	0.6	0.8	0.9	1.0	1.1	1.2	1.3	1.5	1.7	2.2	2.7	3.3
	200	0.1	0.1	0.2	0.2	0.3	0.3	0.4	0.5	0.5	0.6	0.7	0.8	0.9	1.0	1.1	1.3	1.5	1.8	2.3	2.7
	300	0.1	0.1	0.2	0.2	0.2	0.3	0.3	0.4	0.5	0.6	0.7	0.7	0.8	0.9	1.0	1.2	1.3	1.7	2.1	2.5
100	100	0.1	0.1	0.2	0.2	0.3	0.3	0.4	0.5	0.6	0.7	0.8	0.9	1.0	1.1	1.2	1.4	1.6	2.0	2.5	3.0
	200	0.1	0.1	0.1	0.2	0.2	0.3	0.3	0.4	0.4	0.5	0.6	0.7	0.7	0.8	0.9	1.0	1.2	1.5	1.9	2.2
	300	0.1	0.1	0.1	0.2	0.2	0.2	0.3	0.3	0.4	0.5	0.5	0.6	0.7	0.7	0.8	0.9	1.1	1.3	1.7	2.0
150	150	0.1	0.1	0.1	0.2	0.2	0.2	0.3	0.3	0.4	0.5	0.5	0.6	0.7	0.7	0.8	0.9	1.1	1.3	1.7	2.0
	300	—	0.1	0.1	0.1	0.1	0.2	0.2	0.2	0.3	0.3	0.4	0.5	0.5	0.6	0.6	0.7	0.8	1.0	1.2	1.5
200	200	—	0.1	0.1	0.1	0.1	0.2	0.2	0.2	0.3	0.3	0.4	0.5	0.5	0.6	0.6	0.7	0.8	1.0	1.2	1.5
	300		0.1	0.1	0.1	0.1	0.1	0.2	0.2	0.2	0.3	0.3	0.4	0.4	0.5	0.5	0.6	0.7	0.8	1.0	1.2
300	300	—	—	0.1	0.1	0.1	0.1	0.1	0.2	0.2	0.2	0.3	0.3	0.3	0.4	0.4	0.5	0.5	0.6	0.7	0.8
500	500	—	—	—	—	0.1	0.1	0.1	0.1	0.1	0.1	0.2	0.2	0.2	0.2	0.2	0.3	0.3	0.4	0.5	0.6

Reprinted by permission from Chapter 9 of the *Illuminating Engineering Society Handbook*, 5th ed.

TABLE 4. Percent Effective Ceiling or Floor Cavity Reflectance for Various Reflectance Combinations

Per Cent Ceiling or Floor Reflectance	90				80				70			50			30				10		
Per Cent Wall Reflectance	90	70	50	30	80	70	50	30	70	50	30	70	50	30	65	50	30	10	50	30	10
Ceiling or Floor Cavity Ratio																					
0	90	90	90	90	80	80	80	80	70	70	70	50	50	50	30	30	30	30	10	10	10
0.1	90	89	88	87	79	79	78	78	69	69	68	59	49	48	30	30	29	29	10	10	10
0.2	89	88	86	85	79	78	77	76	68	67	66	49	48	47	30	29	29	28	10	10	9
0.3	89	87	85	83	78	77	75	74	68	66	64	49	47	46	30	29	28	27	10	10	9
0.4	88	86	83	81	78	76	74	72	67	65	63	48	46	45	30	29	27	26	11	10	9
0.5	88	85	81	78	77	75	73	70	66	64	61	48	46	44	29	28	27	25	11	10	9
0.6	88	84	80	76	77	75	71	68	65	62	59	47	45	43	29	28	26	25	11	10	9
0.7	88	83	78	74	76	74	70	66	65	61	58	47	44	42	29	28	26	24	11	10	8
0.8	87	82	77	73	75	73	69	65	64	60	56	47	43	41	29	27	25	23	11	10	8
0.9	87	81	76	71	75	72	68	63	63	59	55	46	43	40	29	27	25	22	11	9	8
1.0	86	80	74	69	74	71	66	61	63	58	53	46	42	39	29	27	24	22	11	9	8
1.1	86	79	73	67	74	71	65	60	62	57	52	46	41	38	29	26	24	21	11	9	8
1.2	86	78	72	65	73	70	64	58	61	56	50	45	41	37	29	26	23	20	12	9	7
1.3	85	78	70	64	73	69	63	57	61	55	49	45	40	36	29	26	23	20	12	9	7
1.4	85	77	69	62	72	68	62	55	60	54	48	45	40	35	28	26	22	19	12	9	7
1.5	85	76	68	61	72	68	61	54	59	53	47	44	39	34	28	25	22	18	12	9	7
1.6	85	75	66	59	71	67	60	53	59	52	45	44	39	33	28	25	21	18	12	9	7
1.7	84	74	65	58	71	66	59	52	58	51	44	44	38	32	28	25	21	17	12	9	7
1.8	84	73	64	56	70	65	58	50	57	50	43	43	37	32	28	25	21	17	12	9	6
1.9	84	73	63	55	70	65	57	49	57	49	42	43	37	31	28	25	20	16	12	9	6
2.0	83	72	62	53	69	64	56	48	56	48	41	43	37	30	28	24	20	16	12	9	6
2.1	83	71	61	52	69	63	55	47	56	47	40	43	36	29	28	24	20	16	13	9	6
2.2	83	70	60	51	68	63	54	45	55	46	39	42	36	29	28	24	19	15	13	9	6
2.3	83	69	59	50	68	62	53	44	54	46	38	42	35	28	28	24	19	15	13	9	6
2.4	82	68	58	48	67	61	52	43	54	45	37	42	35	27	28	24	19	14	13	9	6
2.5	82	68	57	47	67	61	51	42	53	44	36	41	34	27	27	23	18	14	13	9	6
2.6	82	67	56	46	66	60	50	41	53	43	35	41	34	26	27	23	18	13	13	9	5
2.7	82	66	55	45	66	60	49	40	52	43	34	41	33	26	27	23	18	13	13	9	5
2.8	81	66	54	44	66	59	48	39	52	42	33	41	33	25	27	23	18	13	13	9	5
2.9	81	65	53	43	65	58	48	38	51	41	33	40	33	25	27	23	17	12	13	9	5
3.0	81	64	52	42	65	58	47	38	51	40	32	40	32	24	27	22	17	12	13	8	5
3.1	80	64	51	41	64	57	46	37	50	40	31	40	32	24	27	22	17	12	13	8	5
3.2	80	63	50	40	64	57	45	36	50	39	30	40	31	23	27	22	16	11	13	8	5
3.3	80	62	49	39	64	56	44	35	49	39	30	39	31	23	27	22	16	11	13	8	5
3.4	80	62	48	38	63	56	44	34	49	38	29	39	31	22	27	22	16	11	13	8	5
3.5	79	61	48	37	63	55	43	33	48	38	29	39	30	22	26	22	16	11	13	8	5
3.6	70	60	17	36	62	51	12	33	18	37	28	30	30	21	26	21	15	10	13	8	5
3.7	79	60	46	35	62	54	42	32	48	37	27	38	30	21	26	21	15	10	13	8	4
3.8	79	59	45	35	62	53	41	31	47	36	27	38	29	21	26	21	15	10	13	8	4
3.9	78	59	45	34	61	53	40	30	47	36	26	38	29	20	26	21	15	10	13	8	4
4.0	78	58	44	33	61	52	40	30	46	35	26	38	29	20	26	21	15	9	13	8	4
4.1	78	57	43	32	60	52	39	29	46	35	25	37	28	20	26	21	14	9	13	8	4
4.2	78	57	43	32	60	51	39	29	46	34	25	37	28	19	26	20	14	9	13	8	4
4.3	78	56	42	31	60	51	38	28	45	34	25	37	28	19	26	20	14	9	13	8	4
4.4	77	56	41	30	59	51	38	28	45	34	24	37	27	19	26	20	14	8	13	8	4
4.5	77	55	41	30	59	50	37	27	45	33	24	37	27	19	25	20	14	8	14	8	4
4.6	77	55	40	29	59	50	37	26	44	33	24	36	27	18	25	20	14	8	14	8	4
4.7	77	54	40	29	58	49	36	26	44	33	23	36	26	18	25	20	13	8	14	8	4
4.8	76	54	39	28	58	49	36	25	44	32	23	36	26	18	25	19	13	8	14	8	4
4.9	76	53	38	28	58	49	35	25	44	32	23	36	26	18	25	19	13	7	14	8	4
5.0	76	53	38	27	57	48	35	25	43	32	22	36	26	17	25	19	13	7	14	8	4

Reprinted by permission from Chapter 9 of the *Illuminating Engineering Society Handbook*, 5th ed.

For cavity dimensions other than those shown here, use the formula

$$CR = \frac{5 \times h_{cc}\,(L + W)}{L \times W}$$

TABLE 5. Factors for Effective Floor Cavity Reflectances Other Than 20 Percent

For 30 percent effective floor cavity reflectance, multiply by appropriate factor below.
For 10 percent effective floor cavity reflectance, divide by appropriate factor below.

Percent Effective Ceiling Cavity Reflectance, ρCC	*80*			*70*			*50*			*10*		
Per Cent Wall Reflectance, ρw	*50*	*30*	*10*	*50*	*30*	*10*	*50*	*30*	*10*	*50*	*30*	*10*
Room Cavity Ratio												
1	1.08	1.08	1.07	1.07	1.06	1.06	1.05	1.04	1.04	1.01	1.01	1.01
2	1.07	1.06	1.05	1.06	1.05	1.04	1.04	1.03	1.03	1.01	1.01	1.01
3	1.05	1.04	1.03	1.05	1.04	1.03	1.03	1.03	1.02	1.01	1.01	1.01
4	1.05	1.03	1.02	1.04	1.03	1.02	1.03	1.02	1.02	1.01	1.01	1.00
5	1.04	1.03	1.02	1.03	1.02	1.02	1.02	1.02	1.01	1.01	1.01	1.00
6	1.03	1.02	1.01	1.03	1.02	1.01	1.02	1.02	1.01	1.01	1.01	1.00
7	1.03	1.02	1.01	1.03	1.02	1.01	1.02	1.01	1.01	1.01	1.01	1.00
8	1.03	1.02	1.01	1.02	1.02	1.01	1.02	1.01	1.01	1.01	1.01	1.00
9	1.02	1.01	1.01	1.02	1.01	1.01	1.02	1.01	1.01	1.01	1.01	1.00
10	1.02	1.01	1.01	1.02	1.01	1.01	1.02	1.01	1.01	1.01	1.01	1.00

Reprinted by permission from Chapter 9 of the *Illuminating Engineering Society Handbook* 5th ed.

TABLE 6. Five Degrees of Dirt Conditions

VERY CLEAN (high-grade offices not near production; laboratories; clean rooms)
Generated dirt: None
Ambient dirt: None (or none enters area)
Removal or filtration: Excellent
Adhesion: None

CLEAN (offices in older buildings or near production; light assembly; inspection)
Generated dirt: Very little
Ambient dirt: Some (almost none enters)
Removal or filtration: Better than average
Adhesion: Slight

MEDIUM (mill offices; paper processing; light machinery)
Generated dirt: Noticeable, but not heavy
Ambient dirt: Some enters area
Removal or filtration: Poorer than average
Adhesion: Enough to be visible after some months

DIRTY (heat-treating; high-speed printing; rubber processing)
Generated dirt: Accumulates rapidly
Ambient dirt: Large amount enters area
Removal or filtration: Only fans or blowers, if any
Adhesion: High—probably due to oil, humidity, or static

VERY DIRTY (similar to "dirty," but luminaires within immediate area of contamination)
Generated dirt: Constant accumulation
Ambient dirt: Almost none excluded
Removal or filtration: None
Adhesion: High

Reprinted by permission from Chapter 9 of the *Illuminating Engineering Society Handbook* 5th ed.

TABLE 7. The LDD Factors for Six Categories and Five Degrees of Dirtiness

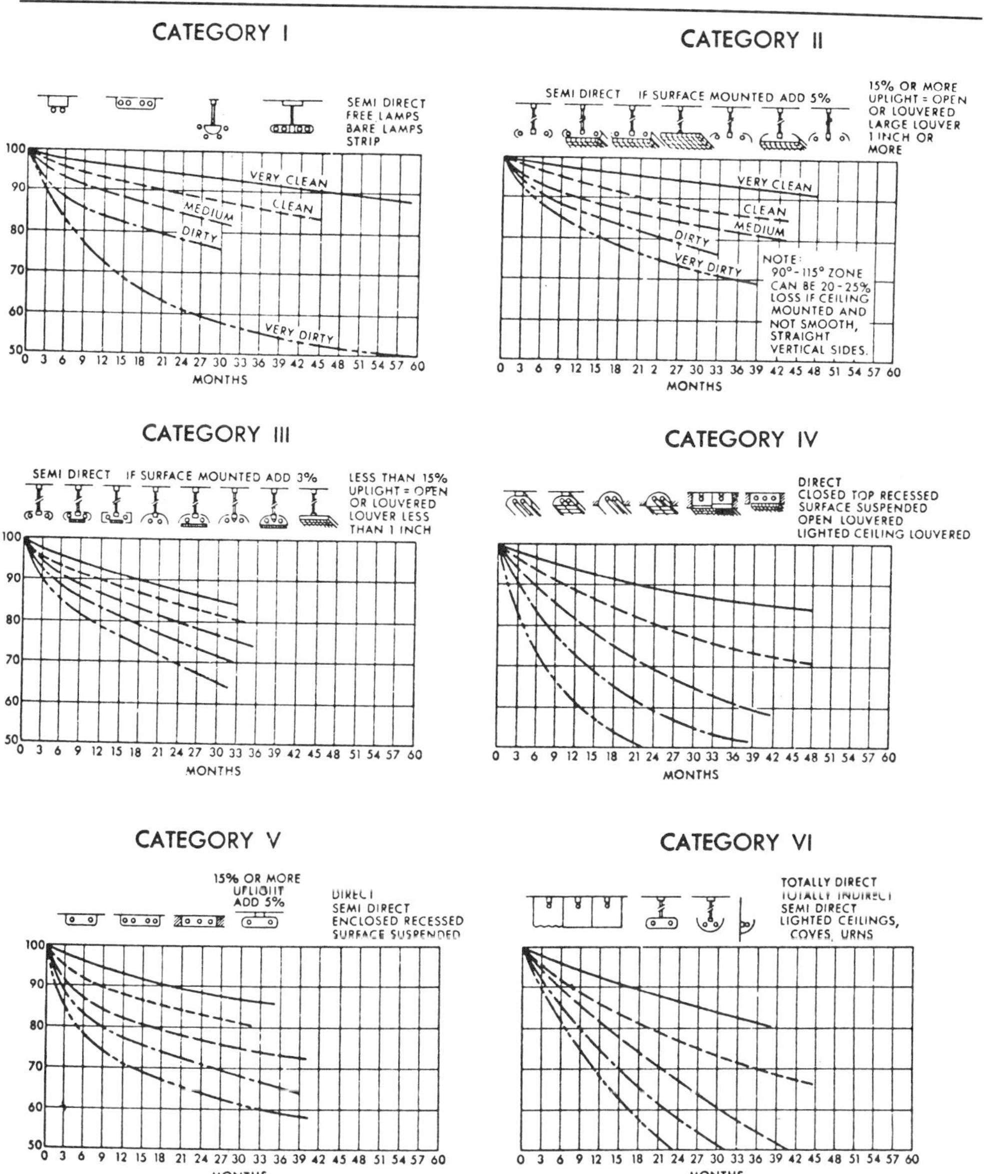

Reprinted with permission from Chapter 9 of the *Illuminating Engineering Society Handbook*, 5th ed.

Index